Butterfly Wing Patterns and Mimicry

Butterfly Wing Patterns and Mimicry

Kenzie Payne

MURPHY & MOORE
www.murphy-moorepublishing.com

www.murphy-moorepublishing.com

MURPHY & MOORE

Cataloging-in-publication Data

Butterfly wing patterns and mimicry / Kenzie Payne.
p. cm.
Includes bibliographical references and index.
ISBN 978-1-63987-731-7
1. Butterflies--Molecular aspects. 2. Butterflies--Physiology.
3. Butterflies--Identification. 4. Genetic polymorphisms. I. Payne, Kenzie.
QL562.2 .B88 2023
572.815 78--dc23

Murphy & Moore Publishing
1 Rockefeller Plaza,
New York City,
NY 10020, USA

ISBN 978-1-63987-731-7

Contents

Preface

Lepidopteran is an order of insects, which includes butterflies and moths. Insects of this order have unique color patterns on their wings. Butterfly wing patterns have identities that can be traced from species to species and typically across genera and families. The evolution and diversification of butterflies can be studied based on homologies among pattern elements and identities. Individuated patterns among the butterflies have emerged from non-individuated precursors that become developmentally autonomous for color pattern formation. This process involves the compartmentalization of the wing areas that in turn leads to the creation of color patterns characterized by repeated elements and emergence of serial homology. Various researches have studied the similarities in the vivid wing patterns of various butterflies and moth species. They have also observed that these patterns usually change rapidly in accordance with the mimicry theory. This theory deals with the organisms which develop a phenotypic similarity to another organism for selective benefits. This theory upholds that phenotypic similarity evolves due to natural selection. Different evaluations, and advanced studies on wing patterns and mimicry have been included in this book. Those in search of information to further their knowledge will be greatly assisted by it.

Various studies have approached the subject by analyzing it with a single perspective, but the present book provides diverse methodologies and techniques to address this field. This book contains theories and applications needed for understanding the subject from different perspectives. The aim is to keep the readers informed about the progresses in the field; therefore, the contributions were carefully examined to compile novel researches by specialists from across the globe.

Indeed, the job of the editor is the most crucial and challenging in compiling all chapters into a single book. In the end, I would extend my sincere thanks to the chapter authors for their profound work. I am also thankful for the support provided by my family and colleagues during the compilation of this book.

Kenzie Payne

Part I
Diversification of NGP

1

Mathematical Model for the Formation of Border Ocelli and Parafocal Elements

H. Frederik Nijhout

Abstract The border ocelli and adjacent parafocal elements are among the most diverse and finely detailed features of butterfly wing patterns. The border ocelli can be circular, elliptical, and heart-shaped or can develop as dots, arcs, or short lines. Parafocal elements are typically shaped like smooth arcs but are also often "V," "W," and "M" shaped. The fusion of a border ocellus with its adjacent parafocal element is a common response to temperature shock and treatment with chemicals such as heparin and tungstate ions. Here I develop a new mathematical model for the formation of border ocelli and parafocal elements. The models are a reaction-diffusion model based on the well-established gradient-threshold mechanisms in embryonic development. The model uses a simple biochemical reaction sequence that is initiated at the wing veins and from there spreads across the field in the manner of a grass-fire. Unlike Turing-style models, this model is insensitive to the size of the field. Like real developmental systems, the model does not have a steady state, but the pattern is "read out" at a point in development, in response to an independent developmental signal such as a pulse of ecdysone secretion, which is known to regulate color pattern in butterflies. The grass-fire model reproduces the sequence of Distal-less expression that determines the position of eyespot foci and also shows how a border ocellus and its neighboring parafocal element can arise from such a single focus. The grass-fire model shows that the apparent fusion of ocellus and parafocal element is probably due to a premature termination of the normal process that separates the two and supports the hypothesis that the parafocal element is the distal band of the border symmetry system.

Keywords Mathematical model • Eyespot • Parafocal element • Grass-fire model • Temperature shock

---

H.F. Nijhout (✉)
Department of Biology, Duke University, Durham, NC 27708, USA
e-mail: hfn@duke.edu

1.1 Introduction

The color patterns of butterflies are extremely diverse, and almost all of the 14,000 or so species can be identified on the basis of their color patterns alone. Adding to this diversity is the fact that dorsal and ventral color patterns are usually entirely different and that many species have polymorphic, sexually dimorphic, and seasonally plastic color patterns. The development and evolution of this diversity of patterns has been of considerable interest, particularly in relation to the genetics and evolution of mimicry (Reed et al. 2011; Nadeau 2016; Baxter et al. 2008; Joron et al. 2006), and the development and evolution of eyespot patterns (Brakefield et al. 1996; Monteiro et al. 1997, 2003; Monteiro 2015; Nijhout 1980).

The organizing principles of color patterns are coming to be increasingly well understood. The diversity of mimicry patterns in *Heliconius* butterflies is due to the variation in only a handful of genes (Nadeau 2016; Kapan et al. 2006), and the specification of color and pattern is now known to be due to a redeployment of many of the genes involved in early embryonic development (Carroll et al. 1994; Martin and Reed 2014; Reed and Serfas 2004; Brunetti et al. 2001).

The developmental mechanism that produces the spatial pattern of pigments that characterizes color patterns is less well understood. It is clear, however, that the wing veins and the wing margin play critical roles in organizing the pattern. This evidence comes, among others, from observations of the color patterns of mutants that lack wing veins and from experimental manipulations that alter the wing margin (e.g., Fig. 1.1 and (Nijhout and Grunert 1988; Koch and Nijhout 2002)).

Fig. 1.1 Color pattern modification in the veinless mutant of *Papilio xuthus* (*right*), compared with the normal pattern (*left*). The longitudinal veins are missing and so are the venous patterns. The submarginal bands are smoothly continuous and parallel to the wing margin, suggesting that the wing margin also plays an important role in color pattern determination

1.2 Eyespots and Parafocal Elements

The color patterns of butterflies are organized as a set of three-symmetry systems (Süffert 1929; Schwanwitsch 1924, 1929; Nijhout 1991). The *basal symmetry system* is often absent or represented only by its distal band. The *central symmetry system* runs in the middle region of the wing and is centered on the discal spot. The *border symmetry system* runs along the distal region of the wing usually paralleling the wing margin (Fig. 1.2). The most complex patterns are typically found in the border symmetry system. The principal elements of the border symmetry system are the border ocelli or eyespots. Although the canonical morphology of an ocellus is a set of concentric circles of contrasting pigments with a well-defined central spot called the focus (Nijhout 1980), circular elements are actually quite uncommon within the larger diversity of butterfly color patterns. More often the shape of the "ocellus" deviates significantly from the circular (heart shaped, dagger shaped, bar shaped) and is often hardly recognizable as homologous to a circular element (Nijhout 1990, 1991).

The proximal and distal bands of the border symmetry system have very different characters. The proximal band, when present, is typically arc shaped, or nearly straight. The distal bands are almost always present and have an exceptionally diverse array of shapes. Because its development and evolution are quite independent of that of the border ocelli, this element has been given a special name: the parafocal element (Nijhout 1990). Süffert (1929) recognized this as the distal band of the border symmetry system but did not give it a special name, and Schwanwitsch (1924) thought it was actually part of the submarginal band system. The results given below in this paper support Süffert's interpretation, as does the recent work of Otaki and colleagues (Dhungel and Otaki 2009; Otaki 2009, 2011).

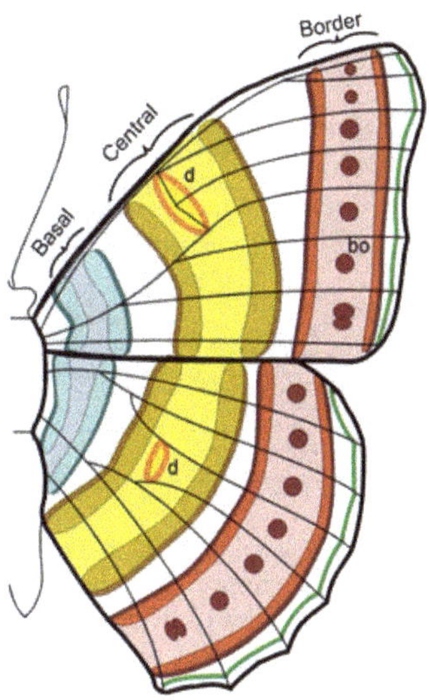

Fig. 1.2 The nymphalid ground plan showing three symmetry systems: basal, central, and border. The border symmetry system has border ocelli (bo) on the compartment midlines. These border ocelli can develop into elaborate eyespots but also into many other shapes. The shape of the distal band of the border symmetry system can also be very diverse, and this band is recognized as the parafocal element

The parafocal elements are developmentally closely related to the border ocelli. Indeed the two are developmentally interdependent in that they appear to arise from a common determination mechanism, although the determinants of their shape are quite different.

1.3 Puzzling Results of Temperature Shock Experiments

A number of investigators have observed that when color pattern aberrations are induced by temperature shock and various chemicals, one of the commonly observed features is a partial or complete fusion of the ocellus and the parafocal element (Otaki 2008; Nijhout 1985, 1991; Nijhout and Grunert 1988). The smooth fusion of these two pattern elements (Fig. 1.3) suggests that that must share a common developmental mechanism. If we interpret the series shown in Fig. 1.3 in reverse order, then it would seem that a single pattern element breaks into two, with the distal one forming the parafocal element and the proximal one the ocellus. None of the current models of color pattern formation can account for this.

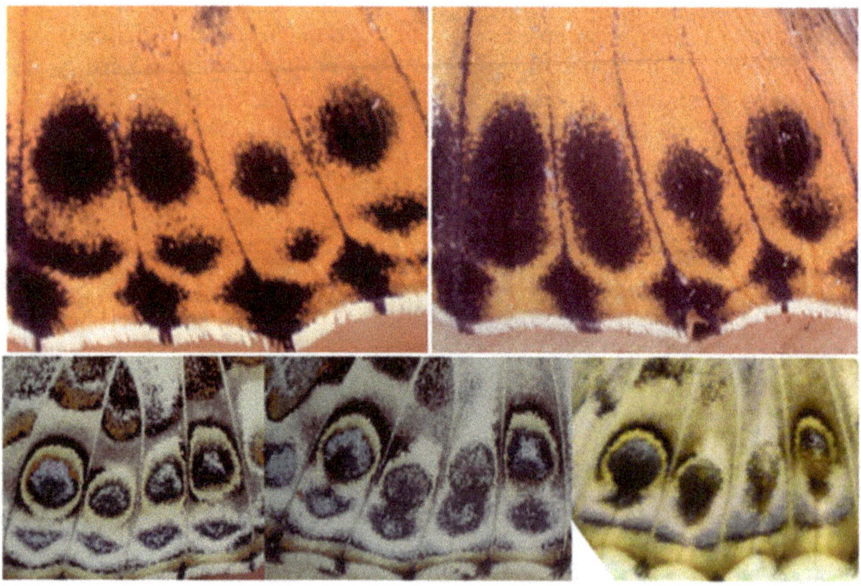

Fig. 1.3 Fusion of ocelli and parafocal elements after temperature shock in *Vanessa cardui*. *Top row*, dorsal surface. *Bottom row*, ventral surface. Normal patterns are on the left in each row. *Bottom row* shows a moderately affected pattern in the middle, and a severely affected pattern in which both pattern elements are completely fused is on the right

1.4 Models of Color Pattern Formation

Previous models for color pattern formation in butterflies have shown that it must be a two-step process. The first step is the establishment of organizing centers, and the second step is the organization of patterns of pigment synthesis by signals produced by these organizing centers. The best known of these organizing centers is the focus, a group of cells that occurs at the center of a canonical eyespot. The foci express both notch and Distal-less, in succession (Carroll et al. 1994; Reed and Serfas 2004), followed by the expression of Spalt and Engrailed in their surrounding, corresponding to the presumptive colored regions of the eyespot (Zhang and Reed 2016; Brunetti et al. 2001).

The mechanism that determines the placement of foci on the wing is still unknown. Foci always occur exactly on the midline of wing compartments delineated by wing veins (i.e., equidistant from the veins). Intervenous stripe patterns (e.g., Fig. 1.6) also occur exactly along the midlines of wing compartments, and in certain papilionids, these stripes break up into spot-like patterns (Nijhout 1991), suggesting a common developmental origin of stripes and spots.

Color pattern determination begins in the wing imaginal disk shortly after the wing venation system is established. The wing imaginal disk is composed of two cell layers, for the dorsal and ventral wing surfaces, respectively. The two cell layers are tightly adhered to each other via a basement membrane. Wing veins develop as tube-like separations between the two layers. The veins are continuous with the hemocoel and allow entry of hemolymph into the developing and growing wing. A special vein called the bordering lacuna (Nijhout 1991) develops around the periphery of the wing imaginal disk and connects the end points of the wing veins (Fig. 1.4).

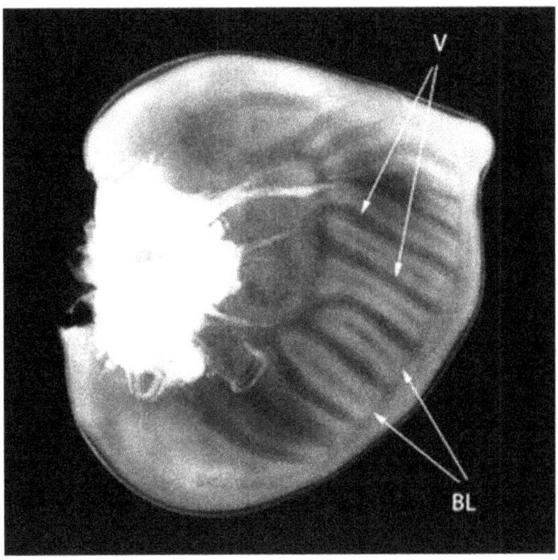

Fig. 1.4 Wing imaginal disk of *Junonia coenia* at the time of color pattern determination. *V* veins, *BL* bordering lacuna. The veins delineate the compartments for pattern formation

The wing veins are bordering lacunae and are the only structural elements in the wing disk when pattern formation begins, and theoretical models of pattern formation assume that these structural elements are the first initiators or organizers for pattern development because they are the only way in which developmental signals can enter the wing (an idea supported by pattern aberrations in veinless mutants (Fig. 1.1). Pattern development including placement of the organizing centers must somehow depend on signals arising from the wing veins and bordering lacuna.

A successful theoretical model for the placement of foci was based on Turing-like reaction diffusion (Turing 1952) using kinetics developed by Meinhardt (1982). The model assumes that a pattern is produced by two chemicals, an autocatalytic activator and an inhibitor, which control each other's synthesis, which can diffuse freely from cell to cell, and in which the inhibitor acts over a larger distance than the activator. Starting with a system at steady state, introducing a small amount of activator from the wing veins, results in a spatial pattern of activator production that rises first as a stripe along the midline between two wing veins in the distal portion of the wing compartment. The end of this midline stripe becomes a particularly strong source of activator production and gradually represses the rest of the stripe, resulting in a stable point-like pattern on the midline resembling the position of a focus. The exact position of the focus as well as the number of foci produced depends on boundary conditions, size of the field, and parameter values of the reaction scheme. This model gained support from the finding that it predicted the spatial sequence of expression of the gene Distal-less, one of the early determinants of color pattern, almost precisely (Fig. 1.5) (Nijhout 1990, 2010).

The spatial pattern of point-like foci and various line-like distributions of the presumptive activator is then used in the second stage of pattern formation to induce the synthesis of specific pigments. A simple diffusion-threshold mechanism using these activator distributions as the origins of new diffusible morphogens proved sufficient to explain almost the entire diversity of color patterns found in the butterflies (Nijhout 1990).

There is, however, a significant problem with this model and, in particular, with the reaction-diffusion mechanism that sets up the initial prepattern of activator distributions. Reaction-diffusion mechanisms are notoriously sensitive to field size and to the exact choice of parameter values and boundary conditions. Even small changes in any of these factors can produce extremely different spatial patterns of activator distribution. Reaction-diffusion mechanisms are particularly sensitive to the size of the field and produce wildly different patterns in fields of different sizes. This seems biologically unrealistic. Biological systems tend to be quite robust to parameter variation and size variation, such as produced by the abundant and often severe genetic and environmental variation to which organisms are subject (Nijhout 2002). In particular, in butterflies, identical patterns often develop in adjoining wing compartments of very different dimensions. Finally, although Turing-style reaction-diffusion mechanisms can be made to produce a wide diversity of realistic patterns, there are, with the possible exception of some fish pigment patterns, no

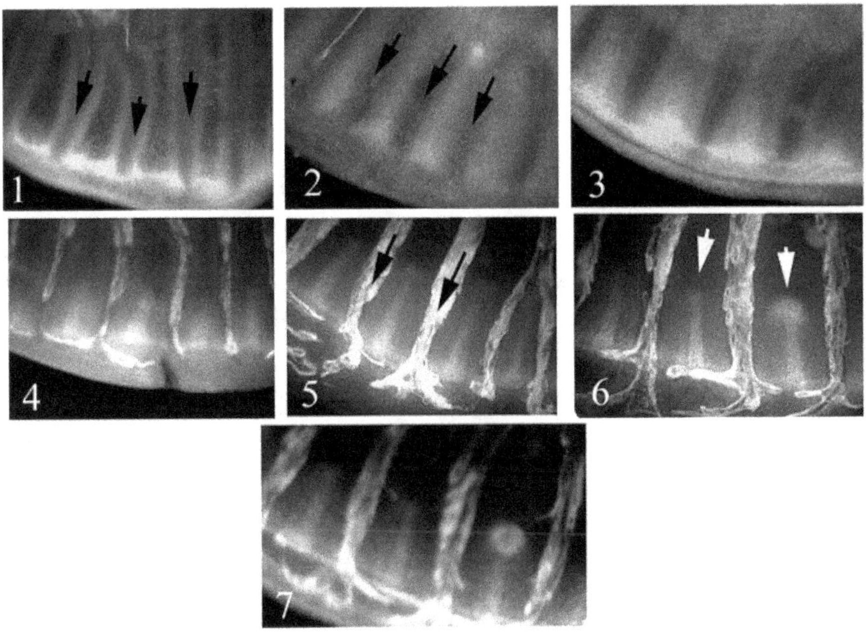

Fig. 1.5 Time series of the development of the expression pattern of Distal-less in the imaginal wing disk of *Junonia coenia*. *Black arrows* indicate the position of wing veins. *White arrows* point to the developing stalks and spots of the Distal-less. Initially Distal-less is expressed along the wing veins and wing margin (*Plate 1*), but then the expression becomes gradually concentrated to the wing compartment midline (*Plates 2–5*). A spot develops at the tip of the midline bar in wing compartment that will develop an ocellus, and the midline bar gradually disappears (*Plates 6–7*)

instances in which they have been experimentally proven to operate during development and in which the activator and inhibitor have been identified (Kondo and Miura 2010).

This has led me to search for a simpler and more robust mechanism that could produce the diversity of color patterns observed. Developmental genetic studies of embryonic development have revealed a broad array of gene regulatory networks that produce dynamically changing spatial patterns of gene expression, in which the product of one gene acts as a transcriptional regulator of one or more other genes. The effect of a gene spreads either by diffusion of the gene product to adjoining cells or by cell-surface signaling interaction among neighboring cells.

These mechanisms for pattern formation are conceptually and physically simple. They are in effect *diffusion-threshold mechanisms*, in which a substance diffuses away from the cells where it is produced and exerts its effect when it rises above a threshold in surrounding cells. These diffusion-threshold mechanisms can be generalized into what I'll call a *grass-fire model.*

1.5 The Grass-Fire Model

The model consists of the simplest possible set of reactions. A molecule we will call fuel is initially distributed across the field and serves as substrate for the first reaction to produce the product P1. P1 in turn serves as the substrate for the production of P2 and so forth. The model is given by:

$$\partial \text{fuel}/\partial t = -\text{k1}*\text{fuel}*\text{P1} + \text{D}_{\text{fuel}}*\nabla^2\text{fuel}$$

$$\partial \text{P1}/\partial t = \text{k1}*\text{fuel}*\text{P1} - \text{k2}*\text{P1} + \text{D}_{\text{P1}}*\nabla^2\text{P1}$$

$$\partial \text{P2}/\partial t = \text{k2}*\text{P1} - \text{k3}*\text{P2} + \text{D}_{\text{P2}}*\nabla^2\text{P2}$$

Initially there is only fuel, and the patterning mechanism is initiated when P1 is introduced at some point in the field, for instance, along the margins of the field. The model resembles a grass-fire with a fire front, initiated at the ignition point where P1 is introduced, that consumes fuel and leaves combustion products behind, some of which can be used in other reactions. In addition to these reactions, we assume that all chemicals can diffuse from areas of high concentration to low concentration. We assume for the present that all reactions are mass action. Thus we have an exceptionally simple reaction-diffusion system.

In the course of time fuel is depleted, as are all subsequent metabolites. This system does not produce a stable end pattern but rather a slowly changing spatial pattern of values of the three variables. In this respect it resembles the early gene expression patterning events in the *Drosophila* embryo in which a successive series of diffusion gradient-threshold events produce a dynamically progressing spatial pattern of gene expression (Tomancak et al. 2002). We assume that an independent event "reads" the spatial pattern of chemicals at some time point in the development. In butterflies this could be the ecdysone signals that initiate a molt or the wandering stage, both of which occur during the period of color pattern formation and also control growth and morphogenesis of the wing imaginal disk.

The nature of fuel, P1 and P2, is undetermined. Any system with mass action kinetics will do, nor are the kinds of kinetics restricted to mass action. Saturation kinetics like Michaelis-Menten and Hill produce the same patterns as mass-action kinetics over a range of parameter values. The reactions could therefore represent a biochemical reaction sequence, a gene activation sequence, a successive activation of signaling cascades, or a combination of these.

1.6 Basic Patterns

We assume the field is a rectangle that represents a compartment in the wing imaginal disk, where the top and long sides are wing veins and the bottom short side is the bordering lacuna. The reactions can be initiated only along these edges. Variation in pattern can come about by a variation in the position of the initiation points (along the entire margin or only near the proximal, middle, or distal ends), the initial distribution of fuel (homogeneous, proximodistal gradient, vein to mid-line gradient), and the distribution of the enzymes or rate constants, that run the reactions (homogeneous, proximodistal gradient, vein-to-midline gradient).

1.7 Venous and Intervenous Patterns

Some of the simplest and most widespread patterns are stripes that run along the midline of a compartment and patterns that run parallel to the wing veins. Figure 1.6 illustrates several examples. The patterns show that the veins do not induce pattern along their entire length. In Fig. 1.6a the pattern is only induced in the mid-region of the vein but not near the proximal and distal ends. There is often a proximodistally graded width of the venous bands suggesting (e.g., Fig. 1.6d–e) that the strength of induction, or the propagation rate of the inductive signal, is graded. These patterns are readily produced by the grass-fire model, as illustrated in Fig. 1.7. A proximodistal gradient of reaction rate constants produces venous bands that taper along the length of the vein (Fig. 1.7c). Intervenous stripes (Fig. 1.7a) can be made if the entire wing vein induces the pattern and both the fuel and reaction rates are homogeneously distributed. Reed and Serfas (2004) have shown that in butterflies without eyespots, but with intervenous stripes, there is a long central midline stripe of notch and Distal-less expression. Notch and Distal-less also specify the position of eyespot foci (see below), thus the patterns of P1 and P2 may simulate the expression of these two peptides.

1.8 Simulation of Notch and Distal-Less Progression

The progression of Distal-less expression (Fig. 1.5), beginning with a short midline stripe of the emerging from the margin, followed by the development of a spot at the apex of the stripe, followed by a regression of the stripe, leaving a spot of Distal-less expression behind, was accurately reproduced by a Turing-style reaction diffusion program (Nijhout 1990). Indeed, it provided strong, albeit circumstantial, support for reaction diffusion as the underlying mechanism of focus formation.

Fig. 1.6 Vein-dependent patterns. *Top row* shows venous patterns of *Anaxita decorata* (**a**) and intervenous patterns of *Pseudacraea lucretia* (**b**) and *Eteone eupolis* (**c**). Bottom row (**d–f**) shows individual variation in *Danaus affinis*. In *Danaus* the *white venous* pattern varies in the extent to which it expands from the wing veins

Reed and Serfas (2004) and Zhang and Reed (2016) have shown that this pattern of Distal-less expression is preceded by an almost identical pattern of notch expression.

The grass-fire model produces both pattern sequences (Fig. 1.8), simply by assuming that only the distal portion of the wing veins acts as initiation sources and that the fuel is distributed in a shallow gradient that is higher near the midline than near the veins. The shape of the focal spot is slightly elongated across the long axis of the wing compartment, just as the expression of notch and Distal-less described by Reed and Serfas (2004) and Zhang and Reed (2016). The pattern of P2 is identical to that of P1 but lags behind a little, and P2 still has a stalk when P1 is already resolved into a spot (Fig. 1.8). Thus the progression of P1 and P2 resemble those of notch and Distal-less, respectively.

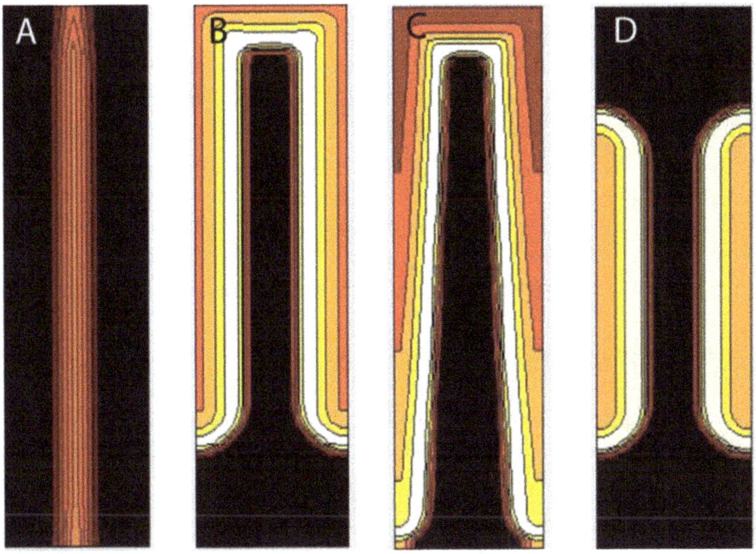

Fig. 1.7 Model simulation for venous and intervenous patterns. In each case the wing veins were used as the initiation points and the "fuel" was either homogeneously distributed or graded slightly from top to bottom (proximal to distal)

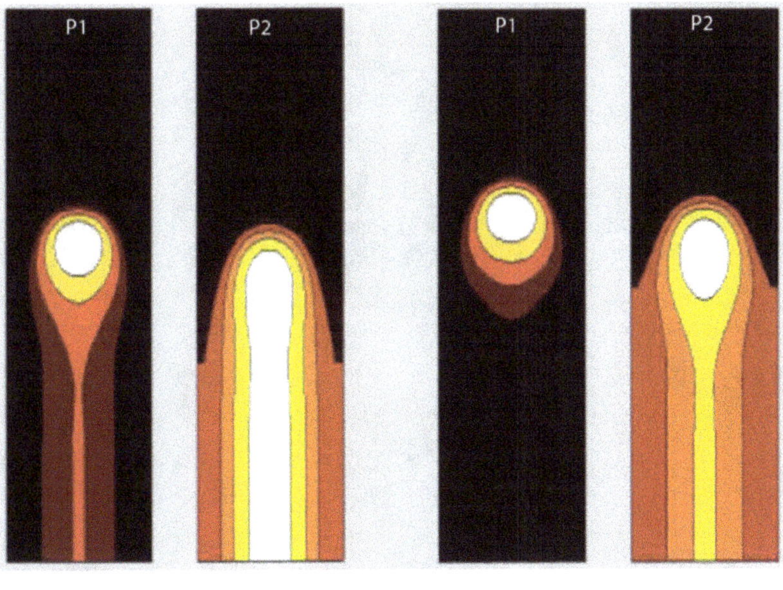

Fig. 1.8 Model simulations of focus formation. Two runs are shown with slightly different initial distributions of "fuel." The distributions of P1 and P2 are shown, which could correspond to the notch and Distal-less, respectively. The two patterns differ in the shape of the lateral gradient of the "stalk," which affects the shape of the parafocal element that will develop

1.9 Shape of the Parafocal Elements

As noted above, once the foci are established, the second step in color pattern formation is a signal that originates from the foci and that specifies a pattern of pigment biosynthesis in their surroundings. We use the grass-fire model for this second step as well, using the focus as the initiation point.

If the grass-fire model is started from a single point source, the pattern produced naturally breaks into two fronts, moving distally and proximally, respectively. If the initial substrate that is used is homogeneously distributed, a circular pattern will form that breaks into two semicircular arcs that move away from the initiation point.

A characteristic feature of the parafocal elements is that they are always symmetrical around the wing compartment midline and are often Λ, V, W, or M shaped (e.g. Fig. 1.9), suggesting a special function of the midline in shaping this element. If the parafocal element is formed by a moving reaction front, then movement near

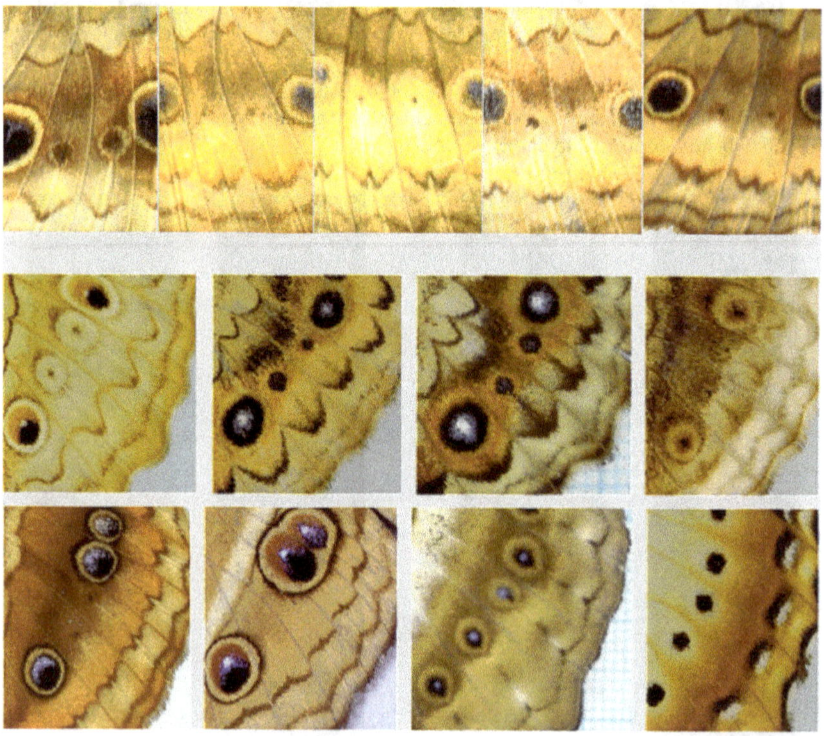

Fig. 1.9 Variation and diversity of parafocal element shapes. *Top row*, individual variation in *Junonia coenia*. *Bottom two rows*, diversity of parafocal elements in selected Junoniini (*middle row J. atlites, J. villida, J. villida, J. oenone. Bottom row J. genoveva, J. almana, Yoma algina, Precis ceryne*)

the midline and/or the veins must be either more rapid or slower than movement elsewhere. One way to accomplish this is by having a required metabolite or precursor to the reaction distributed in a pattern that is symmetrical to the midline. A clear candidate for this is the gradient left behind by the midline pattern that preceded the formation of the focal spot (Fig. 1.8). This midline concentration gradient decays only gradually, and its profile depends on the parameter values and initial fuel distribution.

The hypothesis then is that the shape of the parafocal elements is determined by a gradient left behind by the process that formed the focus. This idea can be tested computationally. Figures 1.10a and 1.11 show a sample of the diversity of parafocal element shapes that can be produced by this model. Although these shapes closely mimic those of real parafocal elements (e.g., Fig. 1.9), the shape of the ocellus is not circular, as would typically be the case.

To produce both perfectly circular eyespots and the right diversity of parafocal element shapes, it is necessary to assume that the focus could be the source of two different signals (one perhaps initiated by notch and the other by Distal-less) that use different substrates. If one signal uses a homogeneously distributed substrate, it will produce a circular eyespot (Fig. 1.10b), and if the other uses the gradient left behind by the focus-forming process, it produces the parafocal element. Interestingly, this second source also produces an arc-shaped pattern on the proximal side of the eyespot (Fig. 1.10c–g). This finding is consistent with Süffert's idea that the parafocal element is the distal band of the border symmetry system: the parafocal element and the proximal arc produced by the second source make up paired bands of the border symmetry system. These model results also support the ideas about the nature of parafocal elements and border symmetry systems proposed by Otaki (Dhungel and Otaki 2009; Otaki 2009, 2011).

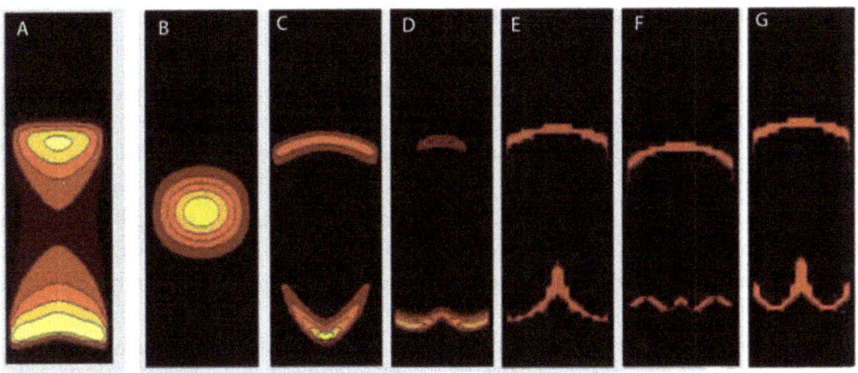

Fig. 1.10 Simulations of pattern generated by focal sources. (**a**) A single source breaks up into an ocellus and a parafocal element, but the ocellus is not circular. (**b–g**) A double source at the focus, one producing the eyespot (**b**) and the other producing the parafocal element and the proximal arc-shaped band of the border symmetry system (**c–g**)

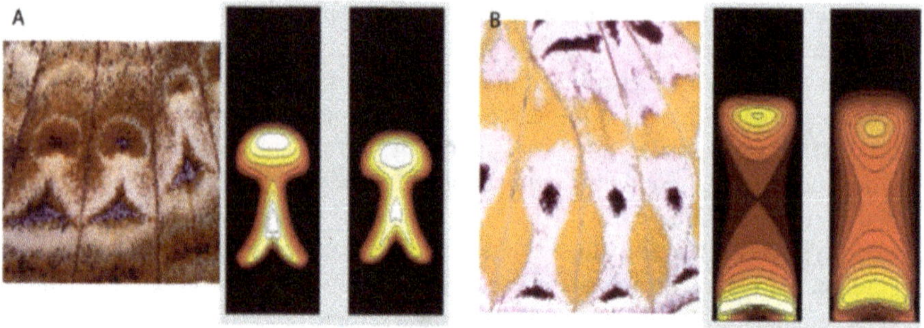

Fig. 1.11 Simulations of patterns that can be generated with a single focal source resembling those of *Vanessa tameamea* (**a**) and *Euryphura concordia* (**b**)

1.10 Fusion and Separation of Ocelli and Parafocal Elements

When the pupae of butterflies are exposed to a temperature shock, many individuals exhibit a fusion between the ocellus and the parafocal element. The degree of fusion is quite variable from individual to individual, and in extreme cases the two fuse into a single pattern element (Fig. 1.3). A possible reason for this effect is that temperature shock freezes the progression of pattern determination, possibly by activating heat shock or stress proteins that stop biosynthetic or transcriptional activity (Mitchell and Lipps 1978; Crews et al. 2016; Welte et al. 1995). The grass-fire model shows that a single pattern element can split into two and that both ocelli and parafocal elements can be produced from a common source.

1.11 Modes of Pattern Evolution

The developing pattern depends on only a few variables: the kinetic parameters of the reactions and the initial gradients of fuel. For all models explored here, these gradients are simple. Beside homogeneous distributions, we used smooth proximodistal gradients or smooth gradients symmetrical to the wing compartment midline, parallel to the wing veins. The latter could be readily set up by diffusion from, or absorption by, the wing veins. Thus the anatomical features of the wing, the wing veins and bordering lacunae, are the only features used to initiate pattern formation.

A significant way in which the proposed patterning mechanism differs from the assumptions of a typical Turing-style reaction-diffusion mechanism is that the system is never at steady state, but the pattern slowly changes over time. The developing pattern becomes fixed, so to speak, by an event such as a pulse of hormone secretion that begins or ends a developmental period, as occurs at several points during insect metamorphosis (Nijhout 1994, 1999; Nijhout et al. 2014). This property is consistent also with the progressive time-varying patterns of gene expression during embryonic development (Tomancak et al. 2002).

This feature also adds a mode of pattern evolution. Pattern evolution could typically occur due to changes in parameter value reaction rates and gradient shapes. But it is also possible that evolutionary changes in the time when a developing pattern is frozen can lead to changes in the final color pattern. This adds a flexible mode of heterochromic evolution.

Moreover, if, as suggested above, the fixation of pattern depends on the timing of hormone secretion, this mechanism could also account for seasonal polyphenisms of butterfly color patterns. Seasonal polyphenisms in color patterns come about through changes in the timing of ecdysone secretion (Rountree and Nijhout 1995; Brakefield et al. 1998; Koch et al. 1996; Koch and Bückmann 1987) and thus may fix the progression of pattern at different stages. On this view, seasonally polyphenic patterns can be thought of as an expression of plastic heterochrony. Once a plastic pattern switch is established, additional adaptive changes in the patterning system can evolve to refine or further alter the pattern.

Acknowledgments This work was supported by grants IOS-0641144, IOS-1121065, and IOS-155734 from the National Science Foundation.

References

Baxter SW, Papa R, Chamberlain N, Humphray SJ, Joron M, Morrison C, Ffrench-Constant RH, Mcmillan WO, Jiggins CD (2008) Convergent evolution in the genetic basis of Mullerian mimicry in *Heliconius* butterflies. Genetics 180:1567–1577

Brakefield P, Gates J, Keys D, Kesbeke F, Wijngaarden P, Monteiro A, French V, Carroll S (1996) Development, plasticity and evolution of butterfly eyespot patterns. Nature 384:236–242

Brakefield P, Kesbeke F, Koch P (1998) The regulation of phenotypic plasticity of eyespots in the butterfly *Bicyclus anynana*. Am Nat 152:853–860

Brunetti CR, Selegue JE, Monteiro A, French V, Brakefield PM, Carroll SB (2001) The generation and diversification of butterfly eyespot color patterns. Curr Biol 11:1578–1585

Carroll S, Gates J, Keys D, Paddock S, Panganiban G, Selegue J, Williams J (1994) Pattern formation and eyespot determination in butterfly wings. Science 265:109–114

Crews SM, Mccleery WT, Hutson MS (2016) Pathway to a phenocopy: heat stress effects in early embryogenesis. Dev Dyn 245:402–413

Dhungel B, Otaki JM (2009) Local pharmacological effects of tungstate on the color-pattern determination of butterfly wings: a possible relationship between the eyespot and parafocal element. Zool Sci 26:758–764

Joron M, Jiggins CD, Papanicolaou A, Mcmillan WO (2006) Heliconius wing patterns: an evo-devo model for understanding phenotypic diversity. Heredity 97:157–167

Kapan DD, Flanagan NS, Tobler A, Papa R, Reed RD, Gonzalez JA, Restrepo MR, Martinez L, Maldonado K, Ritschoff C, Heckel DG, Mcmillan WO (2006) Localization of Mullerian mimicry genes on a dense linkage map of *Heliconius erato*. Genetics 173:735–757

Koch P, Bückmann D (1987) Hormonal control of seasonal morphs by the timing of ecdysteroid release in *Araschnia levana* L. (Nymphalidae: Lepidoptera). J Insect Physiol 33:823–829

Koch PB, Nijhout HF (2002) The role of wing veins in colour pattern development in the butterfly *Papilio xuthus* (Lepidoptera: Papilionidae). Eur J Entomol 99:67–72

Koch P, Brakefield P, Kesbeke F (1996) Ecdysteroids control eyespot size and wing color pattern in the polyphenic butterfly *Bicyclus anynana* (Lepidoptera: Satyridae). J Insect Physiol 43:223–230

Kondo S, Miura T (2010) Reaction–diffusion model as a framework for understanding biological pattern formation. Science 329:1616–1620

Martin A, Reed RD (2014) Wnt signaling underlies evolution and development of the butterfly wing pattern symmetry systems. Dev Biol 395:367–378

Meinhardt H (1982) Models of Biological pattern formation. Academic, London

Mitchell H, Lipps L (1978) Heat shock and phenocopy induction in Drosophila. Cell Adhes Commun 15:907–918

Monteiro A (2015) Origin, development, and evolution of butterfly eyespots. Annu Rev Entomol 60:253–271

Monteiro A, Brakefield PM, Vernon F (1997) Butterfly eyespots: the genetics and development of the color rings. Evolution 51:1207–1216

Monteiro A, Prijs J, Bax M, Hakkaart T, Brakefield PM (2003) Mutants highlight the modular control of butterfly eyespot patterns. Evol Dev 5:180–187

Nadeau NJ (2016) Genes controlling mimetic colour pattern variation in butterflies. Curr Opin Insect Sci 17:24–31

Nijhout HF (1980) Pattern formation on lepidopteran wings: determination of an eyespot. Dev Biol 80:267–274

Nijhout HF (1985) Cautery induced colour patterns in *Precis coenia* (Lepidoptera: Nymphalidae). J Embryol Exp Morphol 86:191–203

Nijhout HF (1990) A comprehensive model for colour pattern formation in butterflies. Proc R Soc Lond B Biol Sci 239:81–113

Nijhout HF (1991) The development and evolution of butterfly wing patterns. Smithsonian Institution Press, Washngton, DC

Nijhout HF (1994) Insect hormones. Princeton, Princeton University Press

Nijhout HF (1999) Control mechanisms of polyphenic development in insects. Bioscience 49:181–192

Nijhout HF (2002) The nature of robustness in development. BioEssays 24:553–563

Nijhout HF (2010) Molecular and physiological basis of colour pattern formation. In: Jérôme C, Stephen JS (eds) Advances in insect physiology. Academic, London

Nijhout HF, Grunert LW (1988) Colour pattern regulation after surgery on the wing disks of *Precis coenia* (Lepidoptera: Nymphalidae). Development 102:337–385

Nijhout HF, Riddiford LM, Mirth C, Shingleton AW, Suzuki Y, Callier V (2014) The developmental control of size in insects. Wiley Interdiscip Rev Dev Biol 3:113–134

Otaki JM (2008) Phenotypic plasticity of wing color patterns revealed by temperature and chemical applications in a nymphalid butterfly Vanessa indica. J Therm Biol 33(2):128–139

Otaki JM (2009) Color-pattern analysis of parafocal elements in butterfly wings. Entomol Sci 12:74–83

Otaki JM (2011) Generation of butterfly wing eyespot patterns: a model for morphological determination of eyespot and parafocal element. Zool Sci 28:817–827

Reed RD, Serfas MS (2004) Butterfly wing pattern evolution is associated with changes in a notch/distal-less temporal pattern formation process. Curr Biol 14:1159–1166

Reed RD, Papa R, Martin A, Hines HM, Counterman BA, Pardo-Diaz C, Jiggins CD, Chamberlain NL, Kronforst MR, Chen R, Halder G, Nijhout HF, Mcmillan WO (2011) *optix* Drives the repeated convergent evolution of butterfly wing pattern mimicry. Science 333:1137–1141

Rountree DB, Nijhout HF (1995) Hormonal control of a seasonal polyphenism in *Precis coenia* (Lepidoptera: Nymphalidae). J Insect Physiol 41:987–992

Schwanwitsch BN (1924) On the ground-plan of wing-pattern in Nymphalids and certain other families of the Rhopaloeerous Lepidoptera. Proc Zool Soc Lond 94:509–528

Schwanwitsch BN (1929) Two schemes of the wing-pattern of butterflies. Z Morphol Okol Tiere 14:36–58

Süffert F (1929) Die Ausbildung der imaginalen Flügelschnittes in der Schmetterlingspuppe. Z Morphol Okol Tiere 14:338–359

Tomancak P, Beaton A, Weiszmann R, Kwan E, Shu S, Lewis SE, Richards S, Ashburner M, Hartenstein V, Celniker SE, Rubin GM 2002 Systematic determination of patterns of gene expression during *Drosophila* embryogenesis. Genome Biol 3, research0088.1-88.14.

Turing AM (1952) The chemical basis of morphogenesis. Philos Trans R Soc Lond Ser B Biol Sci 237:37–72

Welte MA, Duncan I, Lindquist S (1995) The basis for a heat-induced developmental defect: defining crucial lesions. Genes Dev 9:2240–2250

Zhang L, Reed RD (2016) Genome editing in butterflies reveals that spalt promotes and distal-less represses eyespot colour patterns. Nat Commun 7:11769

2

Pattern Elements of Nymphalid Ground Plan Among Satyrinae Tribes

Carla M. Penz

Abstract Based on the most recent nymphalid phylogeny, the Satyrinae can be tentatively organized into the species-rich tribe Satyrini plus a clade that includes the Morphini, Brassolini, Haeterini, Elymniini, Melanitini, Dirini, Zetherini, and Amathusiini. Members of the latter eight tribes have the largest body sizes within Satyrinae and also show extraordinary wing pattern variation. Representatives of these tribes are illustrated herein, and pattern elements of the nymphalid ground plan are identified. Five themes are briefly discussed in light of their pattern diversification: (1) central symmetry system dislocations, (2) variation in ventral hind wing ocelli, (3) the color band between elements *f* and *g*, (4) sexual dimorphism and mimicry, and (5) transparency. Within an ecological and evolutionary standpoint, selected genera are provided as examples to explore wing patterns involved in male mating displays, camouflage, and mimicry.

Keywords Pierellization • Ocelli • Sexual dimorphism • Mimicry • Camouflage • Transparency • Mating behavior

2.1 Introduction

The evolution of adult diurnal activity in Lepidoptera paved the way for the widespread use of color for intra- and interspecific signaling (Grimaldi and Engel 2005; Kemp et al. 2015). Following approximately 90 million years of morphological and species diversification (Wahlberg et al. 2009), butterflies in the family Nymphalidae have played an important role in our understanding of how wing color patterns mediate intraspecific interactions and also the evolution of aposematism, mimicry, and camouflage (Vane-Wright and Ackery 1984; Chai 1990; Nijhout 1991; Rutowski 1991). Whether they target conspecifics or other animals, the evolutionary diversification of butterfly color signals involved impressive modifications of wing pattern elements (WPEs hereafter).

C.M. Penz (✉)
Department of Biological Sciences, University of New Orleans, 2000 Lakeshore Dr., New Orleans, LA 70148, USA
e-mail: cpenz@uno.edu

The characterization of a ground plan that identifies individual pattern components across butterfly wings provided a useful framework for research on development, genetics, and evolution (Schwanwitsch 1924; Süffert 1927; Nijhout 1991 and references therein). Border ocelli are the best studied of all individual WPEs possibly because they are conspicuous and ubiquitous in the family Nymphalidae. The Satyrinae constitutes an excellent group to study variation in the border ocelli alone and also how different WPEs can become integrated to produce particular visual effects.

Most Satyrinae species are small bodied and relatively uniform in appearance, such as members of the tribe Satyrini (85% of the species in the subfamily, Peña and Wahlberg 2008). There are, however, noticeable exceptions. Large-bodied species are grouped in a clade that includes the Brassolini (Fig. 2.1), Morphini (Fig. 2.2), Haeterini (Fig. 2.3), Elymniini (Fig. 2.4), Melanitini (Fig. 2.4), Dirini (Fig. 2.5), Zetherini (Fig. 2.6), and Amathusiini (Fig. 2.7; Wahlberg et al. 2009). Exhibiting remarkable color diversification, these butterflies form the focus of this chapter to provide the first detailed comparison among early satyrine tribes. Representatives were selected for an examination of both ventral and dorsal WPEs (see Nijhout 1991 for terminology), and a list of examined species is given in Appendix. Five themes are briefly described and illustrated and, as much as possible, discussed within the context of the natural history and behavior of the butterflies. More detailed accounts will be presented elsewhere (Penz in prep.).

2.2 Central Symmetry System Dislocations in Forewing and Hind Wing

The term pierellization (Schwanwitsch 1925) refers to the dislocation of elements that pertain to the central symmetry system in such a way that distal elements below vein M_3 align themselves with proximal ones located above such vein. This is visible in the ventral forewings of several species in the Brassolini, Morphini, Haeterini, and Dirini (Figs. 2.1c, e, 2.2d–e, 2.3b and 2.5a), and it varies within and between genera. Taking the genus *Pierella* as an example, the anterior dislocation of element *f* below forewing vein M_3 is found in species with rather plain ventral coloration (e.g., *P. lamia* in Fig. 2.3b; also *luna* and *hortona*, not illustrated). Such dislocation disrupts the interplay between *f* and *g*, which seem to serve as boundaries for a light-colored band that occurs in their congenerics (see below). In *Pierella* species that show ventral forewing pierellization of *f*, elements *f* and *g* are also broadly separated on the ventral hind wing (Fig. 2.3b).

Although pierellization seems to be less common on the ventral hind wing, it occurs in some species that display dead leaf camouflage (e.g., *Caerois gerdrudtus* Fig. 2.2a) or parallel bars (*Morpho marcus*, Fig. 2.2d–e). Some camouflaged species, however, do not show hind wing dislocation of element *f* (e.g., *Amathuxidia amythaon*, Fig. 2.7b), suggesting that ventral camouflage evolved

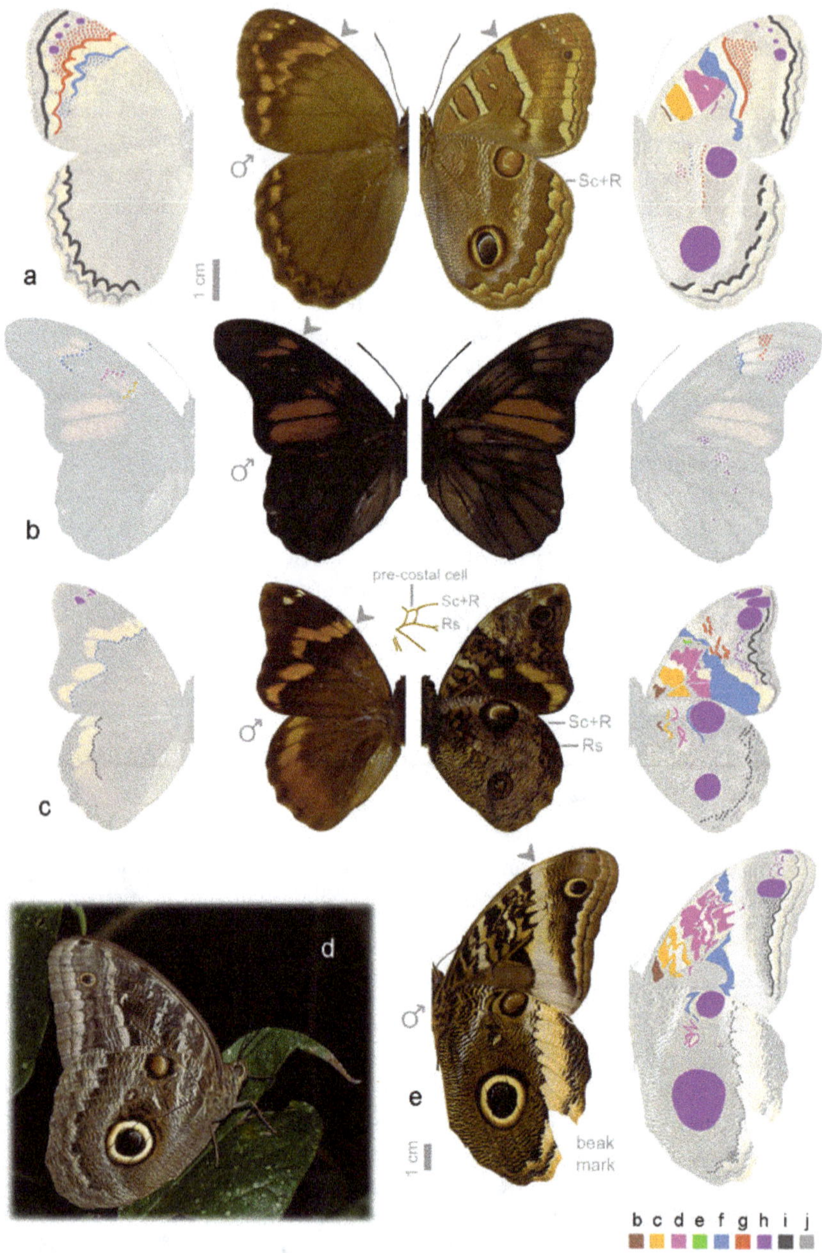

Fig. 2.1 Color-coded wing pattern elements in selected Brassolini. Left side of butterfly image in dorsal view, right side in ventral view. Gray arrows indicate colorful band associated with element *f*. (**a**) *Opoptera syme*. (**b**) *Penetes pamphanis*. (**c**) *Opsiphanes sallei*, note venation detail showing precostal cell present at the base of the hind wing. (**d**) *Caligo illioneus* male perched on leaf (photo by David Powell). (**e**) *Caligo atreus*. All butterflies at the same scale except *C. atreus*

Fig. 2.2 Color-coded wing pattern elements in selected Morphini. Left side of butterfly image in dorsal view, right side in ventral view. Gray arrow indicates colorful band associated with element *f*. (**a**) *Caerois gerdrudtus*. (**b**) *Morpho sulkowskyi*, ventral pattern elements are visible in dorsal view in this semitransparent species. (**c**) *Morpho hecuba*. (**d**) and (**e**) *Morpho marcus*. All butterflies at the same scale except *M. hecuba*

independently multiple times. In the case of *M. marcus*, the comparison of male and female ventral hind wing patterns was helpful to identify the alignment and amalgamation of elements *f* and *d* to produce broad bars (compare Fig. 2.2d–e to *Amathusia phiddippus* in Fig. 2.7e). When *M. marcus* butterflies are at rest, the hind wing bars visibly converge toward an enlarged tornus where the eye-catching parfocal elements seem to be forming a deflection point for predator attack (Fig. 2.2d–e; also present in other species, Fig. 2.2b–c), a pattern that evolved independently in members of the Amathusiini (Fig. 2.7b, e).

2.3 Variation in Ventral Hind Wing Ocelli

Ocelli can take many forms within the Nymphalidae (Nijhout 1991). Species in the eight studied tribes show a broad range of variation, while some species display a complete series at the postmedial area of the wing stereotypical of the nymphalid ground plan (e.g., *Ethope himachala*, Fig. 2.6e; *Faunis eumeus*, Fig. 2.7c); in others the ocelli are markedly reduced (e.g., *Penetes pamphanis*, Fig. 2.1b). Although various types of ocelli are found in members of all tribes, here I limit my discussion to three aspects of the ventral hind wing ocelli: the location of the first ocellus of the series, proximal dislocation of the ocelli, and their use in signaling.

In most members of the eight tribes, the first conspicuous ocellus of the ventral hind wing series is located below vein Rs (Figs. 2.2c, 2.4b–d, 2.5a and 2.7a–b), but there are notable exceptions. In all Brassolini species with well-developed ocelli, the first ocellus is found below Sc + R (Fig. 2.1a, c). All members of Brassolini have a precostal cell (Fig. 2.1c), which increases the distance between Sc + R and Rs, and provides physical space for a well-developed ocellus. Although the function of the precostal cell is unknown, this points to a possible association between wing venation and color pattern in Brassolini. Furthermore, in some Brassolini species this ocellus expands beyond the cell where it originates, suggesting selection for larger size (Fig. 2.1a, c, e). Some members of the Dirini also have a well-developed ocellus below Sc + R, and that of *Paralethe dendrophilus* is particularly large (Fig. 2.5b). In this species the base of Rs is separated from Sc + R, which increases cell height in an analogous way to what is found in Brassolini. Finally, in the transparent Haeterini *Dulcedo*, *Pseudohaetera*, *Haetera*, and *Cithaerias* the first ocellus is located below M_1 (Fig. 2.3e–f), a pattern unique to these taxa.

Border ocelli are usually located in the postmedial area, but dislocations occur in several taxa. Proximal dislocations are more common than distal ones, and the former are associated with a corresponding shift of central symmetry system WPEs. Notable proximal dislocations are found in taxa of Brassolini, Morphini, and Amathusiini (Figs. 2.1, 2.2 and 2.7). In many Brassolini and also *Morpho*, the hind wing ocelli are clearly positioned in the medial area of the wing, which can produce a striking visual effect depending on their size (Penz and Mohammadi 2013; Figs. 2.1d–e and 2.2c). Ocelli dislocations can be uneven with the first, or first and second, ocelli taking a more proximal position than the remaining of the series

Fig. 2.3 Color-coded wing pattern elements in selected Haeterini. Left side of butterfly image in dorsal view, right side in ventral view. Gray arrows indicate colorful band associated with element

(Figs. 2.6e and 2.5b). Finally, the hind wing ocelli are uniquely dislocated distally in the transparent Haeterini genera by being positioned very near the wing margin (Fig. 2.3e–f). The ocellus below M_1 becomes highly visible when these transparent butterflies alight with their wings closed.

The ventral ocellus located at the hind wing tornus has been hypothesized to function as a defense, either a deflection point in the event of a predator attack or a startle mechanism that prevents or delays attacks (DeVries 2002, 2003; Hill and Vaca 2004; Stevens 2005). Although these hypotheses are compelling, my field observations suggest that in some taxa, ventral hind wing ocelli might have an additional function. Males of some *Caligo* species aggregate at leks along forest edges to wait for virgin females (Freitas et al. 1997, Srygley and Penz 1999' Fig. 2.1d). As they fly into the lek, the large ventral ocelli appear to help airborne females locate perched males (pers. obs.), suggesting a potential function in male-female interactions. *Pierella lucia* has two large white ocelli at the hind wing tornus that show perfect dorsoventral correspondence, likely enhancing light reflection (Fig. 2.3a, c). Hill and Vaca (2004) demonstrated that the hind wing tornus of *Pierella lucia* is weaker than surrounding wing areas, thus supporting the deflection hypothesis (see beak marks in Fig. 2.3a). Nonetheless I once observed the complex courtship behavior of this species. While a female was perched on a leaf, a male hovered in her view, beating the forewings only and keeping the hind wings open and motionless. The male clearly displayed the ventral hind wing ocelli to the female as he repeatedly dipped closer and closer to her. Dorsal ocelli have been considered more important during mating displays (e.g., Oliver et al. 2009), but my observations suggest that ventral ocelli may also be used in this context. In the case of both *Caligo* and *Pierella lucia*, it is possible that both natural and sexual selection could be operating concomitantly on the ventral hind wing ocelli. This is perhaps the case in other species as, for example, male *Faunis phaon leucis* that has larger ventral ocelli than the female (Fig. 2.7d; note that dorsal ocelli are absent in *Faunis*).

2.4 The Color Band Between Elements *f* and *g*

Many nymphalid butterflies have a conspicuous, forewing band that constitutes a highly visible component of the dorsal, and sometimes ventral, coloration (e.g., *Melanitis amabilis*, Fig. 2.4d). This band is common among the species studied here

Fig. 2.3 (continued) f. (**a**) *Pierella lucia*, note multiple beak marks on the hind wing tornus (photo by Andrew Neild). (**b**) *Pierella lamia*. (**c**) *Pierella lucia*. (**d**) *Pierella helvina*. (**e**) *Cithaerias aurora*. (**f**) details of the dorsal hind wing of *Haetera piera*: the ventral orange scales in the ocellus are visible dorsally through transparency; element *g* is expressed on the wing membrane. All butterflies are at the same scale

Fig. 2.4 Color-coded wing pattern elements in selected Elyminiini and Melanitini. Left side of butterfly image in dorsal view, right side in ventral view. Gray arrows indicate colorful band associated with element *f*. (**a**) and (**b**) *Elymnias hypermnestra*. (**c**) *Elymnias patna*. (**d**) *Melanitis amabilis*. All butterflies are at the same scale

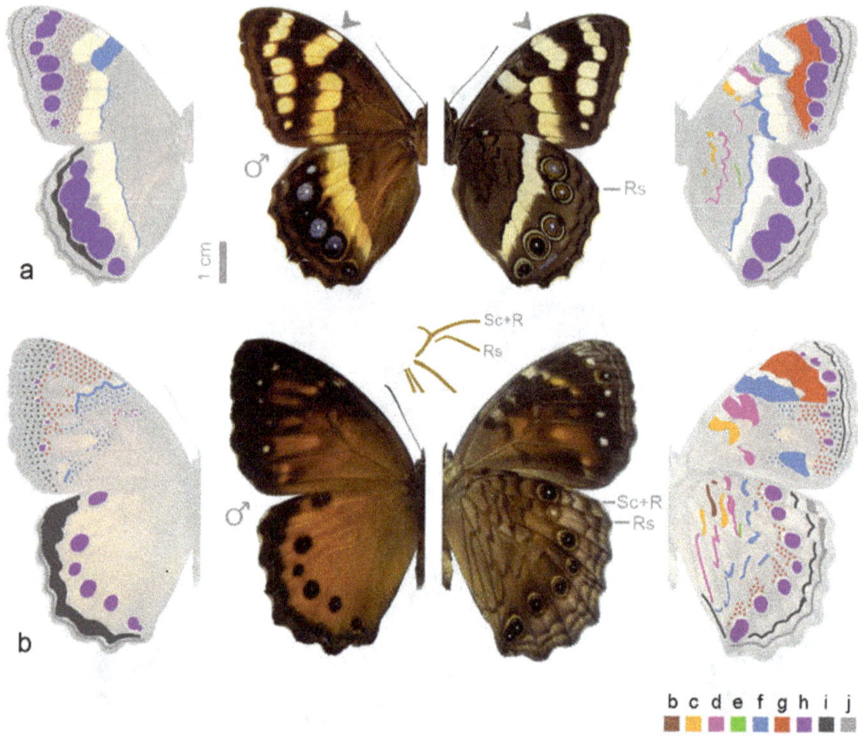

Fig. 2.5 Color-coded wing pattern elements in selected Dirini. Left side of butterfly image in dorsal view, right side in ventral view. Gray arrows indicate colorful band associated with element *f*. (**a**) *Aeropetes tulbaghia*. (**b**) *Paralethe dendrophilus*, note venation detail showing separation of Rs from Sc + R at the base of the hind wing. All butterflies are at the same scale

(see gray arrows in Figs. 2.1, 2.2, 2.3, 2.4, 2.5, 2.6, 2.7, and 2.8). It appears to be associated with element *f* (or bounded between *f* and *g*) and varies between and within the studied tribes. For instance, this band differs noticeably in color, width, and extent of fragmentation between the closely related *Aeropetes tulbaghia* and *Paralethe dendrophilus* (Fig. 2.5a–b). The dorsal forewing band can also vary in orientation (vertical or transverse). A vertical band is found in species where *f* is positioned straight across the medial area of the wing (e.g., Fig. 2.2e). In contrast, a transverse band results from element *f* being slightly diagonal (displaced distally toward the wing tornus, e.g., Fig. 2.7b). Members of the Brassolini, for example, vary in the orientation of this band (compare *Catoblepia* and *Caligo*; Fig. 2.8a–b).

Within the same species and sex, the expression of the band associated with *f* usually differs between the forewing and hind wing and may also show dorso-ventral variation. This is readily apparent in *Pierella helvina* (Fig. 2.3d), where elements *f* and *g* are clearly visible and appear to function as developmental boundaries. Ventrally, the pale-colored band of *P. helvina* is much narrower on the forewing than on the hind wing. Although element *g* forms a continuous line in

Fig. 2.6 Color-coded wing pattern elements in selected Zetherini. Left side of butterfly image in dorsal view, right side in ventral view. Gray arrows indicate colorful band associated with element *f*. (**a**) *Ideopsis vulgaris* (Danaini) model. (**b**) *Penthema lisarda*, hypothesized delimitation of pattern elements based on Nijhout (1991) plus tentative identification of pattern elements (dotted) for species of nonmimetic or intermediate patterns. (**c**) *Neorina hilda*. (**d**) *Penthema adelma*.

the ventral hind wing, it is not expressed dorsally between M_2 and CuA_1, allowing the bright red band to expand distally. For comparison, note that *f* and *g* are also clearly visible on the hind wing of *Pierella lucia* (Fig. 2.3c), where a pale band is expressed ventrally only. The genus *Pierella* constitutes an excellent example of how different WPEs and associated bands can be modified by evolution to give rise to broadly distinctive species-specific patterns (Fig. 2.3b–d).

2.5 Sexual Dimorphism and Mimicry

The species studied here range from sexually monomorphic to slightly or strongly dimorphic, and color pattern divergence implies that selection can operate independently on males and females. When there is little divergence between sexes, both dorsal and ventral WPEs are more conserved in females (Figs. 2.2d–e and 2.4c). In contrast, strong sexual dimorphism can result from simple modifications in few WPEs and the colorful bands associated with them (e.g., *Mielkella singularis*, Penz and Mohammadi 2013) or more complex changes involving a larger number of WPEs (Fig. 2.2d–e).

Strong sexual dimorphism can arise through sexual selection operating on male pattern or natural selection on female pattern (see Kunte 2008 and Oliver and Monteiro 2010 for reviews). Here I confine my discussion to potential natural selection on female pattern. Females could diverge from males to become less conspicuous to potential predators, as might have been the case in five species of *Morpho* (see example in Fig. 2.2e). Furthermore, the evolution of mimetic convergence can be limited to the female sex, although not always the case. Female-limited mimicry has evolved independently in members of various tribes (e.g., Fig. 2.4a–b), and depending on the model, it required simple or complex changes in WPEs. For instance, the convergence of female *Catoblepia orgetorix* with monomorphic *Caligo atreus* (Fig. 2.8a–b) involved a relatively simple set of color pattern modifications. When compared to other species of *Catoblepia*, the band associated with element *f* is dislocated proximally on the dorsal forewing of *C. orgetorix*, its color changed from orange to white, and it acquired purple iridescence. On the dorsal hind wing, the band associated with *i* became wider and changed color from orange to yellow. Mimicry is rare in neotropical Satyrinae, and this example is peculiar as neither *Caligo* nor *Catoblepia* are known to possess chemical defenses.

In contrast, mimicry (female-limited or both sexes) is common in the old-world tribes Zetherini and Elymniini and the Amathusiini genus *Taenaris*. In their case, evolution took two distinctive paths. Figure 2.8c–e shows cross-tribal convergence that resulted from an extreme reduction in the expression of most WPEs plus the

Fig. 2.6 (continued) (**e**) *Ethope himachala*. (**f**) *Ethope noirei*. (**g**) *Zethera pimplea*, note that males of other *Zethera* species have small, dorsal ocelli on both wings. All butterflies are at the same scale

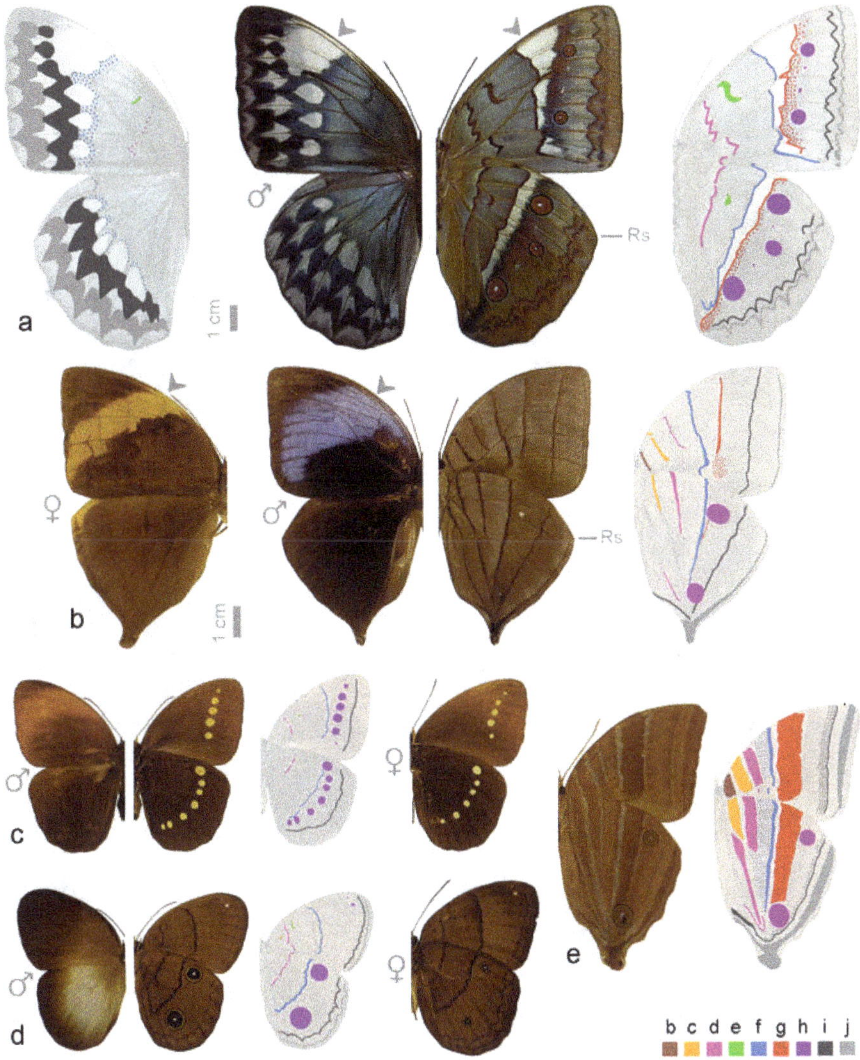

Fig. 2.7 Color-coded wing pattern elements in selected Amathusiini. Left side of butterfly image in dorsal view, right side in ventral view. Gray arrows indicate colorful band associated with element *f*. (**a**) *Stichopthalma godfreyi* (photo by Saito Motoki). (**b**) *Amathuxidia amythaon*. (**c**) *Faunis eumeus*. (**d**) *Faunis phaon leucis*. (**e**) *Amathusia phidippus*. All butterflies are at the same scale except *S. godfreyi*

increase in size of some ocelli to create a similar visual appearance. In other taxa, mimicry involved complex modifications of most WPEs. Nonmimetic and intermediate patterns can help interpret WPE modifications that lead to mimetic convergence of zetherines onto chemically protected danaines (e.g., *Ideopsis vulgaris*, Fig. 2.6a). Figure 2.6c, d, f, and g exemplify a series of such modifications, which

Fig. 2.8 Examples of mimetic convergence. Gray arrows indicate colorful band associated with element *f*. (**a**) nonmimetic male and mimetic female of *Catoblepia orgetorix*. (**b**) *Caligo atreus* model. (**c**) *Taenaris artemis*. (**d**) *Hyanthis hodeva*. (**e**) nonmimetic male and mimetic female of *Elymnias agondas*. All butterflies are at the same scale

can be used to hypothesize the WPE configuration of *Penthema* (Fig. 2.6b; see also Nijhout 1991). Notably, male and female of the sexually dimorphic *Zethera pimplea* have brown and off-white dorsal coloration, but the female pattern is more intricate and Danaini-like than the male (Fig. 2.6g). Although some WPEs can be identified in *Elymnias* species that have complex Danaini-like dorsal patterns, they are generally difficult to interpret (Fig. 2.4a–c).

2.6 Transparency

Layers of scales make butterfly wings generally impenetrable to light. Nevertheless, some members of Satyrinae have evolved partial or complete transparency. In *Morpho sulkowskyi*, the dorsal scale size and pigmentation are reduced to such a degree that the ventral WPEs are visible through the wing (Fig. 2.2b). Partial transparency has evolved in more than one species of *Morpho*, but its function within the context of their natural history is unknown.

Scale cover is dramatically reduced in *Dulcedo*, *Pseudohaetera*, *Haetera*, and *Cithaerias* (Haeterini; Fig. 2.3e–f), and this possibly evolved ca. 29 million years ago (Cespedes et al. 2015). Transparency makes these butterflies nearly invisible in the forest understory and can be considered a defense against predation. Despite the extensive absence of scales, some WPEs are conserved, and this suggests they serve a function in the behavior of these butterflies. For example, their hind wing ocellus below M_1 is highly visible (Fig. 2.3e–f), and it might be involved in signaling. In the forest, male *Cithaerias* that are perched on the ground repeatedly flash their vivid dorsal hind wing colors (pers. obs.), which can likely be seen by other males or potential mates flying nearby.

The interplay between lost versus conserved wing color patterns is an interesting attribute of transparent Haeterini for two reasons. First, some pattern elements are expressed directly onto the wing membrane to form scale-less bands (Fig. 2.3e–f). This shows that the loss of scales does not necessarily lead to a loss of pattern. Membrane-level expression of WPEs can also be seen in areas that have scales, for example, the ocellus in Fig. 2.3f. To my knowledge, *Dulcedo*, *Pseudohaetera*, *Haetera*, and *Cithaerias* are the only butterflies in which WPEs are expressed on the wing membrane. Second, these butterflies show differential dorsoventral regulation of scale formation. For instance, in most transparent Haeterini, the ocellus below M_1 has a complete set of rings on the ventral hind wing surface, but the dorsal one lacks the orange ring (Fig. 2.3f). In *Cithaerias*, colorful scales are present on the dorsal hind wing only, and WPEs expressed at the wing membrane are thus more visible on the ventral surface (Fig. 2.3e). The colorful dorsal vestiture does not seem to correspond to a given WPE, and it spreads across the hind wing surface unaffected by elements *g*, *i*, and *j*. This begs the question of whether these WPEs are expressed on the ventral surface only (C. M. Penz, work in progress).

2.7 Concluding Remarks

The butterflies that form the focus of this chapter provide remarkable examples of color pattern variation. The significant changes in ocelli size and shape observed in *Bicyclus* selection experiments (e.g., Monteiro et al. 1997) suggest that butterflies can undergo rapid adaptive evolution. As a result, lineages might accumulate substantial wing pattern element modifications in relatively short evolutionary time scales. This is consonant with the observation that every tribe studied here includes species with nearly complete to highly reduced wing pattern elements—evolution repeats itself. Convergent appearance resulting from different pattern element modifications could reflect similarities in natural history or microhabitat use, e.g., ventral stripes in species of neotropical *Caerois* and old-world *Amathuxidia* (Figs. 2.2a and 2.7b). Field observations on mating behavior suggest the ventral hind wing ocelli may be used in male-female interactions in species of *Caligo* and *Pierella* (Figs. 2.1d and 2.3a), and this adds a new dimension to previous work. In the tribes studied here, pattern reduction is intriguing because it is accomplished in exceptionally different ways—pattern elements might not be expressed, or the scale vestiture may disappear almost completely (Figs. 2.8c–e and 2.3e–f). Transparency evolved independently in various ecologically and behaviorally distinct groups of Lepidoptera, the Haeterini being an example. How is scale loss adaptive in different taxa, what are the developmental mechanisms involved, and is it reversible? To further our understanding of the role wing coloration plays within the Satyrinae, the work presented here advocates baseline research on two fronts: documentation of pattern variation and field studies aimed at placing wing color diversification in a behavioral and evolutionary context.

Acknowledgments Many thanks to Fred Nijhout and Toshio Sekimura for the invitation to contribute to this volume, the Milwaukee Public Museum (US), Natural History Museum (UK), Florida Museum of Natural History (US), Natural History Museum of Los Angeles County (US), Carnegie Museum of Natural History (US), Smithsonian Institution (US) for specimen loans; Saito Motoki (Japan), Andrew Neild (UK), David Powel (US), Joel Atallah (US), the Nymphalid Systematics Group (Sweden), and Yale Peabody Museum (US) for photographs of preserved or live specimens, and Phil DeVries for comments on the manuscript. This work is dedicated to Neda Mohammadi who, I hope, will forever be fond of butterflies.

Appendix: List of Examined Taxa

Note that most, but not all, tribes within the focal clade are monophyletic (Wahlberg et al. 2009), and the classification used here is therefore tentative and expected to change (e.g., Zetherini). Genera and species are listed in alphabetic order, and those marked with an asterisk were examined from images only.

Brassolini: *Aponarope sutor*; *Bia actorion*, *B. peruana*; *Blepolenis batea*, *B. bassus*; *Brassolis dinizi*, *B. sophorae*; *Caligo atreus*, *C. idomeneus*, *C. martia*,

C. oberthuri; *Caligopsis seleucida*; *Catoblepia berecynthia*, *C. orgetorix*, *C. xanthus*; *Dasyophthalma creusa*, *D. rusina*; *Dynastor darius*; *Eryphanis aesacus*, *E. automedon*, *E. bubocula*; *Mielkella singularis*; *Narope cyllastros*, *N. panniculus*; *Opoptera aorsa*, *O. fruhstorferi*, *O. syme*; *Opsiphanes cassiae*, *O. invirae*, *O. sallei*; *Orobrassolis ornamentalis*; *Penetes pamphanis*; *Selenophanes cassiope*, *S. josephus*, *S. supremus*. See Penz and Mohammadi (2013) for additional species. Morphini: *Antirrhea archaea*, *A. avernus*, *A. philoctetes*; *Caerois chorineus*, *C. gerdrudtus*; *Morpho aega*, *M. anaxibia*, *M. aurora*, *M. catenarius*, *M. cypris*, *M. hecuba*, *M. helenor*, *marcus*, *M. menelaus*, *M. rhetenor*, *M. theseus*. Haeterini: *Cithaerias andromeda*, *C. aurora*, *C. aurorina*, *C. bandusia*, *C. pireta*, *C. pyritosa*, *C. pyropina*; *Dulcedo polita*; *Haetera piera*; *Pierella helvina*, *P. hortona*, *P. hyalinus**, *P. lamia*, *P. lena*, *P. lucia*, *P. luna*, *P. nereis*; *Pseudohaetera mimica*. Elymniini: *Elymnias agondas*, *E. cumaea*, *E. hypermnestra*, *E. nessaea*, *E. patna*; *Elymniopsis bammakoo*. Melanitini: *Melanitis amabilis*, *M. constantia*, *M. leda*. Dirini + *Manataria*: *Aeropetes tulbaghia*; *Dingana dingana**; *Dira clytus**; *Paralethe dendrophilus*; *Torynesis mintha**; *Manataria maculata*. Zetherini: *Ethope diademoides*, *E. himachala*, *E. noirei**; *Hyantis hodeva*; *Morphopsis albertisi*, *M. biakensis*, *M. meeki*, *M. ula*; *Neorina crishna*, *N. hilda*, *N. lowi*, *N. patria*; *Penthema adelma*, *P. darlisa*, *P. formosanum*; *Xanthotaenia busiris*; *Zethera incerta*, *Z. musa*, *Z. musides*, *Z. pimplea*. Amathusiini: *Amathusia binghami*, *A. phidippus*, *A. plateni*; *Amathuxidia amythaon*; *Discophora bambusae*, *D. sondaica*, *D. timora*; *Ensipe cycnus*, *E. euthymius*; *Faunis canens*, *F. eumeus*, *F. menado*, *F. stomphax*, *F. phaon leucis*; *Melanocyma faunula*; *Morphotenaris schoenbergi*; *Stichophthalma camadeva*, *S. godfreyi**, *S. howqua*, *S. louisa*, *S. nourmahal*, *S. sparta*; *Taenaris artemis*, *T. butleri*, *T. catops*, *T. myops*, *T. onolaus*; *Thaumantis diores*, *T. noureddin*, *T. odana*; *Thauria aliris*; *Zeuxidia amethystus*, *Z. aurelius*, *Z. doubledayi*.

References

Cespedes A, Penz CM, DeVries PJ (2015) Cruising the rain forest floor: butterfly wing shape evolution and gliding in ground effect. J Anim Ecol 84:808–816

Chai P (1990) Relationships between visual characteristics of rain forest butterflies and responses of a specialized insectivorous bird. In: Wicksten M (compiler) adaptive coloration in invertebrates. Proceedings of a Symposium sponsored by the American Society of Zoologists. College Station, Texas, pp 31–60

DeVries PJ (2002) Differential wing-toughness among palatable and unpalatable butterflies: direct evidence supports unpalatable theory. Biotropica 34:176–181

DeVries PJ (2003) Tough models versus weak mimics: new horizons in evolving bad taste. J Lep Soc 57:235–238

Freitas AVL, Benson WW, Marini-Filho OJ, Carvalho RM (1997) Territoriality by the dawn's early light: the Neotropical butterfly *Caligo idomeneus* (Nymphalidae: Brassolinae). J Res Lepidoptera 34:14–20

Grimaldi D, Engel MS (2005) Evolution of the insects. Cambridge University Press, Cambridge, MA

Hill RI, Vaca JF (2004) Differential wing strength in *Pierella* butterflies (Nymphalidae, Satyrinae) supports the deflection hypothesis. Biotropica 36:362–370

Kemp DJ, Herberstein ME, Fleishman LJ, Endler JA, Bennett AT, Dyer AG, Hart NS, Marshall J, Whiting MJ (2015) An integrative framework for the appraisal of coloration in nature. Am Nat 185:705–724

Kunte K (2008) Mimetic butterflies support Wallace's model of sexual dimorphism. Proc R Soc B Biol Sci 275:1617–1624

Monteiro A, Brakefield PM, French V (1997) The genetics and development of an eyespot pattern in the butterfly *Bicyclus anynana*: response to selection for eyespot shape. Genetics 146:287–294

Nijhout HF (1991) The development and evolution of butterfly wing patterns. Smithsonian series in comparative evolutionary biology. Smithsonian Institution Press, Washington, DC

Oliver JC, Monteiro A (2010) On the origins of sexual dimorphism in butterflies. Proc R Soc B Biol Sci 278:1981–1988

Oliver JC, Robertson KA, Monteiro A (2009) Accommodating natural and sexual selection in butterfly wing pattern evolution. Proc R Soc B Biol Sci 276:2369–2375

Peña C, Wahlberg N (2008) Pre-historic climate change increased diversification of a group of butterflies. Biol Lett 4:274–278

Penz CM, Mohammadi N (2013) Wing pattern diversity in Brassolini butterflies (Nymphalidae, Satyrinae). Biota Neotrop 13:1–27

Rutowski RL (1991) The evolution of male mate-locating behavior in butterflies. Am Nat 138:1121–1139

Schwanwitsch BN (1924) On the ground-plan of wing-pattern in Nymphalids and certain other families of the Rhopaloeerous Lepidoptera. P Zool Soc London 94:509–528

Schwanwitsch BN (1925) On a remarkable dislocation of the components of the wing pattern in a Satyride genus *Pierella*. Entomologiste 58:226–269

Srygley RB, Penz CM (1999) The lek mating system in Neotropical owl butterflies: *Caligo illioneus* and *C. oileus* (Lepidoptera, Brassolinae). J Insect Behav 12:81–103

Stevens M (2005) The role of eyespots as anti-predator mechanisms, principally demonstrated in the Lepidoptera. Biol Rev 80:573–588

Süffert F (1927) Zur vergleichende Analyse der schmetterlingzeichmung. Biol ZBL 47:385–413

Vane-Wright RI, Ackery PR (1984) The biology of butterflies. Symposium of the Royal Entomological Society of London, Number 11. Academic, Saint Louis

Wahlberg N, Leneveu J, Kodandaramaiah U, Peña C, Nylin S, Freitas AVL, Brower AVZ (2009) Nymphalid butterflies diversify following near demise at the cretaceous/tertiary boundary. Proc R Soc B Biol Sci 276:4295–4302

3

Macroevolutionary Aspects of Lepidopteran Camouflage Patterns

Takao K. Suzuki

Abstract Lepidopteran camouflage patterns offer sophisticated and captivated examples of morphological evolution. Previous studies focused on how and why camouflage patterns are modulated at the microevolutionary level and determined, for instance, the adaptive role of camouflage patterns in avoiding predator attacks. However, less attention has been paid to the macroevolution of camouflage, including the evolutionary paths leading to the origination of leaf mimicry patterns. To understand the deep origins and evolvability of camouflage patterns, a key principle comes from a highly conserved ground plan (termed the nymphalid ground plan; NGP). The ground plan generates a variety of morphological forms, while it maintains its own type. This review introduces several seminal studies that used NGP-known features to reveal the macroevolutionary aspects of lepidopteran camouflage patterns, providing a roadmap for further understanding this biological phenomenon. The following core themes are discussed: (1) how complex camouflage patterns evolved (macroevolutionary pathways), (2) what kind of flexible mechanisms facilitate the origin of such complex patterns (macro-evolvability), and (3) how such complex patterns are tightly integrated through the coupling and uncoupling of ancestral developmental mechanisms (body plan character map). These approaches will provide new research lines for studying the evolution of camouflage patterns and the underlying flexibility of the NGP.

Keywords Crypsis and masquerade • Butterfly and moth • Comparative morphology • Macroevolution • Evolutionary path • Phylogenetic comparative methods • Tinkering • Morphological integration and modules • Morphometrics • Genotype-phenotype map

---

T.K. Suzuki (✉)
Transgenic Silkworm Research Unit, Division of Biotechnology, Institute of Agrobiological Sciences, National Agriculture and Food Research Organization (NARO), 1-2 Oowashi, Tsukuba, Ibaraki 305-8634, Japan
e-mail: homaresuzuki@gmail.com

3.1 Introduction

Complex and sophisticated camouflage patterns have fascinated many biologists (Poulton 1890; Cott 1940; Edmunds 1974; Ruxton et al. 2004; Stevens 2016). Recently, camouflage has been classified into two major types: crypsis (blended into environmental backgrounds to avoid detection by potential predators) and masquerade (special resemblance to natural objects to avoid recognition by potential predators) (Stevens and Merilaita 2009; Merilaita and Stevens 2011; Skelhorn et al. 2010a, b; Skelhorn 2015). Prominent cases of camouflage are found in butterfly and moth wing patterns, including tree bark crypsis in *Biston betularia* (van't Hof et al. 2016), lichen crypsis in *Agriopodes fallax* (Schmidt et al. 2014), leaf vein masquerade in the noctuid moth *Oraesia excavata* (Fig. 3.1a; Suzuki 2013) or in the nymphalid butterflies *Kallima inachus* and *K. paralekta* (Fig. 3.1b; Suzuki et al. 2014), and dried leaf masquerade in *Polygonia c-album* (Wiklund and Tullberg 2004). Most studies focused on the microevolutionary aspects of camouflage generation. For example, research on the industrial melanism shown in peppered moths deciphered both the adaptive significance (Cook et al. 2012) and the genetic basis of cryptic color variation (Cook and Saccheri 2013; van't Hof et al. 2016). Studies on the seasonal polyphenism of the butterflies *Araschnia levana* (Koch and Bückmann 1985), *Bicyclus anynana* (Brakefield and Larsen 1984; Monteiro et al. 2015), and *Polygonia c-aureum* (Fukada and Endo 1966; Endo 1984; Endo et al. 1988) have also uncovered hormonal switches in the generation of the cryptic patterns matching dry or autumnal color environments. In contrast, the macroevolution of camouflage has received little attention. The present review focuses on the comparative morphology of camouflage patterns in butterfly and moth wings and proposes a research roadmap for further advancing our understanding of the generative mechanisms underlying camouflage evolution.

For addressing the macroevolutionary aspects of lepidopteran camouflage, a key principle is that comparison of the anatomy of many species allows the extraction of

Fig. 3.1 Camouflage of moth and butterfly wing patterns. (**a**) *Oraesia excavata*. (**b**) *Kallima inachus* (Figure panel **a** is reproduced with modification from Suzuki (2013). Figure panel **b** is reproduced with modification from Suzuki et al. (2014))

a common theme behind diversity, termed the "body plan" or "ground plan," which refers to the structural composition of organisms based on homologous elements shared among species (Wagner 2014). To date, butterfly and moth (at least within Macrolepidoptera) wing patterns are thought to be based on a highly conserved ground plan (termed the nymphalid ground plan, NGP; Fig. 3.2a; Schwanwitsch 1924; Süffert 1927; Nijhout 1991). The NGP describes the diversification of wing patterns as modifications of an assembly of discrete pattern elements shared among species (Schwanwitsch 1956; Nijhout 1991) and is suggested to be homologous and inherited across species. From the comparative morphology point of view, the essential question is how effective is the NGP scheme in understanding lepidopteran camouflage patterns? Moreover, if certain camouflage patterns are illustrated by the NGP, what information can this scheme provide for understanding

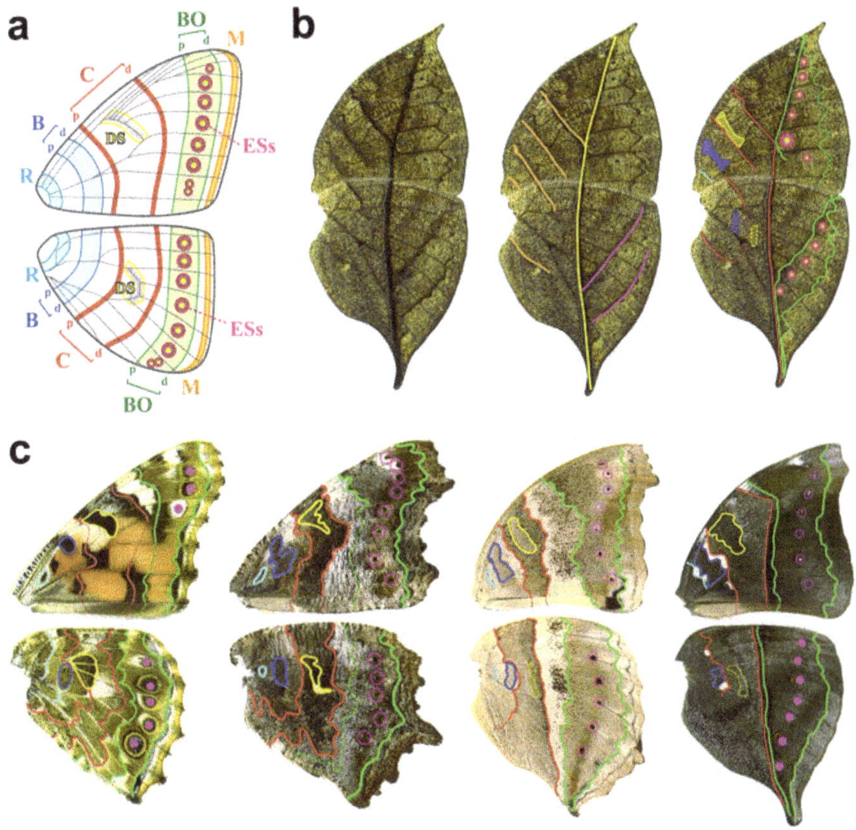

Fig. 3.2 Nymphalid ground plan and the variations generating diversified wing patterns. (**a**) Nymphalid ground plan (NGP). (**b**) Leaf vein-like pattern and the NGP of *Kallima inachus*. (**c**) NGP of *Vanessa cardui*, *Nymphalis vaualbum*, *Yoma sabina*, *Doleschallia bisaltide* (This figure is reproduced with modification from Suzuki et al. (2014))

lepidopteran camouflage patterns and can it contribute to the morphological evolution and organization of such spectacular examples of adaptation to the environment?

The present review introduces several NGP studies that are crucial for revealing the macroevolutionary aspects of lepidopteran camouflage patterns and provide a basis for further understanding this biological phenomenon. First, the foundations for using comparative morphology to identify homologous elements across species are described along with how NGP has led the way to the elaboration of diverse wing pattern configurations. Next, the potential of phylogenetic comparative methods to reveal the sequential evolutionary steps that built up leaflike patterns from nonmimetic ones is discussed. Third, the scheme of the NGP is used for discussing a flexible building logic of leaf mimicry patterns. Fourth, a methodological framework for analyzing the degree of integration and modularity in leaf vein-like pattern is proposed, and arguments favoring the evolutionary origin of *de novo* functional modules are presented. Finally, a research roadmap for further macroevolutionary studies on the origin and diversification of camouflage patterns is proposed.

3.2 Morphological Foundations of the Nymphalid Ground Plan

The concepts of body plan and ground plan are traditionally rooted in comparative morphology (Rieppel 1988). The criteria for identifying structural or positional homologs across different species were summarized by Remane (1952) and are considered a validated procedure in systematic and comparative morphology studies (Williams and Ebach 2008). These criteria consist of three principal rules: (1) similarity of topographical relationships, (2) similarity of special features, and (3) transformational continuity through intermediate ontogeny or phylogeny. The first criterion is logically consistent with Geoffroy St. Hilaire's "*principe des connexions*" (Saint-Hilaire 1818), the second is based on the specific properties of a character of interest, and the third is based on the evolutionary continuity of developmental genetic mechanisms underlying the character of interest. Although the concept of homology is still widely discussed (Patterson 1982; Roth 1988; Wagner 1989, 2007; Brower and Schawaroch 1996; Hall 2000), Remane's criteria remain valuable consensuses that crystallize empirical facts through numerous careful observations of morphological structures. Currently, these criteria provide a powerful tool to decipher the homology of anatomical structures in a broad spectrum of animals and plants (for animals: Nagashima et al. 2009; Hutchinson et al. 2011; Luo 2011; Holland et al. 2013; for plants: Sattler 1984; Buzgo et al. 2004).

The NGP is a scheme for describing homologous elements shared across species and thus should be evaluated within the logical framework of Remane's criteria.

Although Remane's criteria were inherent to NGP studies by Schwanwitsch (1956) and Nijhout and Wray (1986), to my knowledge, there is no explicit citation to Remane's work in NGP studies. Recently, I tackled to apply Remane's criteria to analyze the NGP of *Kallima inachus* and *K. paralekta* leaf vein-like patterns and succeeded in demonstrating that these can be explained by the NGP (Fig. 3.2b; Suzuki et al. 2014), and the results were consistent with Schwanwitsch (1956) analysis and validated the empirical inference proposed by Süffert (1927). The wing patterns of species closely related to *Kallima* spp. can also be explained by the NGP, although these patterns differ from that found in *Kallima* spp. (Fig. 3.2c, only four species were selected; for further details, see Suzuki et al. 2014). Interestingly, these analyses revealed that the differences between the leaf vein-like pattern and the other non-leaf patterns resulted only from differences in the character states of NGP elements. Thus, comparative morphology provides in-depth information about the way of diversification of lepidopteran wing patterns, even in extreme cases such as leaf mimicry.

It is important to mention that the NGP framework has limitations, which are most evident when lepidopteran wing patterns have so dramatically deviated from a stereotypical pattern that they challenge reasonable homology assignments. For example, the wing patterns of some papilionids are intensively fragmented through dislocation and thus difficult to connect to the NGP (Mallet 1991). In the nymphalid butterflies *Heliconius* sp., the NGP has undergone complex rearrangements that culminated in a highly modified state (Mallet 1991), although NGP was previously reported for this genus (Nijhout and Wray 1988). In such cases, less derived species can provide clues on intermediate states and clarify the nature of homologous characters but are prone to misidentifications without a more mechanistic understanding of wing pattern architecture. To further understand the evolutionary trajectories of the NGP, it is necessary to investigate the molecular mechanisms underlying NGP. Previous studies revealed the molecular mechanisms underlying eyespots (*ocelli*), one of the NGP elements in butterfly wings (Carroll et al. 1994; Brakefield et al. 1996; Keys et al. 1999; Brunetti et al. 2001; Beldade and Brakefield 2002; Monteiro et al. 2006; Oliver et al. 2012; Monteiro et al. 2013; Monteiro 2015; Zhang and Reed 2016; Beldade and Peralta 2017). Molecular studies have also uncovered several morphogens (e.g., *Wnt1/wingless*, *WntA*) and transcription factors (e.g., *aristaless2*, *engrailed*) associated with other elements of NGP (Brunetti et al. 2001; Monteiro et al. 2006; Martin and Reed 2010, 2014).

3.3 Evolutionary Path: Gradual Evolutionary Steps Toward Leaf Vein-Like Patterns

The ground plan architecture of lepidopteran wing patterns provides a starting point to investigate the evolutionary paths leading to complex camouflage patterns, but how can these trajectories be analyzed in exquisitely detailed phenotypes?

Character polarity has been used in most studies investigating the evolutionary processes that generate traits (Donoghue 1989; Swofford and Maddison 1992; Wiley and Lieberman 2011), and it refers to the biased phylogenetic placement of certain states of a character of interest (Fig. 3.3a). Clear detection of character polarity indicates a nested hierarchical relationship between traits, whose character states are evolutionarily transformed from ancestral to derived states in a specific temporal order. As shown in Fig. 3.3a, the evolution of trait A follows that of the trait B. However, this approach has a crucial practical limitation: traits of interest often lack a clear character polarity. To cope with this limitation, some statistical methods, collectively termed phylogenetic comparative methods (PCMs), were developed for analyzing traits' evolution (Fig. 3.3b; Harvey and Pagel 1991; Losos and Miles 1994; Garamszegi 2014). In PCMs, statistical testing is incorporated into the examination of phylogenetic information and character states to analyze the evolution of traits (Pagel 1999a). Accordingly, these methods can be used to detect subtle nuances of trait evolution that lack a clear signature of character polarity and thus can be applied in a broad spectrum of scenarios featuring a complex distribution of character states. In such scenarios, PCMs can be used in the reconstruction of traits' ancestral states (Schluter et al. 1997; Pagel 1999b; Pagel et al. 2004) or to infer the temporal order in which traits evolved, within a phylogenetic framework (Pagel 1994; Pagel and Meade 2006).

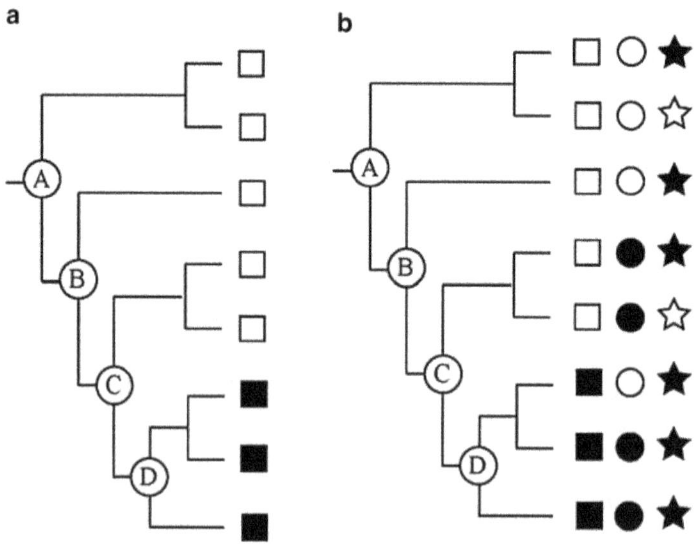

Fig. 3.3 How to infer macroevolutionary paths toward complex traits. (**a**) Simple case of character polarity, in which a trait (*square*) evolved from state 0 (*open square*) to state 1 (*close square*) at the node D of the phylogeny. (**b**) Complex case of character polarity, in which phylogenetic comparative methods were used to estimate the ancestral states of the traits (*squares*, *circles*, and *stars*)

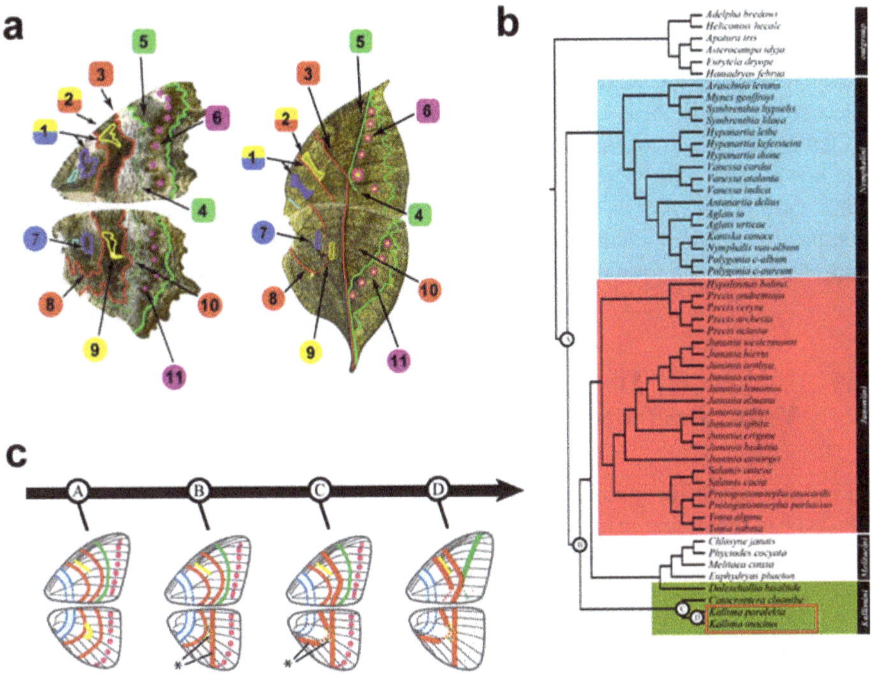

Fig. 3.4 Evolutionary steps that generated Kallima sp. leaf vein-like patterns. (**a**) Decomposition of wing patterns into 11 character states. (**b**, **c**) Reconstructed ancestral character states are represented at four selected nodes (*A*, *B*, *C*, and *D*), which are illustrated as the time required for the evolutionary transformation of wing patterns (from *A* to *D*). In the molecular phylogeny, the genus *Kallima* is evidenced using a *red box* (This figure is reproduced with modification from Suzuki et al. (2014))

The evolution of leaf resemblance in *Kallima* spp. has been a long-term conundrum and remains unresolved. Under a gradualistic view (Darwin 1871; Wallace 1889; Poulton 1890; Weissman 1902; Watson et al. 1936), the leaf mimicry pattern is a product of slow gradual evolution, with natural selection progressively perfecting masquerade forms; under the alternative saltationist view, leaf mimicry pattern evolved via relatively sudden leaps in the morphospace without intermediate forms (Mivart St 1871; Goldschmidt 1945). Despite the enthusiastic debate, no formal assessment of the tempo and mode of evolution in leaf mimicking has been provided so far. Recently, I applied PCMs to gain insight on the evolutionary paths that led to the leaf vein-like pattern in *Kallima* spp. (Fig. 3.4; Suzuki et al. 2014). If overall phenotypes are treated as integrated units, PCM analyses cannot reconstruct the evolutionary history of complex traits, simply by informing how many times the traits evolved (e.g., Mugleston et al. 2013). To avoid this, the butterfly wing patterns including the leaf patterns were decomposed into a set of several subcomponents using the NGP (Fig. 3.4a), which allowed inferring the ancestral states of each component and reconstructing the evolutionary process as the sum of the changes occurring in all components (Suzuki 2017). Thus, tracing ancestral states at various phylogenetic nodes illustrates the sequential transformation of the character states of multiple components that led to the complex traits (Fig. 3.4b). This analysis revealed the successive steps in the evolution of leaf masquerade patterns from nonmimetic wing patterns within a phylogenetic framework (Fig. 3.4c; Suzuki et al. 2014) and provided the first evidence for gradual evolutionary origin of leaf mimicry (Skelhorn 2015). Thus, combining NGP and PCMs information provides an insight into the structural complexity of lepidopteran wing patterns and the possibility to depict the evolutionary paths leading to the formation of complex and detailed patterns (Suzuki 2017).

3.4 Tinkering: The Flexible Building Logic of Leaf Vein-Like Patterns

In addition to the reconstruction of evolutionary paths described above, identifying the NGP of lepidopteran wing patterns will provide resources to assess the different ways to produce leaf vein-like patterns. Regarding this issue, Schwanwitsch (1956) described the NGP of several species presenting leaf patterns such as *Siderone marthesia* (Fig. 3.5a), *Zaretis isidora* (Fig. 3.5b), and *K. inachus* (Fig. 3.2b). According to his scheme, the mode of derivation from the NGP is in most part repeated in these three species. Interestingly, the genera *Siderone* and *Zaretis* (Charaxinae, a subfamily of Nymphalidae) are taxonomically distant from the genus *Kallima* (Nymphalinae), which is also supported by Wahlberg et al. (2009) molecular phylogeny. Because convergence is considered to represent independently evolved features that are both structurally and superficially similar (Stayton

Fig. 3.5 Leaf vein-like variations on the same NGP theme. (**a**) *Siderone marthesia*. (**b**) *Zaretis isidora*. (**c**) *Oraesia excavata*. The NGP of *S. marthesia* and *Z. isidora* is based on Schwanwitsch (1956) (Figure panel c is reproduced with modification from Suzuki (2013))

2015), the similar mode of derivation from the NGP found in Charaxinae and Nymphalinae probably resulted from independent events of convergent evolution.

Does this similar mode of NGP-derived patterns, which seems to indicate that leaf pattern construction modes are quite constrained in butterflies, hold true for more distantly related taxa than Nymphalinae and Charaxinae? To address this question, I here compare the NGP of the leaf vein-like pattern found in the noctuid moth *O. excavata*, one of the most abundant moths in Northeast Asia (Fig. 3.1a), to that of *K. inachus*. Although these leaf vein-like patterns look similar, both consisting of a main vein and two sets of lateral veins, the way in which these two leaf patterns were built from the NGP is quite different (Figs. 3.2b and 3.5c). For example, in *K. inachus* butterflies, the main vein of the leaf pattern is derived from a green element (the proximal band of the border symmetry system) and a red element (the distal band of the central symmetry system), whereas in *O. excavata*, the main leaf vein is derived only from green elements (the border symmetry system). These observations showed that Lepidoptera leaf patterns can evolve through different paths, revealing a higher flexibility than that suggested from the analysis centered on nymphalid butterflies only.

This flexibility in leaf pattern building could be discussed within the concept of tinkering, which was in biology proposed by François Jacob (1977). This concept was described as "a tinkerer who does not know exactly what he is going to produce, but uses whatever he finds around him, whether it be pieces of string, fragments of wood, or old cardboards; in short it works like a tinkerer who uses everything at his disposal to produce some kind of workable object." Based on this statement, the leaf patterns of *Kallima* spp. and *Oraesia* spp. evolved in a tinkering mode of innovation, managing with odds and ends. Additionally, and although it might seem unexpected, the dead leaves of Charaxinae might have achieved the same construction style observed in *Kallima* as a result of tinkering evolution. Strictly speaking, tinkering likely refers to the evolutionary process of building up traits and not just to the traits. Thus, the flexible building logic of Lepidoptera leaf patterns might reflect the tinkering logic of the evolutionary processes behind them.

3.5 Modularity: Developmental Modules of the NGP and a Simple Cryptic Pattern

How a morphological structure is integrated is crucial to understand the genetic and developmental architecture of trait adaptation (Olson and Miller 1958; Cheverud 1996; Klingenberg 2008). The concept of morphological integration postulates that functionally related elements are tightly coupled (Olson and Miller 1958; Cheverud 1996). A special form of integration is modularity, in which units are tightly coupled but can be individually decoupled (Wagner and Altenberg 1996). Modularity results from the regulatory interactions of developmental mechanisms (Klingenberg 2008) and/or from accumulated structural changes shaped by natural

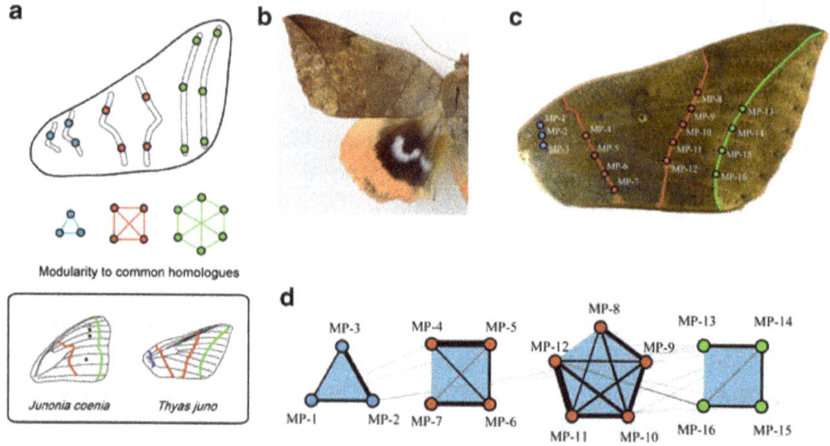

Fig. 3.6 Modularity of the simple cryptic pattern of *Thyas juno*. (**a**) Schematic illustration of divergence strategy in moth and butterfly wing patterns. The modularity of simple patterns corresponds to the original developmental modules of the NGP. (**b**) Forewings and hind wings of *T. juno*. (**c**) Forewings comprise four elements, each corresponding to an NGP element. (**d**) Morphological correlation network of the *T. juno* forewing pattern. In this correlation network, nodes represent measurement points and lines represent the correlations between measurement points (larger correlation coefficients are indicated by *thicker arrow edges* and *darker lines*). The modules detected are illustrated as *light-blue areas* (This figure is reproduced with modification from Suzuki (2013))

selection (Lande 1979; Arnold 1983; Wagner and Altenberg 1996). Previous studies suggested that the NGP is the sum of several developmental modules where each NGP element is genetically and/or developmentally autonomous (Fig. 3.6a; Nijhout 1991, 1994, 2001; Beldade and Brakefield 2002). In fact, the central symmetry system of the NGP appears to be a genetically and phenotypically independent unit (Brakefield 1984; Paulsen and Nijhout 1993; Paulsen 1994, 1996), and eyespots are developmental units formed by factors diffused from foci (Nijhout 1980; French and Brakefield 1995). These considerations strongly suggest that butterfly and moth wing patterns, including camouflage patterns, obey to NGP's rule of modularity.

How are lepidopteran camouflage patterns integrated and modularized? To address this issue, the relatively simple camouflage pattern of the noctuid moth *Thyas juno* was examined (Fig. 3.6b; Suzuki 2013). At rest, this species displays only the cryptic forewings covering the conspicuous hind wings, but, once it detects a potential enemy, the forewings are unfolded and display the warning-colored hind wings. The forewing pattern consists of four elements, each corresponding to an NGP element (Fig. 3.6c). To detect the modules involved in an overall wing pattern, I developed a new analytical method (termed morphological correlation network), which allows analyzing geometric morphometric data by combining graph theory and the statistical physics of spin glass (Suzuki 2013; Esteve-Altava 2016). This approach revealed that the modules involved in *T. juno* wing pattern corresponded to individual NGP symmetry elements, which might reflect the original modular architecture of the NGP (Fig. 3.6d) as supported by previous considerations regarding NGP organization (Nijhout 1991, 1994, 2001; Beldade and Brakefield 2002). Although studying a practical case is limited, at least in relatively simple camouflage patterns, these results supported the hypothesis that the genetic and developmental architectures underlying camouflage patterns reflect the original developmental modules of the NGP (Fig. 3.6a).

3.6 Evolutionary Origin of *De Novo* Modules: Rewiring of the NGP Developmental Modules to Generate Functional Modules

How modules of morphological structures originated is an important question to understand the complex adaptation of phenotypes (Wagner et al. 2007; Klingenberg 2008). A previous conceptual study proposed that modules evolved through the opposite processes of integration (coupling) and parcellation (uncoupling) (Wagner and Altenberg 1996). This conceptual framework seems to be crucial to comprehend the evolution of butterfly and moth wing patterns through modifications of the NGP. Contrasting to the early establishment of the conceptual basis, how *de novo* modules originated still remains poorly understood (Moczek et al. 2015). The question here is how modules of complex camouflage patterns originated within the context of morphological integration and parcellation.

To address this question, the modular architecture of the leaf vein-like pattern of *O. excavata* (Figs. 3.1a and 3.5c) was investigated using the morphological correlation network method (Suzuki 2013). This study revealed that the leaf pattern of *O. excavata* is highly modularized, with each module corresponding to each component of the leaf vein, implying the functional modules (Fig. 3.7b). To examine the extent of the association between these functional modules and the developmental modules of the NGP, the morphological correlation network of the *O. excavata* wing pattern was replotted (Fig. 3.7c). Interestingly, functional modules were generated by the coupling and uncoupling of NGP developmental modules. For example, the functional module of the left lateral vein (i.e., module 2) originated from coupling two distinct modules of the central and border symmetry systems, and the developmental module of the border symmetry system was uncoupled into three functional modules (i.e., modules 2, 3, 4). Thus, this analysis clearly demonstrated that, at least in the evolution of complex camouflage patterns such as leaf masquerade, *de novo* modules originated through the reintegration of NGP developmental modules (Fig. 3.7a).

Unlike the previous studies in which the NGP was considered to comprise autonomous units (Fig. 3.6; Nijhout 1991, 1994, 2001; Beldade and Brakefield 2002), the modules in the *O. excavata* leaf pattern originated through reintegration to new modules (Fig. 3.7). This discrepancy could be due to differences between simple and complex patterns (Figs. 3.6a and 3.7a). Previous studies often

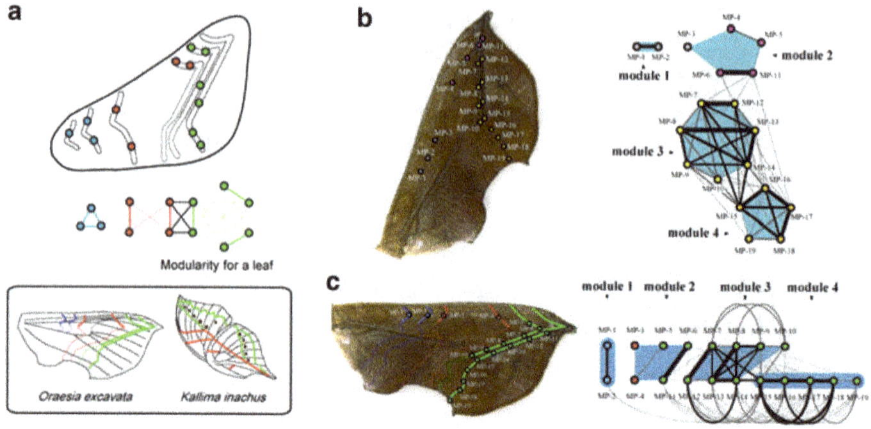

Fig. 3.7 Modularity of the leaf vein-like pattern of *Oraesia excavata*. (**a**) Schematic illustration of divergence strategy in moth and butterfly wing patterns. The modularity of complex patterns evolved through rewiring the original developmental modules of the NGP. (**b**) Forewings of *O. excavata* and its morphological correlation network. In this correlation network, nodes represent measurement points and lines represent correlations between measurement points (larger correlation coefficients are indicated by *thicker arrow edges* and *darker lines*). The modules detected are illustrated as *light-blue* areas. (**c**) Replot of the correlation network of *O. excavata* wing pattern based on the NGP (This figure was reproduced with modification from Suzuki (2013))

emphasized that the genetically and developmentally autonomous units of the NGP allowed further uncoupling pattern elements (e.g., dislocation), and such individualization is thought to allow establishing separate evolutionary trajectories, thereby contributing to the evolvability of lepidopteran wing patterns. In addition to this previous perspective, the present review emphasizes the importance of coupling of pattern elements in wing morphological diversification and proposes a new organizing principle, a "rewiring" strategy (i.e., coupling and uncoupling) of the NGP, in which a combination of decoupling and coupling processes "rewires" the correlations among common parts (Fig. 3.7a; Suzuki 2013).

3.7 Next Research Programs

Quantitative analyses, together with the scheme of the NGP, have begun to set a new path for understanding camouflage patterns of butterfly and moth wings. The NGP provides a foundation for the evolutionary pathways, evolvability, and genetic/developmental architecture underlying the complex and diversified camouflage patterns, through which the ground plan is modified. In this final section, further research programs are discussed.

3.7.1 *Macroevolutionary Pathways Toward Camouflage Patterns*

Diversification based on NGP modifications is not a random process but occurs in a certain sequential order. As shown above, mathematical methods using Bayesian statistics enabled analyzing the evolutionary origin and sequential steps toward the various camouflage patterns (Suzuki et al. 2014; Suzuki 2017). This approach allows to test whether camouflage patterns originated gradually or suddenly and to analyze the evolutionary process through which modifications were accumulated generating camouflage patterns.

Furthermore, comparing multiple evolution processes allows examining evolutionary pathways considering whether processes within them are possible or not. For example, comparing the evolutionary processes involved in butterfly leaf masquerade and lichen cryptic patterns may reveal common/different evolutionary mechanisms between the different camouflage patterns. Similarly, comparing the evolutionary processes of leaf masquerade among distinct taxa may reveal how many pathways are involved in the evolution of lepidopteran leaf patterns and/or addressing the mechanisms allowing the multiple origins of leaf mimicry in Lepidoptera. To date, studies considering macroevolution discussed only the tempo, mode, and trends of evolution (Simpson 1944; Carroll 2001). In addition to these research directions, studying the evolutionary processes and pathways involved in complex and diversified traits is expected to add a new direction in the research field of macroevolution.

3.7.2 *Macro-evolvability of the NGP*

The deep involvement between body plan and evolvability has often been discussed (Vermeij 1973; Riedl 1978; Kirschner and Gerhart 1998; Graham et al. 2000). Regarding evolvability, Vermeij (1973) proposed the concept of versatility, which focuses on the number and range of independent parameters controlling morphological form. As described above, the evolution of the *O. excavata* leaf pattern involved the reintegration of the original developmental modules of the NGP (Fig. 3.7), suggesting that the increase in the number of parameters controlling shape allowed new adaptations, reflecting the versatility of the NGP (Suzuki 2013). In addition, the flexible logic of leaf mimicry patterns suggests a new component (e.g. tinkering) in the evolvability of the ground plan (Fig. 3.2b and 3.5). It has been pointed out that evolvability has various definitions, and Pigliucci proposed its classification in an evolutionary time scale (Pigliucci 2008). Following his definition, I would like to propose the term "macro-evolvability" to define the long time scale evolution that generates various forms through modifications of the ground plan.

Furthermore, one extreme case when examining the macro-evolvability of the ground plan is to determine under which circumstances the ground plan is partially or fully broken. In other words, this approach provides an insight into evolvable limitation of the NGP. Unlike that considered before Darwin's theory of evolution, the ground plan is also subject to natural selection, and therefore some or all of it might be broken with the evolutionary emergence of a specific form derived from the ground plan. Are there possibilities that the NGP was broken? The wing pattern of a mimicry butterfly, *Heliconius* sp., might be considered (Jiggins et al. 2017) a possible example of such a situation. Under this consideration, several questions are raised: How was the NGP deconstructed in *Heliconius* butterflies? What kind of natural selection promotes NGP loss? Does the evolutionary acquisition of Müllerian mimicry affect the loss of the NGP? To address such questions, it will be necessary to combine morphological and molecular studies to verify NGP integrity (Martin et al. 2012; Martin and Reed 2014), because the NGP might be difficult to identify in these butterflies.

3.7.3 Body plan Character Map: Genetic and Developmental Architectures of the NGP

What kind of genetic and developmental architectures underlies the ground plan? In previous studies, this issue was discussed from various perspectives, including the perspectives of transcriptomics (Duboule 1994; Kalinka et al. 2010; Irie and Kuratani 2011, 2014; Quint et al. 2012; Levin et al. 2016) and gene regulatory networks (Davidson and Erwin 2006; Wagner 2007). From the morphological integration and modularity perspective, two major schemes were proposed: the genotype-phenotype map (G-P map; Fig. 3.8a; Wagner and Altenberg 1996) and developmental mapping (D map; Fig. 3.8b; Klingenberg 2008). Both schemes describe how modules of traits were generated through internal interactions, but while the G-P map is based on genetics, the D map is based on ecological

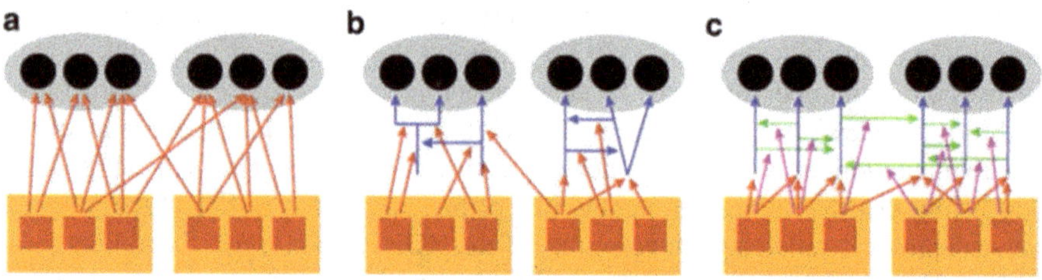

Fig. 3.8 Genetic and developmental architectures of a modularized phenotypic trait. (a) Genotype-phenotype map (G-P map). (b) Developmental map (D map). (c) Body plan character map (BC map). All schemes describe the relationship between genes (*red squares*) and the subcomponents (*black circles*) of a phenotypic trait, when the trait is modularized (*gray circles*). The G-P map describes the construction of modularity through changes in pleiotropic effects (*red arrows*), whereas the D map describes the modulated pathways of the developmental system (*blue arrows*) affected by changes in pleiotropy. The BC map describes the construction of modularity through the coupling (*green arrows*) and uncoupling (*light green arrows*) of the original developmental pathways of the ground plan (blue arrows), where subcomponents (*black circles*) are homologous parts, and each phenotypic trait is the ground plan of interest (Figure panel a was modified from Wagner and Altenberg (1996), and panel b was modified from Klingenberg (2008))

evolutionary developmental biology. These two schemes cover a broad spectrum of biological traits but are less likely to be practical for deciphering a specific genetic and developmental architecture of traits. From the perspective of comparative morphology, a specific scheme to comprehend the complexity and diversification of traits needs to be established.

How can the genetic and developmental architectures that create various forms by modification of the ground plan be depicted? Considering the experimental facts explained above, two major components seem to be involved: one arises from the original developmental modularity of the ground plan and the other from rewiring the developmental modules of the ground plan. In general, the ground plan is a sum of homologous parts, and it is thought that each homologous part constitutes one developmental module because each part is individually identifiable (Wagner 1989, 2014). An example of the component derived from rewiring the developmental modules of the NGP is the functional modules found in *O. excavata* leaf pattern (Suzuki 2013). In the present review, I propose a scheme for integrating the genetic and developmental architecture underlying the variations of a theme of the ground plan, termed Body plan Character Map (BC Map; Fig. 3.8c). This scheme describes the core generation process of the ground plan and the reorganization process that transforms it into various designs, which can only be revealed using the morphological approach described in this study combined with molecular data.

Acknowledgments I thank Toshio Sekimura, Fred Nijhout, and the Chubu University (Japan) for organizing the 2016 conference and stimulating the writing of this review. I would also like to thank Professor Hideki Sezutsu for his valuable comments on this manuscript and Professor Shigeru Kuratani for his inspired comments on comparative morphology discussion and importance of the macroevolutionary views. This manuscript was rigorously reviewed by two referees, Shigeyuki Koshikawa and Arnaud Martin.

References

Arnold SJ (1983) Morphology, performance and fitness. Am Zool 23:347–361

Beldade P, Brakefield PM (2002) The genetics and evo-devo of butterfly wing patterns. Nat Rev Genet 3:442–452

Beldade P, Peralta CM (2017) Developmental and evolutionary mechanisms shaping butterfly eyespots. Curr Opin Insect Sci 19:22–29

Brakefield PM (1984) The ecological genetics of quantitative characters of *Maniola jurtina* and other butterflies. In: Vane-Wright RI, Ackery PR (eds) Symposia of the royal entomological society, the biology of butterflies. Academic Press, London, pp 167–190

Brakefield PM, Larsen TB (1984) The evolutionary significance of dry and wet season forms in tropical butterflies. Biol J Linn Soc 22:1–22

Brakefield PM, Gates J, Keys D, Kesbeke F, Wijngaarden PJ, Monteiro A, French V, Carroll SB (1996) Development, plasticity and evolution of butterfly eyespot patterns. Nature 384:236–242

Brower AVZ, Schawaroch V (1996) Three steps of homology assessment. Cladistics 12:265–272

Brunetti CR, Selegue JE, Monteiro A, French V, Brakefield PM, Carroll SB (2001) The generation and diversification of butterfly eyespot color patterns. Curr Biol 11:1578–1585

Buzgo M, Soltis DE, Soltis PS, Ma H (2004) Towards a comprehensive integration of morphological and genetic studies of floral development. Trends Plant Sci 9:164–173

Carroll SB (2001) Chance and necessity: the evolution of morphological complexity and diversity. Nature 409:1102–1109

Carroll SB, Gates J, Keys D, Paddock SW, Panganiban GF, Selegue JE, Williams JA (1994) Pattern formation and eyespot determination in butterfly wings. Science 265:109–114

Cheverud JM (1996) Developmental integration and the evolution of pleiotropy. Am Zool 36:44–50

Cook LM, Saccheri IJ (2013) The peppered moth and industrial melanism: evolution of a natural selection case study. Heredity 110:207–212

Cook LM, Grant BS, Saccheri IJ, Mallet J (2012) Selective bird predation on the peppered moth: the last experiment of Michael Majerus. Biol Lett. doi:10.1098/rsbl.2011.1136

Cott HB (1940) Adaptive coloration in animals. Methuen and Co., London

Darwin C (1871) The descent of man. John Murray, London

Davidson EH, Erwin DH (2006) Gene regulatory networks and the evolution of animal body plans. Science 311:796–800

Donoghue MJ (1989) Phylogenies and the analysis of evolutionary sequences, with examples from seed plants. Evolution 43:1137–1156

Duboule D (1994) Temporal collinearity and the phylotypic progression: a basis for the stability of a vertebrate Bauplan and the evolution of morphologies through heterochrony. Development (Suppl):135–142

Edmunds M (1974) Defence in animals. Longman, New York

Endo K (1984) Neuroendocrine regulation of the development of seasonal forms of the Asian comma butterfly *Polygonia c-aureum*. Dev Growth Diff 26:217–222

Endo K, Masaki T, Kumagai K (1988) Neuroendocrine regulation of the development of seasonal morphs in the Asian comma butterfly, *Polygonia c-aureum* L. difference in activity of summer-morph-producing hormone from brain extracts of the long-day and short-day pupae. Zool Sci 5:145–152

Esteve-Altava B (2016) In search of morphological modules: a systematic review. Biol Rev. doi:10.1111/brv.12284

French V, Brakefield PM (1995) Eyespot development of butterfly wings: the focal signal. Dev Biol 168:112–123

Fukada S, Endo K (1966) Hormonal control of the development of seasonal forms in the butterfly *Polygonia c-aureum* L. Proc Jpn Acad 42:1082–1987

Garamszegi LZ (ed) (2014) Modern phylogenetic comparative methods and their application in evolutionary biology, concepts and practice. Springer, Heidelberg

Goldschmidt RB (1945) Mimetic polymorphism, a controversial chapter of Darwin. Q Rev Biol 20:205–230

Graham LE, Cook ME, Busse JS (2000) The origin of plants: body plan changes contributing to a major evolutionary radiation. Proc Natl Acad Sci U S A 97:4535–4540

Hall B (ed) (2000) Homology: the hierarchical basis of comparative biology. Academic Press, San Diego

Harvey PH, Pagel M (1991) The comparative method in evolutionary biology. Oxford University Press, Oxford

Holland LZ, Carvalho JE, Escriva H, Laudet V, Schubert M, Shimeld SM, Yu J-K (2013) Evolution of bilaterian central nervous systems: a single origin? EvoDevo 4:27

Hutchinson JR, Delmer C, Miller CE, Hildebrandt T, Pitsillides AA, Boyde A (2011) From flat foot to fat foot: structure, ontogeny, function, and evolution of elephant "sixth toes.". Science 334:1699–1703

Irie N, Kuratani S (2011) Comparative transcriptome analysis reveals vertebrate phylotypic period during organogenesis. Nat Commun 2. doi:10.1038/ncomms1248

Irie N, Kuratani S (2014) The developmental hourglass model: a predictor of the basic body plan? Development 141:4649–4655

Jacob F (1977) Evolution and tinkering. Science 196:1161–1166

Jiggins CD, Wallbank RWR, Hanly JJ (2017) Waiting in the wings: what can we learn about gene co-option from the diversification of butterfly wing patterns? Phil Trans R Soc B 372:20150485. doi:10.1098/rstb.2015.0485

Kalinka AT, Varga KM, Gerrard DT, Preibisch S, Corcoran DL, Jarrells J, Ohler U, Bergman CM, Tomancak P (2010) Gene expression divergence recapitulates the developmental hourglass model. Nature 468:811–814

Keys DN, Lewis DL, Selegue JE, Pearson BJ, Goodrich LV, Johnson RL, Gates J, Scott MP, Carroll SB (1999) Recruitment of a hedgehog regulatory circuit in butterfly eyespot evolution. Science 283:532–534

Kirschner M, Gerhart J (1998) Evolvability. Proc Natl Acad Sci U S A 95:8420–8427

Klingenberg CP (2008) Morphological integration and developmental modularity. Ann Rev Ecol Evol Syst 39:115–132

Koch PB, Bückmann D (1985) The seasonal dimorphism of *Araschnia levana* L. (Nymphalidae) in relation to hormonal controlled development. Verb Dt Zool Ges 78:260

Lande R (1979) Quantitative genetic analysis of multivariate evolution, applied to brain: body size allometry. Evolution 33:402–416

Levin M, Anavy L, Cole AG, Winter E, Mostov N, Khair S, Senderovich N, Kovalev E, Silver DH, Feder M, Fernandez-Valverde SL, Nakanishi N, Simmons D, Simakov O, Larsson T, Liu S-Y, Jerafi-Vider A, Yaniv K, Ryan JF, Martindale MQ, Rink JC, Arendt D, Degnan SM, Degnan BM, Hashimshony T, Yanai I (2016) The mid-developmental transition and the evolution of animal body plans. Nature 531:637–641

Losos JB, Miles DB (1994) Adaptation, constraint, and the comparative method: phylogenetic issues and methods. In: Wainwright PC, Reilly S (eds) Ecological morphology: integrative organismal biology. University of Chicago Press, Chicago, pp 60–98

Luo Z-X (2011) Developmental patterns in Mesozoic evolution of mammal ears. Ann Rev Ecol Evol Syst 42:355–380

Mallet J (1991) Variations on a theme? Nature 354:368

Martin A, Reed RD (2010) Wingless and aristaless2 define a developmental ground plan for moth and butterfly wing pattern evolution. Mol Biol Evol 27:2864–2878

Martin A, Reed RD (2014) Wnt signaling underlies evolution and development of the butterfly wing pattern symmetry systems. Dev Biol 395:367–378

Martin A, Papa R, Nadeau NJ, Hill RI, Counterman BA, Halder G, Jiggins CD, Kronforst MR, Long AD, McMillan WO, Reed RD (2012) Diversification of complex butterfly wing patterns by repeated regulatory evolution of a *Wnt* ligand. Proc Natl Acad Sci U S A 109:12632–12637

Merilaita S, Stevens M (2011) Crypsis through background matching. In: Stevens M, Merilaita S (eds) Animal camouflage, mechanisms and function. Cambridge University Press, Cambridge, pp 17–33

Mivart St GJ (1871) On the genesis of species. Macmillan, London

Moczek AP, Sears KE, Stollewerk A, Wittkopp PJ, Diggle P, Dworkin I, Ledon-Rettig C, Matus DQ, Roth S, Abouheif E, Brown FD, Chiu C-H, Cohen CS, Tomaso AWD, Gilbert SF, Hall B, Love AC, Lyons DC, Sanger TJ, Smith J, Specht C, Vallejo-Marin M, Extavour CG (2015) The significance and scope of evolutionary developmental biology: a vision for the 21st century. Evol Dev 17:198–219

Monteiro A (2015) Origin, development, and evolution of butterfly eyespots. Annu Rev Entomol 60:253–271

Monteiro A, Glaser G, Stockslager S, Glansdorp N, Ramos D (2006) Comparative insights into questions of lepidopteran wing pattern homology. BMC Dev Biol 6:52

Monteiro A, Chen B, Ramos DM, Oliver JC, Tong X, Guo M, Wang W-K, Fazzino L, Kamal F (2013) Distal-less regulates eyespot patterns and melanization in *Bicyclus* butterflies. J Exp Zool B 320:321–331

Monteiro A, Tong X, Bear A, Liew SF, Bhardwaj S, Wasik BR, Dinwiddie A, Bastianelli C, Cheong WF, Wenk MR, Cao H, Prudic KL (2015) Differential expression of ecdysone receptor leads to variation in phenotypic plasticity across serial homologs. PLoS Genet 11:e1005529
Mugleston JD, Song H, Whiting MF (2013) A century of paraphyly: a molecular phylogeny of katydids (Orthoptera: Tettigoniidae) supports multiple origins of leaf-like wings. Mol Phylo Evol 69:1120–1134
Nagashima H, Sugahara F, Takechi M, Ericsson R, Kawashima-Ohya Y, Narita Y, Kuratani S (2009) Evolution of the turtle body plan by the folding and creation of new muscle connections. Science 325:193–196
Nijhout HF (1980) Pattern formation on lepidopteran wings: determination of an eyespot. Dev Biol 80:267–274
Nijhout HF (1991) The development and evolution of butterfly wing patterns. Smithsonian Institution Press, Washington, DC
Nijhout HF (1994) Symmetry systems and compartments in lepidopteran wings: the evolution of a patterning mechanism. Development (Suppl):225–233
Nijhout HF (2001) Elements of butterfly wing patterns. J Exp Zool 291:213–295
Nijhout HF, Wray GA (1986) Homologies in the colour patterns of the genus *Charaxes* (Lepidoptera: Nymphalidae). Biol J Linn Soc 28:387–410
Nijhout HF, Wray GA (1988) Homologies in the colour patterns of the genus *Heliconius* (Lepidoptera: Nymphalidae). Linn Soc Biol J 33:345–365
Oliver JC, Tong XL, Gall LF, Piel WH, Monteiro A (2012) A single origin for nymphalid butterfly eyespots followed by widespread loss of associated gene expression. PLoS Genet 8:e1002893
Olson EC, Miller RL (1958) Morphological integration. University of Chicago Press, Chicago
Pagel M (1994) Detecting correlated evolution on phylogenies: a general method for the comparative analysis of discrete characters. Proc R Soc B 255:37–45
Pagel M (1999a) Inferring the historical patterns of biological evolution. Nature 401:877–884
Pagel M (1999b) The maximum likelihood approach to reconstructing ancestral character states of discrete characters on phylogenies. Syst Biol 48:612–622
Pagel M, Meade A (2006) Bayesian analysis of correlated evolution of discrete characters by reversible-jump Markov chain Monte Carlo. Am Nat 167:808–825
Pagel M, Meade A, Barker D (2004) Bayesian estimation of ancestral character states on phylogenies. Syst Biol 53:673–684
Patterson C (1982) Morphological characters and homology. In: Joysey KA, Friday AE (eds) Problems of phylogenetic reconstruction. Academic Press, London, pp 21–74
Paulsen SM (1994) Quantitative genetics of butterfly wing color patterns. Dev Genet 15:79–91
Paulsen SM (1996) Quantitative genetics of the wing color pattern in the buckeye butterfly (*Precis coenia* and *Precis evarete*): evidence against the constancy of g. Evolution 50:1585–1597
Paulsen SM, Nijhout HF (1993) Phenotypic correlation structure among elements of the color pattern in *Precis coenia* (Lepidoptera: Nymphalidae). Evolution 47:593–618
Pigliucci M (2008) Is evolvability evolvable? Nat Rev Genet 9:75–82
Poulton EB (1890) The colours of animals: their meaning and use, especially considered in the case of insects. Kegan Paul, Trench, Trübner and Co., Ltd., London
Quint M, Drost H-G, Gabel A, Ullrich KK, Bönn U, Grosse I (2012) A transcriptomic hourglass in plant embryogenesis. Nature 490:98–101
Remane A (1952) Die Grundlagen des Naturlichen Systems, der Vergleichenden Anatomie und der Phylogenetik. Theoretische Morphologie und Systematik I. Geest & Portig K.-G., Leipzig
Riedl R (1978) Order in living organisms: a systems analysis of evolution. Wiley, New York
Rieppel O (1988) Fundamentals of comparative biology. Birkhauser Verlag, Basel
Roth VL (1988) The biological basis of homology. In: Humphries CJ (ed) Ontogeny and systematics. Columbia University Press, New York, pp 1–26
Ruxton GD, Sherratt TN, Speed MP (2004) Avoiding attack: the evolutionary ecology of Crypsis, warning signals and mimicry. Oxford University Press, Oxford
Saint-Hilaire EG (1818) Philosophie Anatomique, Tome Premiere. J. B. Baillière, Paris

Sattler R (1984) Homology-a continuing challenge. Syst Botany 9:382–394
Schluter D, Price T, Mooers AØ, Ludwig D (1997) Likelihood of ancestor states in adaptive radiation. Evolution 51:1699–1711
Schmidt BC, Wagner DL, Zacharczenko BV, Zahiri R, Anweiler GG (2014) Polyphyly of lichen-cryptic dagger moths: synonymy of *Agriopodes* Hampson and description of a new basal acronictine genus, *Chloronycta*, gen. n. (Lepidoptera, Noctuidae). Zookeys 421:115–137
Schwanwitsch BN (1924) On the ground-plan of wing-pattern in Nymphalids and certain other families of the Rhopalocerous Lepidoptera. Proc Zool Soc Lond B 34:509–528
Schwanwitsch BN (1956) Color-pattern in Lepidoptera. Entomologeskoe Obozrenie 35:530–546
Simpson GG (1944) Tempo and mode in evolution. Columbia University Press, New York
Skelhorn J (2015) Masquerade. Curr Biol 25:R643–R644
Skelhorn J, Rowland HM, Ruxton GD (2010a) The evolution and ecology of masquerade. Biol J Linn Soc 99:1–8
Skelhorn J, Rowland HM, Speed MP, Ruxton GD (2010b) Masquerade: camouflage without crypsis. Science 327:51
Stayton CT (2015) The definition, recognition, and interpretation of convergent evolution, and two new measures for quantifying and assessing the significance of convergence. Evolution 69:2140–2153
Stevens M (2016) Cheats and deceits: how animals and plants exploit and mislead. Oxford University Press, Oxford
Stevens M, Merilaita S (2009) Animal camouflage: current issues and new perspectives. Phil Trans R Soc B 364:423–427
Süffert F (1927) Zur vergleichenden analyse der schmetterlingszeichnumg. Biol Zentralblatt 47:385–413
Suzuki TK (2013) Modularity of a leaf moth-wing pattern and a versatile characteristic of the wing-pattern ground plan. BMC Evol Biol 13:158
Suzuki TK (2017) On the origin of complex adaptive traits: progress since the Darwin vs. Mivart debate. J Exp Zool B 328:304–320
Suzuki TK, Tomita S, Sezutsu H (2014) Gradual and contingent evolutionary emergence of leaf mimicry in butterfly wings. BMC Evol Biol 14:229
Swofford DL, Maddison WP (1992) Parsimony, character-state reconstructions, and evolutionary inferences. In: Mayden RL (ed) Systematics, historical ecology, & north american freshwater fishes. Stanford University Press, Palo Alto, pp 186–223
van't Hof AE, Campagne P, Rigden DJ, Yung CJ, Lingley J, Quail MA, Hall N, Darby AC, Saccheri IJ (2016) The industrial melanism mutation in British peppered moths is a transposable element. Nature 534:102–105
Vermeij GJ (1973) Adaptation, versatility and evolution. Syst Zool 22:466–477
Wagner GP (1989) The biological homology concept. Ann Rev Ecol Syst 20:51–69
Wagner GP (2007) The developmental genetics of homology. Nat Rev Genet 8:473–479
Wagner GP (2014) Homology, genes, and evolutionary innovation. Princeton University Press, Princeton
Wagner GP, Altenberg L (1996) Complex adaptations and the evolution of evolvability. Evolution 50:967–976
Wagner GP, Pavlicev M, Cheverud JM (2007) The road to modularity. Nat Rev Genet 8:921–931
Wahlberg N, Leneveu J, Kodandaramaiah U, Peña C, Nylin S, Freitas AVL, Brower AVZ (2009) Nymphalid butterflies diversify following near demise at the cretaceous/tertiary boundary. Proc R Soc B 276:4295–4302
Wallace AR (1889) Darwinism: an exploitation of the theory of natural selection with some of its applications. MacMillan & Co., London
Watson DMS, Timofeeff-Ressovsky NW, Salisbury EJ, Turrill WB, Jenkin TJ, Ruggles Gates R, Fisher RA, Diver C, Hale Carpenter GD, Haldane JBS, MacBrid EW, Salaman RN (1936) A discussion on the present state of the theory of natural selection. Proc R Soc B 121:43–73
Weissman A (1902) The evolution theory. Edward Arnold, London

Wiklund C, Tullberg BS (2004) Seasonal polyphenism and leaf mimicry in the comma butterfly. Anim Behav 68:621–627

Wiley EO, Lieberman BS (2011) Phylogenetics: theory and practice of phylogenetic systematics, 2nd edn. Wiley, Hoboken

Williams DM, Ebach MC (2008) Foundations of systematics and biogeography. Springer, New York

Zhang L, Reed RD (2016) Genome editing in butterflies reveals that spalt promotes and distal-less represses eyespot colour patterns. Nat Commun 7:11769

4

Significance of Ligand-Coding Genes for Morphological Evolution

Arnaud Martin and Virginie Courtier-Orgogozo

Abstract What types of genetic changes underlie evolution? Secreted signaling molecules (*syn.* ligands) can induce cells to switch states and thus largely contribute to the emergence of complex forms in multicellular organisms. It has been proposed that morphological evolution should preferentially involve changes in developmental toolkit genes such as signaling pathway components or transcription factors. However, this hypothesis has never been formally confronted to the bulk of accumulated experimental evidence. Here we examine the importance of ligand-coding genes for morphological evolution in animals. We use Gephebase (http://www.gephebase.org), a database of genotype-phenotype relationships for evolutionary changes, and survey the genetic studies that mapped signaling genes as causative loci of morphological variation. To date, 19 signaling genes represent 20% of the cases where an animal morphological change has been mapped to a gene (80/391). This includes the signaling gene *Agouti*, which harbors multiple cis-regulatory alleles linked to color variation in vertebrates, contrasting with the effects of coding variation in its target, the melanocortin receptor MC1R. In sticklebacks, genetic mapping approaches have identified 4 signaling genes out of 14 loci associated with lake adaptations. Finally, in butterflies, a total of 18 allelic variants of the *WntA* Wnt-family ligand cause color pattern adaptations related to wing mimicry, both within and between species. We discuss possible hypotheses explaining these cases of natural replication (genetic parallelism) and conclude that signaling ligand loci are an important source of sequence variation underlying morphological change in nature.

Keywords Signaling ligands • Genotype-phenotype relationships • Mutational target • cis-Regulatory alleles • Gephebase

---

A. Martin (✉)
Department of Biological Sciences, The George Washington University, Washington, DC, USA
e-mail: arnaud@gwu.edu

V. Courtier-Orgogozo
Institut Jacques Monod, CNRS, UMR 7592, Université Paris Diderot, Paris, France

A key aim of developmental biology is to describe the molecular mechanisms underlying pattern formation, i.e., how gene expression patterns are established and how cell differentiation is orchestrated over time. Since the discovery of embryonic induction, which revealed that secreted molecules are capable of instructing and organizing cells in surrounding tissues (Waddington 1940; Spemann and Mangold 2001), cell-cell signaling has become a sine qua non mechanism of pattern formation in many (if not most) developmental systems (Meyerowitz 2002; Rogers and Schier 2011; Urdy 2012; Kicheva and Briscoe 2015). Experimental manipulations of extracellular signals can impact tissue patterning at a distance (Salazar-Ciudad 2006; Nahmad Bensusan 2011; Perrimon et al. 2012; Urdy et al. 2016). It follows that to understand how spatial information is deployed in differentiating tissues, it is critical to characterize the signals that mediate intercellular communication. A handful of genes coding extracellular proteins that act as signaling molecules between neighboring cells have been identified in animals (Nichols et al. 2006; Rokas 2008a; Perrimon et al. 2012): Wnt, TGF-beta, Hedgehog, Notch, EGF, RTK ligands, and TNFs, among other families. These signaling ligands are widely conserved and show highly regulated expression patterns (Salvador-Martínez and Salazar-Ciudad 2015).

In the 2000s it was proposed that the construction of multicellular organisms relies on a small set of conserved genes, referred to as the developmental genetic toolkit (DGT), which comprises a few hundred genes from a few dozen gene families involved in two major processes: cell differentiation and cell-cell communication (Carroll et al. 2005; Floyd and Bowman 2007; Rokas 2008b; Erwin 2009). On the other side, genes that are not part of the DGT were attached to vital routine functions such as metabolism, protein synthesis, or cell division. According to the DGT view, spatial information emerges from an interplay between genetic factors involved in signal transduction and transcriptional control. An inevitable consequence is that morphological evolution should be based, to a large extent, on reusing these toolkit components, and it follows that mutations in the DGT genes themselves should cause evolution of form (Carroll et al. 2005; Carroll 2008). Such proposition was formulated at the beginning of the twenty-first century, while few genes underlying morphological evolution had been identified – less than 50 cases in 2001 (Martin and Orgogozo 2013). As of today, the hypothesis that animal morphological evolution is mainly caused by mutations in DGT genes can now be tested further based on micro-evo-devo studies (Nunes et al. 2013) and the analyses of genotype-phenotype variation in nature (Orgogozo et al. 2015; Stern 2011). Here we investigate one aspect of the DGT view, the importance of genes encoding secreted signaling proteins in driving morphological evolution. We examine whether ligand-coding genes are preferential targets for the generation of morphological evolution. In addition, we confront existing data to predictions that the corresponding allelic variation should be (1) potentially adaptive (Barrett and Hoekstra 2011; Pardo-Diaz et al. 2015), (2) replicated over various phylogenetic levels (Gompel and Prud'homme 2009; Kopp 2009; Martin and Orgogozo 2013), and (3) cis-regulatory rather than coding (Prud'homme et al. 2007; Carroll 2008; Stern and Orgogozo 2008; Liao et al. 2010).

4.1 Gephebase: The Database of Genotype-Phenotype Variations

Experimental studies based on the manipulation of gene function in the laboratory – for instance, based on reverse genetics or on a mutant screen followed by forward genetics mapping – describe the overall architecture of the genotype-phenotype map in a given organism. However, the genetic causes of evolutionary change in nature do not necessarily equate to the mutations studied in the laboratory: evolutionary-relevant mutations may represent a particular subset of all possible mutations. To identify the genetic causes of natural differences between individuals, populations, and species, one can perform forward genetics studies that compare two naturally occurring phenotypic states – in general, using linkage mapping of quantitative trait loci or Mendelian genes or association mapping (Stern 2000). The so-called "loci of evolution" or "quantitative trait gene (QTG)" studies identify pairs of alleles linked to a specific phenotypic difference (Orgogozo et al. 2015), for instance between an ancestral and a derived state. These loci are typically genomic targets of selection when the variation is of adaptive or domesticating potential. Due to experimental limitations, the dataset is biased toward large-effect loci and thus misses a large fraction of what constitutes the total genetic template of evolution (Rockman 2012). Nevertheless, we think that it is crucial to gather the findings of this research program under the banner of a resource that would integrate, for comparative and meta-analytical purposes, our growing knowledge of genotype-phenotype relationships. To facilitate the curation and analysis of the relevant literature [see (Stern and Orgogozo 2008; Streisfeld and Rausher 2011; Martin and Orgogozo 2013) for previous examples], we have created Gephebase (http://www.gephebase.org), a database of genotype-phenotype relationships underlying natural and domesticated variation across Eukaryotes. Here, we use Gephebase to reflect on the importance of signaling ligand genes for morphological evolution in animals.

4.2 Method: Construction of Gephebase and Identification of Signaling Genes

Gephebase is a quality-controlled, manually curated database of published associations between genes and phenotypes in Eukaryotes – containing a total of 1400 entries as of December 31, 2016. For now, genes responsible for human disease and for aberrant mutant phenotypes in laboratory model organisms are excluded and can be found in other databases (OMIM, OMIA, FlyBase, etc.). QTL mapping studies whose resolution did not reach the level of the nucleotide or of the transcriptional unit are also excluded. In Gephebase, each genotype-phenotype association is attributed to only one type of experimental evidence among three possibilities: "association mapping," "linkage mapping," or "candidate gene." This choice is made by Gephebase curators based on the best evidence available for a given genotype-phenotype relationship. Gene-to-phenotype associations identified by linkage mapping with resolutions below 500 kb have priority in the dataset (see Supplementary Materials in Martin and Orgogozo 2013). Association mapping studies are included based on individual judgment, with a strong bias toward SNP-to-phenotype associations that have been confirmed in reverse genetic studies. In other words, Gephebase intends to be more stringent than a compilation of statistically significant SNPs, and attempts to select studies where a given genotype-phenotype association is relatively well supported or understood.

Gephebase presents itself as a collection of entries, where each entry corresponds to an allelic difference at a given gene, either between two closely related species or between two individuals, its associated phenotypic change, and the relevant publications. As of today, the database contains a total of 391 entries related to animal morphological changes: 174 for domesticated or artificially selected traits, 172 for intraspecific trait variations, and 45 for interspecific changes (Table S1, available at http://virginiecourtier.wordpress.com/publications/. We identify 80 cases of natural morphological evolution and domestication in animals (out of 391) that involve 21 different ligand genes (Table 4.1; Table S2, available at http://virginiecourtier.wordpress.com/publications/).

To estimate the proportion of genes encoding signaling ligands in genomes (Table 4.2), we used the BioMart portal from Ensembl (Smedley et al. 2015). All the genes, which have both the following Gene Ontology (GO) annotations, "receptor binding" (Molecular Function, GO:0005102) and "extracellular region" (Cellular Component, GO:0005576), were considered as ligand genes. To count the number of genes with two GO annotations, we used BioMart to extract text files containing Ensembl Gene ID for each GO and each species. We then counted the number of genes having both GO in each species with the following Linux command: *comm -1 -2 <(sort human-GO0005102.txt) <(sort human-GO0005576.txt) | head -n -1 | wc -l* (note that the title line had to be excluded from the count).

Box 4.1: Definitions

Admixture Mapping: a method capitalizing on the current gene flow between two or more previously isolated populations to associate genetic loci to phenotypic traits. Admixture mapping is a form of association mapping.

Association Mapping: a forward genetics method for gephe identification based on a genome-wide statistical association between genetic variants and phenotypic traits, generally in a large cohort of unrelated individuals.

Candidate Gene Approach: a reverse genetics method that tests if a locus defined a priori, based on our current biological knowledge, underlies variation in a phenotype of interest. *Example*: opsin photoreceptor genes are typical candidate genes for differences in color vision.

(continued)

Box 4.1 (continued)

Forward genetics: set of methods used to identify the genetic cause(s) of a given phenotypic trait ("from the phenotype to genes").

Genetic hotspot: a group of orthologous loci that have been associated multiple times to phenotypic variation due to independent mutational events in each lineage (Martin and Orgogozo 2013).

Gephe (neologism for genotype-phenotype relationship; pronounced *jay-fee*): an abstract entity composed of three elements: a variation at a genetic locus (two alleles), its associated phenotypic change (two distinct phenotypic states, e.g., an ancestral and a derived state), and their relationship (Orgogozo et al. 2015). A gephe is usually defined for a given genetic background and environment.

Haplotype: a set of closely linked alleles found on the same chromosome, which is inherited as a single piece.

Heterotopy: change that occurred during evolution in the location of a particular molecular event within the developing organism.

Linkage Mapping: a forward genetics method for gephe identification based on chromosome shuffling and crossing-overs, using the progeny of a hybrid cross. This includes the mapping of quantitative trait loci (QTL) and Mendelian loci.

Mendelian Gene: a segregating genetic unit which is detected through phenotypic differences associated with different alleles at the same locus (Orgogozo et al. 2016).

Morphospace: an abstract representation of all possible morphologies and shapes of an organism.

Orthologous Loci: pieces of DNA that share ancestry because of a speciation event and that are thus found in different species.

Parallel Evolution: here defined as independent repeated sequence variation at a same locus, underlying variation in a similar phenotypic trait (Stern 2013). For other definitions, see (Scotland 2011).

Phenologue: a similar phenotype caused by a conserved genetic mechanism in distant lineages (McGary et al. 2010; Lehner 2013). Used here as the phenotypic counterpart of a gephe involving several cases of parallel evolution.

Quantitative Trait Locus: a portion of DNA (the locus) that is associated with variation in a quantitative phenotypic trait.

Reverse Genetics: set of methods used to alter a given gene in order to characterize its function ("from genes to phenotypes").

Table 4.1 Twenty-one ligand genes with known gephe variations identified in Gephebase – 19 of which related to cases of morphological evolution (accessed December 31, 2016)

Ligand gene	Trait variation	Nb. of natural gephes (intra-/interspecific) in	Nb. of artificially selected gephes (domesticated) in	Comments	Reference PubMed ID
Agouti (ASIP)	Pigmentation	Mammals/birds (**11**)	Mammals/birds (**14**)	See Fig. 4.1 and Gephebase	See Gephebase
BMP2	Fertility (females) + comb morphology (males)	–	Chicken (**1**)	Pleiotropic and joint effect with *HAO1*	22956912 24655072
BMP3	Craniofacial skeleton	–	Dog (**1**)	Uncertain (possible role of *PRKG2*)	22876193
BMP6	Tooth number	Stickleback fish (**1**)	–	cis-Regulatory allele	25205810 26062935 25732776
BMP15	Fertility (not a morphological trait)	–	Sheep (**4**)	Coding alleles	10888873 23637641
CBD103	Pigmentation	Wolf/coyote (**1**)	Dog (same allele)	Coding allele	19197024 17947548
EDA	Armor plates + schooling behavior	Stickleback fish (**1**)	–	cis-Regulatory allele	15790847 22481358 25629660
EDN3	Pigmentation	–	Chicken (**1**)	Large duplication	22216010 25344733 25741364
FGF5	Hair length	–	Cat (**1**), dog (**1**), donkey (**2**)	Coding alleles	See Gephebase
GDF5	Body size	Human (**1**)	–	Allele unknown (SNP association)	18193045
GDF6	Skeletal traits	Stickleback fish (**1**); human (**1**)	–	cis-Regulatory alleles	26774823

GDF9	Fertility (not a morphological trait)	–	Sheep (**2**)	Coding alleles	19713444 20528846
IGF1	Body size	–	Dog (**1**)	cis-Regulatory allele	17412960
IGF2	Muscle and fat content	–	Pig (**1**)	cis-Regulatory allele	14574411
KITLG	Pigmentation	Stickleback fish (**1**); human (**2**)	Cattle (**1**)	cis-Regulatory alleles	See text
Myostatin (GDF8)	Muscular growth	–	Mammals (**14**)	Coding (12) + cis-reg. (2) alleles	See Gephebase
Rspo2	Hair length	–	Dog (**1**)	cis-Regulatory allele	19713490
scabrous	Bristle number	Fruit fly (**2**)	–	Distinct alleles affect thorax and abdomen	7992053
upd-like	Wing size	Jewel wasp (**1**)	–	cis-Regulatory allele QTL fractionation	22363002
wingless (Wnt1)	Pigment patterns (wing, larva)	Fruit fly (**1**)	Silkworm (**1**)	cis-Regulatory complex alleles	23673642 26034272
WntA	Pigment patterns (wing)	Butterflies (10)	–	cis-Regulatory alleles	See text

Table 4.2 Proportion of signaling ligand-encoding genes in the genome of several model species

	Homo sapiens (GRCh38.p7)	*Mus musculus* (GRCm38.p4)	*G. aculeatus* (BROADS1)	*D. melanogaster* (BDGP6)	*C. elegans* (WBcel235)
Nb. of protein-coding genes	22,285	22,222	20,787	18	20,362
Nb. of genes with GO (molecular function) = "receptor binding"	1592	1435	247	200	198
Nb. of genes with GO (molecular function) = "extracellular region"	4814	4225	334	1016	562
Nb. of genes with GO (molecular function) = "extracellular region" and "receptor binding"	930	771	115	105	86
Proportion of signaling ligand genes	4.17%	3.47%	0.55%	0.75%	0.42%

Ligand-encoding genes are defined as the protein-coding genes associated with both "receptor binding" and "extracellular region" Gene Ontology terms

4.3 A Few Select Genes for Body-Wide Switches in Melanin Production in Tetrapods

Among 294 Gephebase morphology entries for tetrapods (Gephebase search term "Tetrapoda," including mammals and reptiles sensu *largo*), 206 genotype-phenotype relationships relate to pigment variation, including 193 entries identifying components of the melanocyte differentiation pathway. Both sampling and ascertainment biases explain this unusual enrichment. First, pigmentation shows a bulk of variation accessible to breeders and natural selection altogether (Protas and Patel 2008; Linderholm and Larson 2013). In combination with the fact that coloration variation often involves few genes, these features have made pigmentation a favorite target for exploring genotype-phenotype relationships (sampling bias). Second, there is predictability in the genetic basis of melanin pigment variation, as illustrated by the fact that the melanocortin 1 receptor (MC1R), a major regulator of melanocyte activation, is the most represented gene in Gephebase with 84 entries (6% of all 1400 entries). Interestingly, 80% of *MC1R* gephes (67/84) were identified by a candidate gene approach. This pattern illustrates well a latent ascertainment bias in the study of vertebrate pigment variation: when interested in the genetic basis of a color variation involving shifts in melanin types (mammalian coat, bird plumage, etc.), it has become a knee-jerk reflex for biologists to look for amino acid changes in MC1R, in particular in domains that had been functionally characterized. As a matter of fact, all of the 67 MC1R gephes based on a candidate gene approach involve mutations affecting the gene-coding region. Thus, both the phenotypic diversity of vertebrate pigmentation traits and their simple genetic basis explain the overrepresentation of MC1R to a large extent. This said, the fact that the remaining 20% of *MC1R* entries were identified by linkage or association mapping validates the idea that *MC1R* is a bona fide driver of color variation in vertebrates. As an explanation for this trend, it is likely that the MC1R protein hosts tuning sites that can modulate pigmentation without affecting other traits and that its mutations can show a dominant effect prone to a rapid adaptive spread (Mundy 2005; Kopp 2009; Kronforst et al. 2012; Reissmann and Ludwig 2013; Wolf Horrell et al. 2016). Other components of the melanocyte activation cascade also form gephes involved in natural and artificial selection of coloration traits (Fig. 4.1). This includes downstream targets of MC1R signal transduction such as the transcription factor gene *MITF* and the melanogenic genes *TYR*, *TYRP1*, and *Pmel17*, all involved in the biogenesis of eumelanosomes.

Upstream of MC1R, two signaling molecules that interact with receptor function are known as allelic sources of color variation in vertebrates. In particular, the antagonist ligand *Agouti/ASIP* is a genetic hotspot for pigment variation with a total of 28 entries in Gephebase. This includes numerous cases where this gene was identified by linkage or association mapping, both in natural and domesticated contexts (Fig. 4.1a–c), making *Agouti* one of the most commonly mapped genes in our dataset. Coding alleles of *Agouti* are recessive loss-of-function mutations

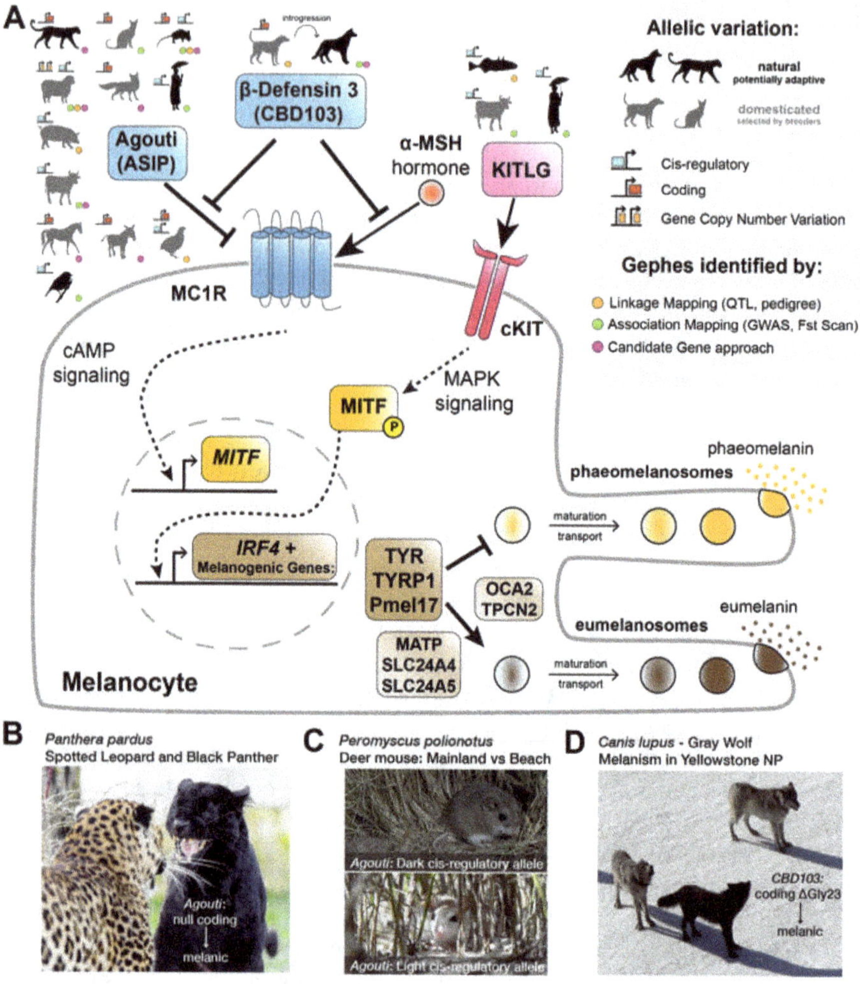

Fig. 4.1 Alleles of secreted ligands associated to pigment variation in vertebrates. (**a**) The MC1R and cKIT signaling pathways each activate a signal transduction regulatory cascade converging on the MITF transcription factor that modulates the expression of melanogenic genes and ultimately activates the maturation and transport of dark eumelanin in melanosomes. Agouti and β-defensin3 are secreted extracellular modulators of MC1R, and KITLG is the agonist ligand of cKIT. Allelic variation at these three genes is associated to pigment variation in vertebrates. (**b**) Black panthers are leopards that carry a null mutation in *Agouti*. (**c**) Adaptive pigment variation in deer mice (*Peromyscus* spp.) has repeatedly involved sequence modifications at the *Agouti* locus. For instance, distinct populations of *P. polionotus* adapted to dark (mainland Florida; *top panel*) and light (coastal Florida; *bottom panel*) color backgrounds via cis-regulatory variants that modulate *Agouti* skin expression. (**d**) Black wolves can be seen at increasing frequencies in packs of the Yellowstone National Park (USA). The melanic allele corresponds to a single amino acid deletion, which was originally selected in domestic dogs and later introgressed in wild in North American wolves and coyotes by hybridization. *Photo credits* – (**b**) Emmanuel Keller,
(**c**) Roger Barbour, (**d**) Doug Smith (Public Domain)

resulting in melanic phenotypes. This contrasts with the melanic gain-of-function coding alleles of MC1R which are dominant, a difference in allelic effects that is used to infer the genetic basis of melanism (Eizirik et al. 2003). The *Agouti* ligand inhibits the basal activation of the MC1R pathway. In an *Agouti*-null context, MC1R is hyper-activated by its active ligand, the pituitary melanocortin hormone α-MSH, which triggers a melanocyte regulatory cascade that culminates with eumelanin production. It has been proposed that wild-type Agouti can become an agonist of MC1R melanic variants (McRobie et al. 2014), suggesting that certain gain-of-function MC1R alleles reverse the responsiveness of the receptor to the Agouti ligand itself. In addition to Agouti, the β-defensin 3/CBD103 peptide is secreted by skin epithelia, strongly binds to MC1R, and was shown to be responsible for melanism in dogs (Candille et al. 2007). In certain melanic dog breeds, one amino acid deletion in β-defensin 3/CBD103 results in dominant melanism, possibly by blocking the inhibitory activity of Agouti or by losing its blocking of α-MSH stimulatory binding (Nix et al. 2013). Of note, the $CBD103^{\Delta G23}$ melanic allele is revealing a complex history that blurs the boundary between wild and domesticated. First, based on ancient DNA studies, it probably originated through domestication from a possible wolflike gene pool as early as 10,000 years ago (Ollivier et al. 2013), introgressing into modern dog breeds. Second, it propagated back in the wild, resulting in relatively recent segregation of melanic phenotypes in North American gray wolves, North American coyotes, and Italian gray wolves (Anderson et al. 2009). The melanic allele shows signatures of positive selection, but it remains unclear if this is due to a fitness effect of the melanic coat or, alternatively, to the antimicrobial properties of β-defensin 3. A few other cases of organism-wide color changes have been found to be positively selected (Vignieri et al. 2010; Barrett and Hoekstra 2011; Laurent et al. 2016).

In conclusion, mutations in MC1R and Agouti account for 54% (112/206) of the gephes dealing with tetrapod pigmentation variation in our current dataset. Such an overrepresentation cannot be explained by experimental bias alone and suggests that MC1R and Agouti are preferential targets for pigmentation evolution in tetrapods.

4.4 cis-Regulatory Evolution Drives Regional Specific Color Shifts

While ligand- and receptor-coding changes likely modulate the strength of signaling, and, thus, pigment synthesis in melanocytes, such changes are likely to affect all the body regions where these genes are expressed. In contrast, region-specific changes in coat, skin, or plumage coloration are more likely to involve cis-regulatory mutations. In a previous meta-analysis of the gephe literature, it was established empirically that localized morphological changes almost always involve cis-regulatory rather than coding variation (Stern and Orgogozo 2008).

Agouti is a hotspot of cis-regulatory evolution for pigment pattern modification and provides one of the most spectacular examples of QTL fractionation. Deer mice display extensive pigment variation matching the color of their environment (Manceau et al. 2010). Fine mapping of this variation revealed that not only the *Agouti* locus is the major driver of pigment variation (Manceau et al. 2011) but also this genetic region decomposes itself into multiple noncoding sub-loci, each tightly associated with parts of the total phenotype (Linnen et al. 2013). Various regulatory elements are involved in directing the expression of three alternative isoforms into different body regions (Mallarino et al. 2016). Each adaptive allele is a complex haplotype that is inherited as a package that underwent multiple local changes. This is of major importance to understand how small leaps in the morphospace occur, as it illustrates the principle that genetic hotspots, in addition to providing a somewhat predictable basis for phenotypic evolution between species, can also accumulate mutations that collectively result in large-effect variation within a single lineage (Stam and Laurie 1996; McGregor et al. 2007; Rebeiz et al. 2011; Martin and Orgogozo 2013; Linnen et al. 2013; Noon et al. 2016).

Thus, the studies of vertebrate pigment variation suggest that a receptor (*MC1R*) and its inverse agonist (*Agouti/ASIP*) are key regulators of melanocyte differentiation, driving adaptive variation in natural contexts as well as novel color features available to farmers and breeders. Coding evolution in either component results in body-wide color shifts, while cis-regulatory evolution of *Agouti*, by tuning the spatial deployment of an MC1R switch-off, permits subtle changes in morphology. The *Agouti/MC1R* axis is not a typical developmental pathway and plays little role during ontogenesis (e.g., see Gene Ontology annotations in Gephebase). In contrast, the *endothelin-3* ligand/*endothelin-receptor B* (EDN3/EDNRB) signaling axis has pleiotropic roles in the differentiation and migration of neural crest cells, and mutations in both EDN3 and EDNRB have been found to cause pigmentation changes in domesticated chicken, cattle, and horse (Santschi et al. 1998; Dorshorst et al. 2011; Qanbari et al. 2014). So far, only domesticated alleles of EDN3/EDNRB that may be under unrealistic selective regimes have been mapped. Thus, while it represents perhaps a genuine DGT component, it remains ambiguous if endothelin pathway genes can be a mutational target of evolution in a natural context. To truly assess the role of signaling ligand genes in morphological evolution, it is useful to focus on radiating lineages that allow a trait-by-trait dissection by forward genetics (i.e., taking advantage of natural variation between closely related lineages – populations and sister species) and, sometimes, natural experiments of replicated evolution (Kopp 2009; Powell and Mariscal 2015). In the next sections, we focus primarily on stickleback fishes and *Heliconius* butterflies, for which numerous linkage mapping efforts have been uncovering the genetic basis of several morphological adaptations.

4.5 Recent Stickleback Fish Adaptations Repeatedly Recruited Ligand Alleles

Three-spined sticklebacks (*Gasterosteus aculeatus*) are a species of marine fishes that repeatedly colonized freshwater environments following the retreat of the Pleistocene glaciers. Adapting to these novel niches involved numerous morphological, physiological, and behavioral modifications all available to genetic dissection by QTL mapping and population scans. Among the 14 gephes that have been mapped in sticklebacks (*Pitx1*, *TSHBeta2*, *KCNH4*, *KITLG*, *EDA*, *GDF6*, *BMP6*, *PRKCD*, *SOD3*, *KCNH4*, *ATP6V0A1*, *ATP1A1*, *Mucin*, *IGK*), 4 involve a secreted ligand gene. Analysis of well-annotated genomes indicates that secreted ligand genes represent less than 5% of the total number of genes within an animal genome (Table 4.2). The proportion of ligand gephes in sticklebacks (28%) is thus higher than expected with the null hypothesis that mutations responsible for phenotypic evolution occur randomly at any gene within a genome (chi^2 test: chi$^2 > 20$; $p < 10^{-5}$).

A single large-effect locus was identified as driving melanin pigment reduction in freshwater populations (Fig. 4.2a). Contrary to expectations, this trait mapped neither to the MC1R pathway nor at its downstream targets, but at the *Kit-ligand* (*KITLG*) locus (Miller et al. 2007; Jones et al. 2012), which encode the secreted signaling component of a parallel pathway (Fig. 4.1). KITLG is the ligand of the KIT receptor, which triggers a MAPK tyrosine kinase transduction cascade that modulates the differentiation and activity of melanocytes (Wehrle-Haller 2003). While the *KIT* receptor has been identified in a total of 17 color-related gephes, it is only linked to domesticated alleles in the cattle, pig, horse, donkey, domesticated fox, and domestic cat (see Advanced Search "Gene name and synonyms" = "KIT" at www.gephebase.org for a complete list). In contrast, cis-regulatory alleles of *KITLG* have been shown to underlie natural pigment variation not only in stickleback fishes but also in humans (Miller et al. 2007; Guenther et al. 2014). An Ala193Asp mutation in *KITLG* has also been shown to cause piebald coat color phenotypes in cattle breeds (Seitz et al. 1999; Qanbari et al. 2014). Of note, cis-regulatory *KITLG* variation may provide tissue-specific effects that limit its potential deleterious pleiotropic effects on cancer risks, as observed in other variant forms of this locus in humans (Karyadi et al. 2013; Litchfield et al. 2016).

Another locus, encoding the bone morphogenetic protein 6 (*BMP6*) ligand, was found to cause tooth gain in freshwater stickleback population (Cleves et al. 2014; Erickson et al. 2015) (Fig. 4.2b). The causal change is cis-regulatory and downregulates *BMP6* expression, late during oral development (see Cleves et al. 2014 correction). Surprisingly, genetic mapping of a second freshwater population revealed that another genomic locus has driven a similar phenotypic output (Ellis

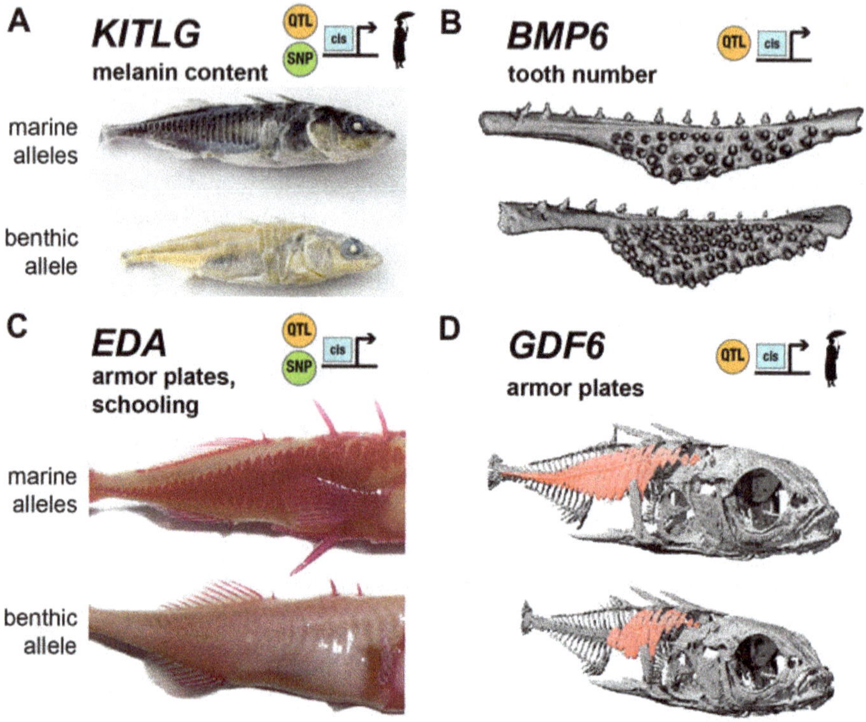

Fig. 4.2 Secreted ligand loci involved in marine-to-lake adaptations in sticklebacks. (**a**) A *KITLG* cis-regulatory variant causes reduced melanization in lake populations (*bottom*) compared to marine alleles (*top*). (**b**) MicroCT images of the tooth plates of a marine vs. a lake-adapted ecotype. The freshwater cis-regulatory *BMP6*-derived allele causes increased tooth area and density. (**c–d**) Armor plates are lateral bony structures, here stained by Alizarin Red (**c**) and false-colored in MicroCT rendering (**d**, *pink*), which were repeatedly reduced or lost in freshwater populations. cis-Regulatory alleles of *EDA* and GDF6 cause distinct effects on plate distribution, number, and size (*Photo credits* – (**a**) Frank Chan and David Kingsley, (**b**) Craig Miller and David Kingsley, (**c**) Nicholas Ellis and Craig Miller, (**d**) Catherine Guenther, Vahan Indjeian, and David Kingsley)

et al. 2015). BMP ligands belong to the TGF-β family, are shared by all bilaterian animals, and play important roles for the regulation of development (De Robertis 2008). Compilation of current data suggests that mutations in TGF-β family genes are often involved in the tinkering of reproductive and skeletal traits during evolution and domestication. Several BMP alleles have been associated to increased fertility in domestic sheeps (*BMP15* and its paralog *GDF9*) (Monestier et al. 2014) and to fecundity and bone allocation in chicken (*BMP2*) (Johnsson et al. 2012). Genetic studies of craniofacial diversity mapped a QTL interval containing the *BMP4* gene in cichlid fishes (Albertson et al. 2005) and found a strong association between a single amino acid change in BMP3 and brachycephalic (short-skulled) dog breeds (Schoenebeck et al. 2012).

Body armor loss, via the reduction of lateral bony plates, has been a recurring adaptation to freshwater in sticklebacks. Two major loci have been characterized. The tumor necrosis factor superfamily gene *Ectodysplasin A* (*EDA*) harbors cis-regulatory variation existing at low frequency in the marine population that has been repeatedly recruited in continental populations to drive plate number reduction (Colosimo et al. 2005; Jones et al. 2012; O'Brown et al. 2015). The same locus also triggers a change in schooling behavior, as fishes from lake habitats have lost the ability to precisely align their body axis when swimming in a group, an effect that is reversed by transgenic overexpression of EDA (Greenwood et al. 2016). In addition, a combination of QTL mapping and genome scan has identified a freshwater-specific allele at the *growth/differentiation factor 6* (*GDF6*) locus, which results in a gain of expression of that gene in the developing epithelium and, ultimately, in a reduction of lateral plate size (Indjeian et al. 2016). Like for *KITLG*, this case also opened a window into human evolution as it was found that a *GDF6* hindlimb-specific enhancer was lost in the human lineage, with skeletal modifications obtained in mice that suggest a potential role in the evolution of bipedalism (Indjeian et al. 2016).

Forward genetics efforts in sticklebacks thus show that ligand genes belonging to classical developmental pathways are an important source of morphological variation of adaptive relevance. Noticeably, all the stickleback gephes described here are cis-regulatory, in accordance with the prediction that tinkering of developmental genes is more likely to involve cis-regulatory changes than coding mutations (Carroll 2008; Stern and Orgogozo 2008). Next, we focus on how accumulated changes in signaling ligand loci have enlarged the landscape of possible morphologies in insect wings.

4.6 The Wnt Beneath My Wings

There are few case studies that characterize adaptive variation for a same set of traits both within and between species. Butterflies of the *Heliconius* genus provide a rich phylogenetic template for such micro-evo-devo studies (Papa et al. 2008; Supple et al. 2014; Kronforst and Papa 2015; Merrill et al. 2015). They display a range of highly variable wing color pattern phenotypes involved in Müllerian mimicry (the collaborative display of similar morphologies to predators from multiple unpalatable species) and sexual selection that are amenable to hybrid crosses followed by linkage mapping. In addition, their natural hybrid zones form a system of choice for high-resolution admixture mapping, looking for SNP-phenotype associations and the smoking guns of selection that are the handful of Mendelian loci that keep adjacent populations phenotypically distinct in the face of constant gene flow and recombination. The Wnt-family signaling ligand WntA has emerged as a key genetic driver of wing pattern evolution in butterflies. Originally discovered as a Mendelian locus responsible for discrete shifts in pattern shapes in the *Heliconius erato* mimetic radiation, this gene shows striking

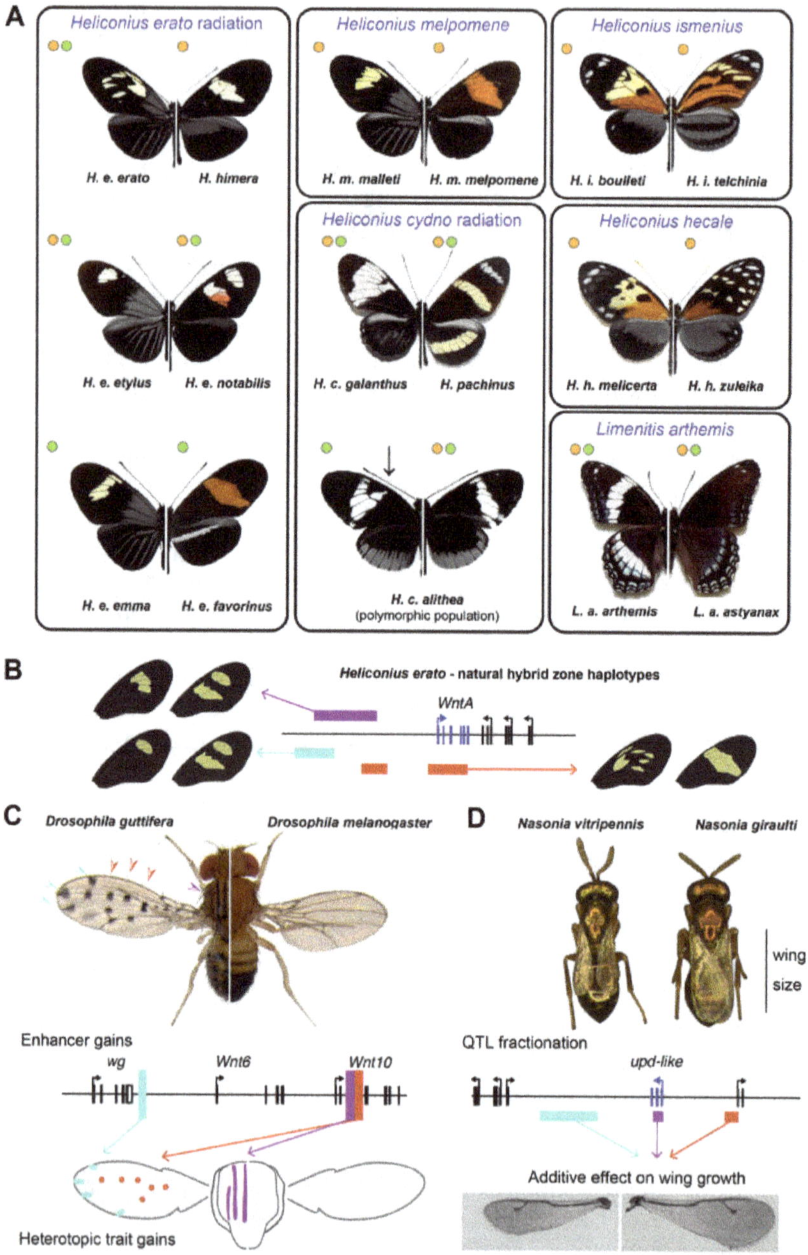

Fig. 4.3 Mapped cis-regulatory alleles of *WntA*, a genetic hotspot of wing pattern shape variation. (**a**) A total of 18 *WntA* cis-regulatory variants have been identified by linkage mapping (*orange dots*) and admixture mapping in natural hybrid zones (*green dots*). Each allele is associated with spatial shifts in *WntA* expression that drive pattern shape variations, in particular, in the median

expression differences in larval wing disks that correlate tightly with the position of presumptive color elements and defines the black contours of forewing color patterns (Martin et al. 2012). Both linkage and admixture mapping approaches have revealed that a versatile pool of *WntA* alleles underlie marked phenotypic differences in at least six geographic races of *H. erato* (Fig. 4.3a, b). Following this discovery, additional mapping efforts discovered that *WntA* variants control pattern variation in four other *Heliconius* species, as well as in *Limenitis arthemis*, a species which diverged from the *Heliconius* genus 65 million years ago (Fig. 4.3a) (Gallant et al. 2014; Huber et al. 2015). All the mapped *WntA* alleles not only underlie phenotypic divergence within species but also convergence between sympatric morphs that evolved in distinct species, thus providing clear examples of adaptive tinkering and repeated evolution of similar patterns. As expected, the causative changes are not found in the *WntA* coding exons, which show little variation in amino acid sequence, but in the adjoining regulatory loci that control *WntA* expression during wing development. The role of *WntA* cis-regulatory mutations may very well extend to much broader phylogenetic levels, as *WntA* expression, which shows spectacular shifts in expression in all the butterfly species assessed so far, always correlates with color pattern features (Martin and Reed 2014). With a total of 18 alleles in 7 species, all associated with wing color pattern variation, *WntA* can be seen as a genuine genetic hotspot of adaptation (Martin and Orgogozo 2013) and a case model for linking regulatory sequence variation, pattern formation, and morphological evolution at multiple time scales.

4.7 Ligand Gene Modularity Allows Interspecific Differences

The current data suggest that the *WntA* locus contains multiple control regions and haplotypes, each being able to reconfigurate part of *WntA* expression and the overall organization of wing patterns. Association mapping reveals at least three adjacent haplotype regions with distinct patterning effects in *H. erato* (Fig. 4.3b) and a single 1.8 kb indel perfectly associated to a polymorphic variant in a sympatric *H. cydno alithea* population (Gallant et al. 2014; Van Belleghem et al. 2017). This said, the

Fig. 4.3 (continued) region of butterfly forewings. Each half-butterfly corresponds to a natural morph. *WntA*-independent color patterns were manually masked and shaded in *gray* to better highlight the wing pattern areas influenced by *WntA*. (**b**) Fractionation of the *H. erato WntA locus* at several haplotypic blocks, each perfectly associated with pattern shape variation across three natural hybrid zones (Van Belleghem et al. 2017). (**c**) Three novel cis-regulatory regions underlie the evolution of novel pigmentation traits in *D. guttifera*. (**d**) Fine QTL mapping of wing size variation in male *Nasonia* wasps identifies three intervals responsible for the differential spatio-temporal recruitment of the *upd-like* growth factor (*Photo credit (use with permission)* – (**c**) Nicolas Gompel and Shigeyuki Koshikawa and (**d**) David Loehlin)

functional dissection of these genetic elements is reaching a technical limitation at this moment due to the inability to test for the function of each cis-regulatory region in butterflies, and we must gain insight into the evolution of ligand gene expression in analog models to explore the logic of cis-regulatory control. Interestingly, detailed analyses of the cis-regulatory region of another *Wnt* locus, this time encompassing *wingless* (syn. *Wnt1*; *wg*) and its tandem paralogs *Wnt6* and *Wnt10* (Fig. 4.3c), show that three novel, tissue-specific cis-regulatory elements drive *wingless* expression and underlie novel color patterns on the wings and thorax of *Drosophila guttifera* fruit flies (Werner et al. 2010; Koshikawa et al. 2015). While these studies lack the phylogenetic resolution and replication observed in butterflies, they provide one of the most detailed mechanistic accounts of truly novel traits, where the deployment of Wnt expression in three different body regions is driven by independent cis-regulatory changes. Of note, *wg* is also associated to color patterns and wing contours in both flies and butterflies (Macdonald et al. 2010; Martin and Reed 2010; Koshikawa et al. 2015), and a redeployment of this gene to new body regions is likely to drive the evolution of new patterns as well, as it seemed to have occurred during the evolution of larval cuticle patterns in Lepidoptera (Yamaguchi et al. 2013). We note that while Koshikawa et al. did not detect any pattern-related *Wnt6* and *Wnt10* expression in *D. guttifera* developing wings (Koshikawa et al. 2015; S. Koshikawa, personal communication), these two paralogs are co-deployed with *wg* in butterflies where they may underlie a more complex architecture, with partially redundant ligand activities (Martin and Reed 2014). Beyond their obvious parallels (wing pigmentation traits; *Wnt* loci), the butterfly and *D. guttifera* data collectively depict a modular landscape of pattern evolution where acquisitions and modifications of cis-regulatory elements allow a fine-tuning of color patterns (Koshikawa 2015).

Another case study provides further support for linking gene regulatory region modularity at a ligand locus and interspecific variation (Loehlin and Werren 2012). Using two *Nasonia* wasp sister species, Loehlin and Werren mapped a male wing size variation QTL to the JAK/STAT pathway ligand gene *unpaired-like* (*upd-like*) and, by a genetic tour de force, were able to genetically break down this locus into three regulatory intervals, each with complementary effects on wing size. In fact, each mapped interval affects various complementary spatiotemporal expression patterns of *upd-like*, ultimately affecting wing growth. Thus, whether the phenotypic output is a growth trait (the *upd-like* case) or a color pattern (the *WntA* and *wg* cases), we have empirical evidence that morphological evolvability depends in these cases on the capacity to modify an expression pattern. In a nutshell, the different case studies linking insect wing variation and ligand genes highlight the importance of modular cis-regulatory architecture in the tinkering of anatomy.

4.8 How, When, and Why Ligand Genes Are Likely Drivers of Pattern Variation, or Not

Our cumulative knowledge of evolutionary genetics foreshadows a relative predictability in the genetic mechanisms that drive phenotypic change (Stern and Orgogozo 2009; Martin and Orgogozo 2013; Orgogozo 2015): by laying out what seems to be common mechanisms or trends in the generation of novelty, we can formulate post hoc expectations that can be generalized over broad taxonomic ranges. The cases of Wnt-based color pattern variation discussed above, *WntA* in nymphalid butterflies and *wg* in *D. guttifera*, both provide a useful model framework for understanding the molecular logic of pattern evolution due to their relative simplicity, as they take place in the two-dimensional canvas of the insect wing epithelium. To the best of our knowledge, these patterning systems are uncoupled from tissue growth, which prevents the complex dynamics found in many other morphological contexts (Salazar-Ciudad 2006; Salazar-Ciudad 2009; Urdy et al. 2016). As simplified spatial output of cellular differentiation, color patterns can be used as a proxy for more complex morphologies, providing fundamental insights that can be applied across all animals. A simple ascertainment emerges from the fly and butterfly data: cis-regulatory evolution of pattern-inducing signaling genes has repeatedly driven the evolution of new patterns and derived pattern shapes. We can elaborate upon a simple gradient model of positional information (Wolpert 1969) generating threshold-dependent pattern boundaries (Fig. 4.4a), to derive five types of ligand gene signaling that can produce morphological outcomes (Fig. 4.4b–f). Since cis-regulatory variation modulates gene expression in time and space, it can affect tissue patterning in multiple ways, and its effect on a ligand gene can be sufficient to induce a new pattern (Fig. 4.4b) or simply change its shape (Fig. 4.4c). In addition, cis-regulatory acquisition of localized repressors can dislocate a pattern and thus affect both pattern number and shape (Fig. 4.4d). Pattern size can also be affected by quantitative or temporal changes in the expression of a secreted factor, without requiring a change in the number of source cells, or, alternatively, by *trans*-interactions upstream of the ligand that would affect its secretion and transport (Fig. 4.4e). Finally, modification in the tissue responsiveness to the signal or its concentration or time-dependent interpretation may modulate the pattern thresholds (e.g., color composition) without affecting the overall size and shape of the pattern (Fig. 4.4f).

These distinct dimensions of pattern variation can be used to generate hypotheses on the molecular targets underlying a given phenotypic state. Below we illustrate this principle, building upon a set of observations made on the variable checkerspot (*Euphydryas chalcedona*). *E. chalcedona* checkerspots display a set of orange patterns outlined by black scales that are each expressing *WntA* or *wg/Wnt6/Wnt10* (Martin and Reed 2014). Each of these patterns can be contracted or expanded by an injection of dextran sulfate or heparin, respectively (Fig. 4.4g). These two sulfated polysaccharide compounds possess a high molecular weight, which restrict them to the extracellular space, and injections are only effective when

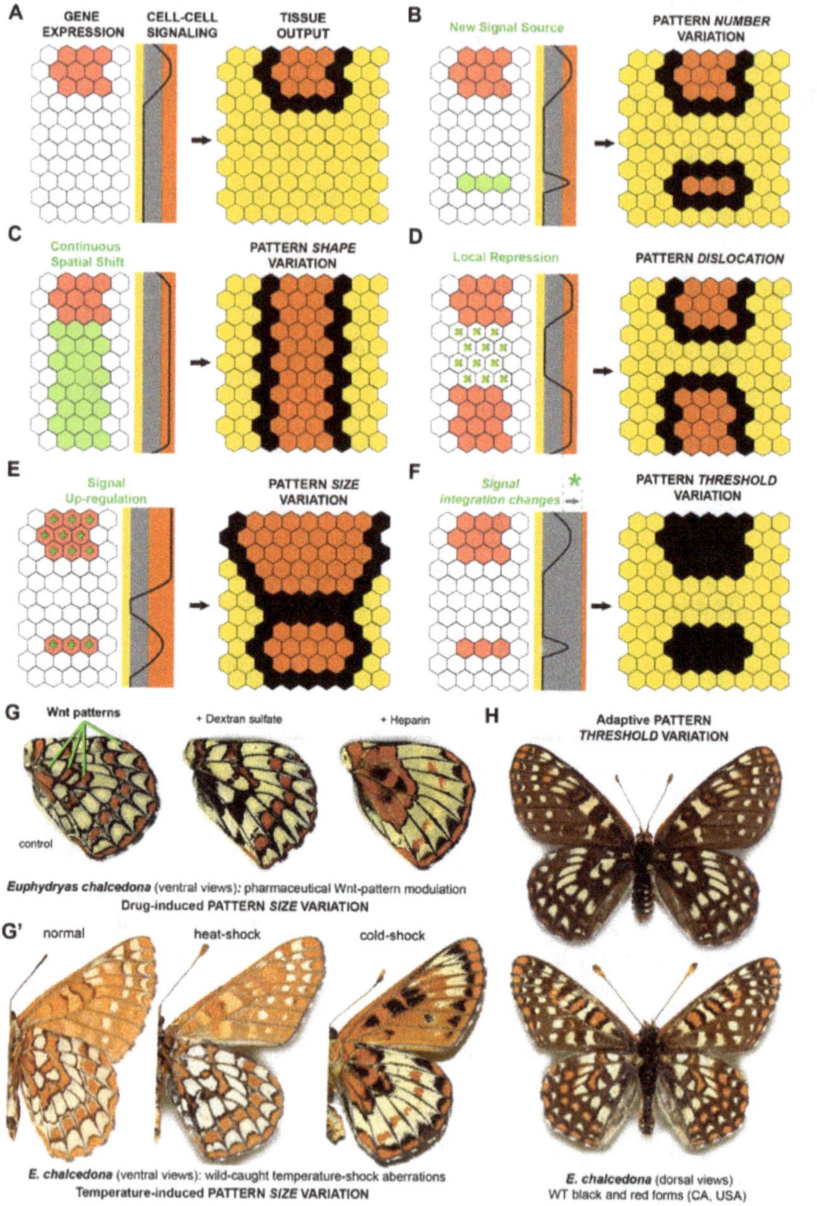

Fig. 4.4 Distinct aspects of pattern variation may rely on different modes of ligand gene modification. (**a**) A three-step model of pattern formation. Ligand-expressing cells (*red hexagons*) deploy a signal that is interpreted by neighboring cells in a concentration-dependent manner, resulting in a three-state output (*yellow*, low signal; *black*, intermediate; *orange*, high). (**b**) Discrete gain of a novel ligand gene expression domain can generate novel pattern elements. (**c**) Continuous spatial modulation of ligand expression can generate new pattern shapes. (**d**) Local

performed within 24 h after pupation, revealing a short time window for pattern formation (Serfas and Carroll 2005; Martin and Reed 2014). Finally, both heparin and endogenous, heparin-like heparan sulfate proteoglycans (HSPGs) are known to bind Wnt ligands in the extracellular space, where they are of critical importance for signal secretion, stability, and transport (Lin 2004). These observations provide a simple alternative mechanism for modifying pattern size: rather than affecting signal strength directly, variation at genes involved in HSPG synthesis could also modulate the spread of Wnt ligands. Similarly, temperature shocks experienced during early pupal life create analogous pattern aberrations (Fig. 4.4g'), suggesting that specific physiological conditions are critical for normal patterning and that, here again, a broad range of molecular mechanisms taking place during cell-cell signaling (e.g., signal secretion, transport, reception, and degradation) could affect pattern size. The variable checkerspot takes its name from the extensive color pattern variations (Bowers et al. 1985; Long et al. 2014b) that can be observed between populations (Fig. 4.4h). Can we predict whether a ligand locus is involved in driving the difference between these Wnt-positive black vs. red/black patterns? Based on the framework developed above, we believe this is in fact an unlikely scenario. Indeed, the variation involves little differences in pattern shape or number and instead consists in color composition differences. A difference in signal sensitivity rather than signal strength between the two forms is more likely to explain the phenomenon, resulting in a threshold trait variation (see Allen et al. 2008 for a discussion of pattern size vs. color composition). We thus predict that this polymorphism could map to a Wnt-pathway gene or to a gene that can modify the output of the Wnt signaling pathway and that this gene should be active during the extracellular signaling phase or shortly thereafter. Alternatively, the threshold traits could also depend on signal temporal dynamics (Sorre et al. 2014). To be formally tested, these competing hypotheses will require linkage or association mapping between natural morphs and illustrate how our current knowledge can guide a different set of predictions, based on the type of observed trait variation.

Fig. 4.4 (continued) loss of ligand expression can result in pattern dislocation. (**e**) Upregulation of a ligand gene can generate enlarged patterns. (**f**) Pattern composition may vary based on modifications of the signal interpretation process, downstream of the ligand gene itself (without affecting its expression or protein). (**g**) Sulfated polysaccharide injections in the variable checkerspot butterfly, performed within 24 h after pupation, affect the size of Wnt-positive patterns. Dextran sulfate results in Wnt pattern contractions, while heparin results in Wnt gain-of-function effects that expand the same patterns. Both compounds illustrate how genetic modulations of the extracellular environment can modulate pattern size. (**g'**) Temperature shocks during early pupal life result in pattern distortions (similar to G panel), indicating a sensitivity of the signaling step to physiological conditions. (**h**) The variable checkerspot is named after its color pattern polymorphism, involved in adaptive mimicry (Bowers et al. 1985; Long et al. 2014b). Differences in *red patterns* may be due to changes in genes modulating Wnt signal, rather than at a Wnt gene locus itself (see **f**)

4.9 Synthesis: Variations of Morphological Relevance in Ligand-Coding Genes Are cis-Regulatory, Complex, and Multiallelic

We have seen in this review that cis-regulatory alleles in ligand genes can drive morphological evolution in nature. Four cases stand out by the level of scrutiny at which they have been examined, as their experimental dissection shines by exceptional levels of phylogenetic replication or genetic resolution: *Agouti* (*Peromyscus maniculatus* – Nebraska Sandhills: light and dark alleles), *WntA* (*Heliconius* spp. and *Limenitis arthemis* butterflies: wing pattern shape variation), *wg* (*Drosophila guttifera*: acquisition of novel pigmented patterns), and *upd-like* (*Nasonia* spp.: wing size differences). Based on the data at hand, we propose a set of hypotheses that can now be confronted to future experimentation:

1. *Ligand cis-regulatory variation underlies heterotopies.* The four loci above provide clear illustrations of the principle that a local modification of morphology (heterotopy) is likely to be based on cis-regulatory variation. Due to their direct role in cell fate induction, ligand genes can be expressed in new places to influence developmental patterning and eventually anatomical phenotypes.
2. *Gene expression shifts require the accumulation of multiple changes clustered into complex alleles.* Fine mapping of the *Agouti* and *upd-like* loci reveals multiple sub-genic regions which independently contribute to the total phenotype (Loehlin and Werren 2012; Linnen et al. 2013). The same is true for *WntA* in a recent hybrid zone study (Fig. 4.3b), where three noncoding regions were each associated to pattern variation in distinct subareas of the butterfly wing, with their combination constituting the complete phenotype (Van Belleghem et al. 2017). Finally, the *wg* study reveals the modular evolution of three tissue-specific enhancers that collectively explain the pigmentation features of *D. guttifera* (Koshikawa et al. 2015; Koshikawa 2015). These four cases are conceptually similar and show that cis-regulatory evolution relies on the accumulation of multiple changes to generate large effects on ligand expression and final morphology.
3. *Parallel evolution is pervasive, even across distant lineages.* The repeated finding of the same orthologous gene causing similar visible trait changes across distinct lineages may be expected under the candidate gene approach, as a result of ascertainment bias. The replicated identification of coding alleles of *MC1R* and *Agouti* is of that order. However, when independent experiments happen to pinpoint the same locus by taking a linkage or association mapping approach, then we can firmly infer that gene reuse underlies a phenomenon of evolutionary repetition (Martin and Orgogozo 2013; Orgogozo et al. 2015). We have seen that cis-regulatory alleles of *Agouti* have been repeatedly mapped in several populations and species of *Peromyscus* deer mice as well as in humans. The stickleback *KITLG* cis-regulatory changes were mirrored by other cis-regulatory variants driving both skin and hair color variation in human populations (Miller et al. 2007; Guenther et al. 2014). Finally, the *WntA* locus was mapped as a

hotspot of wing pattern evolution in five *Heliconius* species as well as in a clade distant by about 65MY (Gallant et al. 2014). This implies that for a given phenotypic trait, the genetic basis of phenotypic variation may be relatively predictable in a post hoc fashion.

4. *Multiallelism could precede the aggregation of complex alleles.* The identification of multiallelism (syn. polyallelism, genetic heterogeneity) by forward genetic approaches is difficult in spite of their suspected importance in human disease (McClellan and King 2010). Indeed, detecting multiallelism requires a multiple-parent QTL scheme, and this has only been recently implemented in a handful of model organisms (Huang et al. 2011; Long et al. 2014a). Furthermore, GWAS studies typically underestimate the contributions of mixed alleles (Thornton et al. 2013). Several studies have nonetheless found that the pool of cis-regulatory variation influencing gene expression levels is multiallelic, to an overwhelming extent (Gruber and Long 2009; Zhang et al. 2011; King et al. 2014). Does this observation hold up for the spatial shift alleles considered here? As it turns out, replicated mapping within the *H. erato* and *H. cydno* radiations has identified six and four noncoding *WntA* alleles underlying ten distinct wing color shapes in these two species groups, respectively (Martin et al. 2012; Papa et al. 2013; Gallant et al. 2014). *WntA* thus exemplifies how repeated cis-regulatory modification of a ligand gene can replicate both within and between species, spanning a phylogenetic spectrum ranging from recently evolved populations (Van Belleghem et al. 2017) to distant lineages (Gallant et al. 2014). Importantly, this multiallelism probably acts as a prerequisite for the formation of complex alleles, as it is likely that adjacent regulatory regions evolve by recombination between blocks that exist as standing variation, rather than solely by cumulative de novo mutations on the same DNA molecule (Rebeiz et al. 2011; Martin and Orgogozo 2013). A chimeric, polyallelic origin can explain the cis-regulatory evolution of *optix* (Wallbank et al. 2015) (see also chapter by CD Jiggins in this volume), a transcription factor locus that, like *WntA*, shows extensive parallelism and multiallelism in the *Heliconius* genus (Reed et al. 2011; Papa et al. 2013; Martin et al. 2014; Kronforst and Papa 2015; Zhang et al. 2016). We expect that further examples of phenotypic radiations will uncover a multiallelic basis, as recently proposed in cichlid fishes (Roberts et al. 2016). The fact that we observe genetic heterogeneity shows that multiple variants can swarm in a gene pool and may thus provide the bricks of change to build novel cis-regulatory activities. We suggest that the large *Agouti*, *upd-like*, *WntA*, and *wg* haplotypes were agglomerated by recombination between multiple alleles segregating in ancestral populations (Martin and Orgogozo 2013).

4.10 Conclusion

In less than a decade, the DGT hypothesis has found validation in the forward genetics literature, where investigations that focused on a morphological difference (without a strong initial bias on the underlying genetics) eventually identified genetic toolkit loci. This is particularly true for signaling genes: four out of seven morphological gephes in sticklebacks involve secreted signaling ligands, and 18 *WntA* alleles have been associated to wing pattern variation in butterflies. We hope that the continuous compilation of the genetic basis of phenotypic evolution into Gephebase will facilitate similarly minded questions of broad interest and perhaps yield to broader insights and meta-analytical thinking in evolutionary genetics.

Acknowledgments We thank Toshio Sekimura, Fred Nijhout, and the Chubu University (Japan) for organizing the 2016 conference and stimulating the writing of this review, and Takao Suzuki and Shigeyuki Koshikawa for their comments and suggestions. We are indebted to the team of Atout Libre (France) for developing the software and website behind the Gephebase database, as well as to Stéphane Prigent, Laurent Arnoult, and the 22 participants of the Loci of Evolution Meta-Analysis Workshop (Paris, September 2016) for their contributions to the design and curation of this resource. The development of Gephebase is funded by a John Templeton Foundation grant to AM and VO (JTF award #43903).

References

Albertson RC, Streelman JT, Kocher TD, Yelick PC (2005) Integration and evolution of the cichlid mandible: the molecular basis of alternate feeding strategies. Proc Natl Acad Sci U S A 102:16287–16292

Allen CE, Beldade P, Zwaan BJ, Brakefield PM (2008) Differences in the selection response of serially repeated color pattern characters: standing variation, development, and evolution. BMC Evol Biol 8:94

Anderson TM, Candille SI, Musiani M et al (2009) Molecular and evolutionary history of melanism in North American gray wolves. Science 323:1339–1343

Barrett RD, Hoekstra HE (2011) Molecular spandrels: tests of adaptation at the genetic level. Nat Rev Genet 12:767–780

Bowers MD, Brown IL, Wheye D (1985) Bird predation as a selective agent in a butterfly population. Evolution:93–103

Candille SI, Kaelin CB, Cattanach BM et al (2007) A β-defensin mutation causes black coat color in domestic dogs. Science 318:1418–1423

Carroll SB (2008) Evo-devo and an expanding evolutionary synthesis: a genetic theory of morphological evolution. Cell 134:25–36

Carroll SB, Grenier JK, Weatherbee SD (2005) From DNA to diversity: molecular genetics and the evolution of animal design. Wiley, Somerset

Cleves PA, Ellis NA, Jimenez MT et al (2014) Evolved tooth gain in sticklebacks is associated with a cis-regulatory allele of Bmp6. Proc Natl Acad Sci 111:13912–13917

Colosimo PF, Hosemann KE, Balabhadra S et al (2005) Widespread parallel evolution in sticklebacks by repeated fixation of ectodysplasin alleles. Science 307:1928–1933

De Robertis E (2008) Evo-devo: variations on ancestral themes. Cell 132:185–195

Dorshorst B, Molin A-M, Rubin C-J et al (2011) A complex genomic rearrangement involving the endothelin 3 locus causes dermal hyperpigmentation in the chicken. PLoS Genet 7:e1002412

Eizirik E, Yuhki N, Johnson WE et al (2003) Molecular genetics and evolution of melanism in the cat family. Curr Biol 13:448–453

Ellis NA, Glazer AM, Donde NN et al (2015) Distinct developmental genetic mechanisms underlie convergently evolved tooth gain in sticklebacks. Development 142:2442–2451

Erickson PA, Cleves PA, Ellis NA et al (2015) A 190 base pair, TGF-β responsive tooth and fin enhancer is required for stickleback Bmp6 expression. Dev Biol 401:310–323
Erwin DH (2009) Early origin of the bilaterian developmental toolkit. Philos Trans R Soc Lond Ser B Biol Sci 364:2253–2261
Floyd SK, Bowman JL (2007) The ancestral developmental tool kit of land plants. Int J Plant Sci 168:1–35
Gallant JR, Imhoff VE, Martin A et al (2014) Ancient homology underlies adaptive mimetic diversity across butterflies. Nat Commun 5:4817
Gompel N, Prud'homme B (2009) The causes of repeated genetic evolution. Dev Biol 332:36–47
Greenwood AK, Mills MG, Wark AR et al (2016) Evolution of schooling behavior in threespine sticklebacks is shaped by the eda gene. Genetics 203:677–681
Gruber JD, Long AD (2009) Cis-regulatory variation is typically polyallelic in Drosophila. Genetics 181:661–670
Guenther CA, Tasic B, Luo L et al (2014) A molecular basis for classic blond hair color in Europeans. Nat Genet 46:748–752
Huang X, Paulo M-J, Boer M et al (2011) Analysis of natural allelic variation in Arabidopsis using a multiparent recombinant inbred line population. Proc Natl Acad Sci 108:4488–4493
Huber B, Whibley A, Poul Y et al (2015) Conservatism and novelty in the genetic architecture of adaptation in Heliconius butterflies. Heredity 114:515–524
Indjeian VB, Kingman GA, Jones FC et al (2016) Evolving new skeletal traits by cis-regulatory changes in bone morphogenetic proteins. Cell 164:45–56
Johnsson M, Gustafson I, Rubin C-J et al (2012) A sexual ornament in chickens is affected by pleiotropic alleles at HAO1 and BMP2, selected during domestication. PLoS Genet 8: e1002914
Jones FC, Chan YF, Schmutz J et al (2012) A genome-wide SNP genotyping array reveals patterns of global and repeated species-pair divergence in sticklebacks. Curr Biol 22:83–90
Karyadi DM, Karlins E, Decker B et al (2013) A copy number variant at the KITLG locus likely confers risk for canine squamous cell carcinoma of the digit. PLoS Genet 9:e1003409
Kicheva A, Briscoe J (2015) Developmental pattern formation in phases. Trends Cell Biol 25:579–591
King EG, Sanderson BJ, McNeil CL et al (2014) Genetic dissection of the *Drosophila melanogaster* female head transcriptome reveals widespread allelic heterogeneity. PLoS Genet 10:e1004322
Kopp A (2009) Metamodels and phylogenetic replication: a systematic approach to the evolution of developmental pathways. Evolution 63:2771–2789
Koshikawa S (2015) Enhancer modularity and the evolution of new traits. Fly (Austin) 9:155–159
Koshikawa S, Giorgianni MW, Vaccaro K et al (2015) Gain of cis-regulatory activities underlies novel domains of wingless gene expression in Drosophila. Proc Natl Acad Sci 112:7524–7529
Kronforst MR, Papa R (2015) The functional basis of wing patterning in Heliconius butterflies: the molecules behind mimicry. Genetics 200:1–19
Kronforst MR, Barsh GS, Kopp A et al (2012) Unraveling the thread of nature's tapestry: the genetics of diversity and convergence in animal pigmentation. Pigment Cell Melanoma Res 25:411–433
Laurent S, Pfeifer SP, Settles ML et al (2016) The population genomics of rapid adaptation: disentangling signatures of selection and demography in white sands lizards. Mol Ecol 25:306–323
Lehner B (2013) Genotype to phenotype: lessons from model organisms for human genetics. Nat Rev Genet 14:168–178
Liao B-Y, Weng M-P, Zhang J (2010) Contrasting genetic paths to morphological and physiological evolution. Proc Natl Acad Sci 107:7353–7358
Lin X (2004) Functions of heparan sulfate proteoglycans in cell signaling during development. Development 131:6009–6021

Linderholm A, Larson G (2013) The role of humans in facilitating and sustaining coat colour variation in domestic animals. Semin Cell Dev Biol 24:587–593

Linnen CR, Poh Y-P, Peterson BK et al (2013) Adaptive evolution of multiple traits through multiple mutations at a single gene. Science 339:1312–1316

Litchfield K, Levy M, Huddart RA et al (2016) The genomic landscape of testicular germ cell tumours: from susceptibility to treatment. Nat Rev Urol 13:409–419

Loehlin DW, Werren JH (2012) Evolution of shape by multiple regulatory changes to a growth gene. Science 335:943–947

Long AD, Macdonald SJ, King EG (2014a) Dissecting complex traits using the Drosophila synthetic population resource. Trends Genet 30:488–495

Long EC, Hahn TP, Shapiro AM (2014b) Variation in wing pattern and palatability in a female-limited polymorphic mimicry system. Ecol Evol 4:4543–4552

Macdonald WP, Martin A, Reed RD (2010) Butterfly wings shaped by a molecular cookie cutter: evolutionary radiation of lepidopteran wing shapes associated with a derived cut/wingless wing margin boundary system. Evol Dev 12:296–304

Mallarino R, Linden TA, Linnen CR, Hoekstra HE (2016) The role of isoforms in the evolution of cryptic coloration in Peromyscus mice. Molecular Ecology 26(1):245–258

Manceau M, Domingues VS, Linnen CR et al (2010) Convergence in pigmentation at multiple levels: mutations, genes and function. Philos Trans R Soc Lond Ser B Biol Sci 365:2439–2450

Manceau M, Domingues VS, Mallarino R, Hoekstra HE (2011) The developmental role of Agouti in color pattern evolution. Science 331:1062–1065

Martin A, Orgogozo V (2013) The loci of repeated evolution: a catalog of genetic hotspots of phenotypic variation. Evolution 67:1235–1250. doi:10.1111/evo.12081

Martin A, Reed RD (2010) Wingless and aristaless2 define a developmental ground plan for moth and butterfly wing pattern evolution. Mol Biol Evol 27:2864–2878

Martin A, Reed RD (2014) Wnt signaling underlies evolution and development of the butterfly wing pattern symmetry systems. Dev Biol 395:367–378. doi:10.1016/j.ydbio.2014.08.031

Martin A, Papa R, Nadeau NJ et al (2012) Diversification of complex butterfly wing patterns by repeated regulatory evolution of a Wnt ligand. Proc Natl Acad Sci U S A 109:12632–12637. doi:10.1073/pnas.1204800109

Martin A, McCulloch KJ, Patel NH et al (2014) Multiple recent co-options of Optix associated with novel traits in adaptive butterfly wing radiations. EvoDevo 5:1–14. doi:10.1186/2041-9139-5-7

McClellan J, King M-C (2010) Genetic heterogeneity in human disease. Cell 141:210–217

McGary KL, Park TJ, Woods JO et al (2010) Systematic discovery of nonobvious human disease models through orthologous phenotypes. Proc Natl Acad Sci 107:6544–6549

McGregor AP, Orgogozo V, Delon I et al (2007) Morphological evolution through multiple cis-regulatory mutations at a single gene. Nature 448:587–590

McRobie HR, King LM, Fanutti C et al (2014) Agouti signalling protein is an inverse agonist to the wildtype and agonist to the melanic variant of the melanocortin-1 receptor in the grey squirrel (*Sciurus carolinensis*). FEBS Lett 588:2335–2343

Merrill R, Dasmahapatra K, Davey J et al (2015) The diversification of Heliconius butterflies: what have we learned in 150 years? J Evol Biol 28:1417–1438

Meyerowitz EM (2002) Plants compared to animals: the broadest comparative study of development. Science 295:1482–1485

Miller CT, Beleza S, Pollen AA et al (2007) Cis-regulatory changes in kit ligand expression and parallel evolution of pigmentation in sticklebacks and humans. Cell 131:1179–1189

Monestier O, Servin B, Auclair S et al (2014) Evolutionary origin of bone morphogenetic protein 15 and growth and differentiation factor 9 and differential selective pressure between mono- and polyovulating species. Biol Reprod 91:83

Mundy NI (2005) A window on the genetics of evolution: MC1R and plumage colouration in birds. Proc R Soc Lond B Biol Sci 272:1633–1640

Nahmad Bensusan M (2011) Interpretation and scaling of positional information during development. Dissertation (Ph.D.), California Institute of Technology
Nichols SA, Dirks W, Pearse JS, King N (2006) Early evolution of animal cell signaling and adhesion genes. Proc Natl Acad Sci 103:12451–12456
Nix MA, Kaelin CB, Ta T et al (2013) Molecular and functional analysis of human β-defensin 3 action at melanocortin receptors. Chem Biol 20:784–795
Noon EP-B, Davis FP, Stern DL (2016) Evolved repression overcomes enhancer robustness. Dev Cell 39(5):572–584
Nunes MD, Arif S, Schlötterer C, McGregor AP (2013) A perspective on micro-evo-devo: progress and potential. Genetics 195:625–634.
O'Brown NM, Summers BR, Jones FC et al (2015) A recurrent regulatory change underlying altered expression and wnt response of the stickleback armor plates gene EDA. elife 4:e05290
Ollivier M, Tresset A, Hitte C et al (2013) Evidence of coat color variation sheds new light on ancient canids. PLoS One 8:e75110
Orgogozo V (2015) Replaying the tape of life in the twenty-first century. Interface Focus 5:20150057
Orgogozo V, Morizot B, Martin A (2015) The differential view of genotype–phenotype relationships. Front Genet 6:179
Orgogozo V, Peluffo A, Morizot B (2016) Chapter one-the "Mendelian Gene" and the "Molecular Gene": two relevant concepts of genetic units. Curr Top Dev Biol 119:1–26
Papa R, Martin A, Reed RD (2008) Genomic hotspots of adaptation in butterfly wing pattern evolution. Curr Opin Genet Dev 18:559–564
Papa R, Kapan DD, Counterman BA et al (2013) Multi-allelic major effect genes interact with minor effect QTLs to control adaptive color pattern variation in *Heliconius erato*. PLoS One 8: e57033
Pardo-Diaz C, Salazar C, Jiggins CD (2015) Towards the identification of the loci of adaptive evolution. Methods Ecol Evol 6:445–464
Perrimon N, Pitsouli C, Shilo B-Z (2012) Signaling mechanisms controlling cell fate and embryonic patterning. Cold Spring Harb Perspect Biol 4:a005975
Powell R, Mariscal C (2015) Convergent evolution as natural experiment: the tape of life reconsidered. Interface Focus 5:20150040
Protas ME, Patel NH (2008) Evolution of coloration patterns. Annu Rev Cell Dev Biol 24:425–446
Prud'homme B, Gompel N, Carroll SB (2007) Emerging principles of regulatory evolution. Proc Natl Acad Sci 104:8605–8612
Qanbari S, Pausch H, Jansen S et al (2014) Classic selective sweeps revealed by massive sequencing in cattle. PLoS Genet 10:e1004148
Rebeiz M, Jikomes N, Kassner VA, Carroll SB (2011) Evolutionary origin of a novel gene expression pattern through co-option of the latent activities of existing regulatory sequences. Proc Natl Acad Sci 108:10036–10043
Reed RD, Papa R, Martin A et al (2011) Optix drives the repeated convergent evolution of butterfly wing pattern mimicry. Science 333:1137–1141
Reissmann M, Ludwig A (2013) Pleiotropic effects of coat colour-associated mutations in humans, mice and other mammals. Semin Cell Dev Biol 24(6–7):576–586
Roberts RB, Moore EC, Kocher TD (2016) An allelic series at pax7a is associated with color polymorphism diversity in Lake Malawi cichlid fish. Mol Ecol 26(10):2615–2639
Rockman MV (2012) The QTN program and the alleles that matter for evolution: all that's gold does not glitter. Evolution 66:1–17
Rogers KW, Schier AF (2011) Morphogen gradients: from generation to interpretation. Annu Rev Cell Dev Biol 27:377–407
Rokas A (2008a) The molecular origins of multicellular transitions. Curr Opin Genet Dev 18:472–478
Rokas A (2008b) The origins of multicellularity and the early history of the genetic toolkit for animal development. Annu Rev Genet 42:235–251

Salazar Ciudad I (2006) On the origins of morphological disparity and its diverse developmental bases. BioEssays 28:1112–1122

Salazar-Ciudad I (2009) Looking at the origin of phenotypic variation from pattern formation gene networks. J Biosci 34:573–587

Salvador-Martínez I, Salazar-Ciudad I (2015) How complexity increases in development: an analysis of the spatial–temporal dynamics of 1218 genes in *Drosophila melanogaster*. Dev Biol 405:328–339

Santschi EM, Purdy AK, Valberg SJ et al (1998) Endothelin receptor B polymorphism associated with lethal white foal syndrome in horses. Mamm Genome 9:306–309

Schoenebeck JJ, Hutchinson SA, Byers A et al (2012) Variation of BMP3 contributes to dog breed skull diversity. PLoS Genet 8:e1002849

Scotland RW (2011) What is parallelism? Evol Dev 13:214–227

Seitz JJ, Schmutz SM, Thue TD, Buchanan FC (1999) A missense mutation in the bovine MGF gene is associated with the roan phenotype in belgian blue and shorthorn cattle. Mamm Genome 10:710–712

Serfas MS, Carroll SB (2005) Pharmacologic approaches to butterfly wing patterning: sulfated polysaccharides mimic or antagonize cold shock and alter the interpretation of gradients of positional information. Dev Biol 287:416–424

Smedley D, Haider S, Durinck S et al (2015) The BioMart community portal: an innovative alternative to large, centralized data repositories. Nucleic Acids Res 43:gkv350

Sorre B, Warmflash A, Brivanlou AH, Siggia ED (2014) Encoding of temporal signals by the TGF-β pathway and implications for embryonic patterning. Dev Cell 30:334–342

Spemann H, Mangold H (2001) Über induktion von embryoanlagen durch implantation artfremder organisatoren. Roux'Arch. Entwicklungsmech 1924; 100: 599–638 (translated and reprinted). Int J Dev Biol 45:13–38

Stam LF, Laurie CC (1996) Molecular dissection of a major gene effect on a quantitative trait: the level of alcohol dehydrogenase expression in *Drosophila melanogaster*. Genetics 144:1559–1564

Stern DL (2000) Perspective: evolutionary developmental biology and the problem of variation. Evolution 54:1079–1091

Stern DL (2011) Evolution, development, & the predictable genome. Roberts and Co. Publishers, Greenwood Village

Stern DL (2013) The genetic causes of convergent evolution. Nat Rev Genet 14:751–764

Stern DL, Orgogozo V (2008) The loci of evolution: how predictable is genetic evolution? Evolution 62:2155–2177

Stern DL, Orgogozo V (2009) Is genetic evolution predictable? Science 323:746–751

Streisfeld MA, Rausher MD (2011) Population genetics, pleiotropy, and the preferential fixation of mutations during adaptive evolution. Evolution 65:629–642

Supple M, Papa R, Counterman B, McMillan WO (2014) The genomics of an adaptive radiation: insights across the Heliconius speciation continuum. In: Ecological Genomics. Springer, pp 249–271

Thornton KR, Foran AJ, Long AD (2013) Properties and modeling of GWAS when complex disease risk is due to non-complementing, deleterious mutations in genes of large effect. PLoS Genet 9:e1003258

Urdy S (2012) On the evolution of morphogenetic models: mechano-chemical interactions and an integrated view of cell differentiation, growth, pattern formation and morphogenesis. Biol Rev 87:786–803

Urdy S, Goudemand N, Pantalacci S (2016) Chapter seven-looking beyond the genes: the interplay between signaling pathways and mechanics in the shaping and diversification of epithelial tissues. Curr Top Dev Biol 119:227–290

Van Belleghem SM, Rastas P, Papanicolaou A et al (2017) Complex modular architecture around a simple toolkit of wing pattern genes. Nat Ecol Evol 1:0052. doi:10.1038/s41559-016-0052

Vignieri SN, Larson JG, Hoekstra HE (2010) The selective advantage of crypsis in mice. Evolution 64:2153–2158
Waddington CH (1940) Organisers and genes. Cambridge University Press, Cambridge
Wallbank RW, Baxter SW, Pardo-Díaz C et al (2015) Evolutionary novelty in a butterfly wing pattern through enhancer shuffling. PLoS Biol 14(1):e1002353
Wehrle-Haller B (2003) The role of Kit-ligand in melanocyte development and epidermal homeostasis. Pigment Cell Res 16:287–296
Werner T, Koshikawa S, Williams TM, Carroll SB (2010) Generation of a novel wing colour pattern by the Wingless morphogen. Nature 464:1143–1148
Wolf Horrell EM, Boulanger MC, D'orazio JA (2016) Melanocortin 1 receptor: structure, function and regulation. Front Genet 7:95
Wolpert L (1969) Positional information and the spatial pattern of cellular differentiation. J Theor Biol 25:1–47
Yamaguchi J, Banno Y, Mita K et al (2013) Periodic Wnt1 expression in response to ecdysteroid generates twin-spot markings on caterpillars. Nat Commun 4:1857
Zhang X, Cal AJ, Borevitz JO (2011) Genetic architecture of regulatory variation in *Arabidopsis thaliana*. Genome Res 21:725–733
Zhang W, Dasmahapatra KK, Mallet J et al (2016) Genome-wide introgression among distantly related Heliconius butterfly species. Genome Biol 17:1

Part II
A Comprehensive Evaluation of Evolution and Eyespots

5

Wing Pattern Plasticity and Environmental Variation

Antónia Monteiro

Abstract Phenotypic plasticity refers to the ability of a genotype to develop into different phenotypes in response to environmental cues. In many instances, this ability is an evolved adaptation to enable organisms to adapt to predictable but variable environments in time or space (West-Eberhard MJ, Developmental plasticity and evolution. Oxford Unversity Press, New York, p 794, 2003; Stearns SC, BioScience 39(7):436–445, 1989; Bradshaw AD, Evolutionary significance of phenotypic plasticity in plants. In: Caspari EW (ed) Adv Genet 13. Academic, New York, pp 115–155, 1956; de Jong G, New Phytol 166(1):101–117, 2005; Moran NA, Am Nat 139(5):971–989, 1992). While much research has focused on the ecological and adaptive significance of the alternative phenotypes produced under different environments, relatively little is still known about the proximate physiological and molecular mechanism translating environmental variation to phenotypic variation and how these mechanisms may have evolved (Beldade P, Mateus ARA, Keller RA, Mol Ecol 20(7):1347–1363, 2011).

Here I provide a review of the literature that has explored how environmental variation, in particular seasonal variation, impacts eyespot size in African satyrid butterflies of the genus *Bicyclus*. Plasticity in eyespot size is undeniably the most conspicuous effect of seasonal variation on the appearance of *Bicyclus* species, and perhaps because of this, its ecological and physiological bases have been under investigation since 1984 (Brakefield PM, Reitsma N, Ecol Entomol 16:291–303, 1991; Brakefield PM, Larsen TB, Biol J Linn Soc 22:1–12, 1984). Much subsequent research on members of this genus, and in particular on the model species *Bicyclus anynana*, uncovered, however, many other morphological, behavioral, physiological, and life history traits that are equally impacted by seasons and, in particular, by rearing temperature (Bear A, Monteiro A, Plos One 8(5), 2013; Dion E, Monteiro A, Yew JY, Scientific Reports 6:39002, 2016; Fischer K, Brakefield PM, Zwaan BJ, Ecology 84(12):3138–3147, 2003a; de Jong MA, Kesbeke F,

A. Monteiro (✉)
Department of Biological Sciences, National University of Singapore, 117543 Singapore, Singapore

Yale-NUS College, 138609 Singapore, Singapore
e-mail: antonia.monteiro@nus.edu.sg

Brakefield PM, Zwaan BJ, Climate Res 43(1–2):91–102, 2010; Mateus ARA, Marques-Pita M, Oostra V, Lafuente E, Brakefield PM, Zwaan BJ, et al. Bmc Biology 12, 2014; Windig JJ, Brakefield PM, Reitsma N, Wilson JGM, Ecol Entomol 19:285–298, 1994; Fischer K, Eenhoorn E, Bot AN, Brakefield PM, Zwaan BJ, Proc R Soc B 270(1528):2051–2056, 2003b; Everett A, Tong XL, Briscoe AD, Monteiro A, BMC Evol Biol 12:232, 2012; Prudic KL, Jeon C, Cao H, Monteiro A, Science 331(6013):73–75, 2011; Westerman E, Monteiro A, Plos One 11(2), 2016; Macias-Munoz A, Smith G, Monteiro A, Briscoe AD, Mol Biol Evol 33(1):79–92, 2016). This review, however, focuses solely on eyespots, the original trait that initiated explorations of phenotypic plasticity in this butterfly genus.

Keywords Plasticity • Eyespots • 20-Hydroxyecdysone • Hormone manipulations • Cucurbitacin • Temperature • Developmental plasticity • Sexual ornaments

5.1 Introduction

Insects have relatively short lives, and this promotes the evolution of seasonal forms or polyphenisms. A short life means that insects can live all their lives within a particular season, in regions of the world that have seasons. This also means that cohorts that emerge in different seasons (spring or summer or wet or dry seasons) will encounter very different biotic and abiotic environments. These environments often exert different selection pressures on the appearance of these insects in order to enhance their survival and reproduction in the respective season. The evolution of adaptive phenotypic plasticity is then a natural response to these predictable, recurrent, but alternate environments that different cohorts of insects experience at different times of the year. This type of plasticity is called a seasonal polyphenism and is especially notable in the highly conspicuous wing patterns of butterflies that inhabit seasonal environments (Brakefield and Larsen 1984; Nijhout 1999, 2003).

One type of wing pattern in butterflies that is especially sensitive to seasonality is the eyespot pattern. Eyespots found in the exposed surfaces of the wings (most of the ventral wing surfaces) are often large in the wet season (WS) and small in the dry season (DS) in the African tropics (Brakefield and Larsen 1984), as well as in many other regions of the world (Fig. 5.1). The ecological significance of this plasticity has been explored with a variety of experiments in the field (Brakefield and Frankino 2009; Ho et al. 2016) and in the lab (Lyytinen et al. 2003, 2004; Prudic et al. 2015; Olofsson et al. 2013; Vlieger and Brakefield 2007). The consensus, so far, is that small cryptic eyespots are an adaptation of the butterfly to avoid being detected by vertebrate predators, who predominate in the DS (Lyytinen et al. 2003), whereas the more conspicuous eyespots are an adaptation to deflect the attacks of invertebrate predators, such as mantids, who predominate in the WS (Prudic et al. 2015).

Fig. 5.1 Patterns of plasticity in *Bicyclus anynana* butterflies. Main image depicts a DS female (*left*) mating with a WS male. Eyespots described in this review are named M1 (*white arrow*) and Cu1 (*blue arrow*). The ventral wing surfaces are often exposed to predators with the exception of the Cu1 forewing eyespot, which is often hidden by the hindwing. The *right panels* depict the hidden (*dorsal*) surfaces of a DS female (*top*) and a DS male displaying sexual dimorphism in their Cu1 eyespots

Butterfly eyespots that are found in hidden (mostly dorsal) surfaces have different patterns of plasticity altogether because these eyespots serve different functions in each of the seasons. These eyespots are used in sexual signaling by both sexes (Prudic et al. 2011; Robertson and Monteiro 2005; Costanzo and Monteiro 2007) (Fig. 5.1). Males use these eyespots to signal to females in the WS, and females use the same eyespots to signal to males in the DS. This leads to patterns of size plasticity that are congruent with those from ventral surface eyespots for males (large in WS males and small in DS males) but not for females. DS females, in particular, have abnormally large dorsal eyespots, which they use for sexual signaling to males in this season (Fig. 5.1), which are at odds with the small size of their ventral exposed counterparts. Females, thus, don't display size plasticity in these eyespots – they are large in both seasons. The patterns of sexual selection operating on dorsal eyespots lead to sexual size dimorphism in dorsal Cu1 eyespots in the DS (Fig. 5.1), as well as a male-specific pattern of plasticity for these eyespots (Bhardwaj et al. 2017).

The review that follows looks critically at the literature that has investigated the environmental, physiological, and molecular mechanisms that regulate eyespot size plasticity in both dorsal and ventral eyespots. In addition, the evolution of phenotypic plasticity in eyespot size is also reviewed.

5.2 Physiological Mechanisms of Eyespot Plasticity

Bicyclus anynana is found from Ethiopia to South Africa (Condamin 1973) and has evolved along a range of climates, but the original lab population of *Bicyclus anynana* stems from Malawi, a country with strong seasonality. The arrival of the dry season in Malawi is primarily cued by decreasing temperatures, whereas the arrival of the wet season is cued by increasing temperatures (Brakefield and Reitsma 1991). Lab-rearing experiments, where photoperiod and thermoperiods were varied, confirmed that average temperature and fluctuations in night- and daytime temperature were the most important determinants of eyespot size plasticity in this species (Brakefield and Mazzotta 1995). Food plant quality, however, also affected eyespot size plasticity (Kooi 1995).

Once environmental cues with significant effects on the induction of plasticity were identified, the next investigations probed how and when these cues interacted with the gene regulatory networks that differentiate the eyespot patterns to modify their output in a plastic manner. In particular, these investigations focused on the mechanisms whereby average daily temperature induced the wet and the dry seasonal forms in *B. anynana*.

The first consideration was whether temperature only exerted its effects on wing pattern development during specific developmental windows or critical temperature-sensitive stages. Early work in this system used temperature-shift experiments to identify the critical period during eyespot development that was sensitive to rearing temperature and able to modify the final size of eyespots (Kooi and Brakefield 1999). These experiments used a variety of shifts differing in length of time that the animals were kept at each of the two alternative temperatures (17 and 27 °C) and times of initiation of the shift. Kooi and Brakefield (1999) concluded that the most important period of sensitivity that led to changes in the size of two of the ventral eyespots (forewing M1 and hindwing Cu1 eyespots) was the final 5th larval instar. Furthermore, while they found that temperatures experienced during the first 24 hrs of pupal development still impacted eyespot size, they concluded that temperatures experienced during this period could not shift a WS wing pattern into a DS pattern and vice versa (Kooi and Brakefield 1999).

More recent work replicated these experiments, with narrower window temperature shifts, and confirmed that the late larval period, in particular, the wandering stage of development, when the larvae stop eating and start looking for a place to pupate, was the most temperature-sensitive stage for the determination of ventral eyespot size plasticity of Cu1 ventral hindwing eyespots (Monteiro et al. 2015). These experiments also highlighted that forewing and hindwing ventral Cu1 eyespots in females responded differently to temperature. Forewing Cu1 eyespots, which are normally hidden by the hindwing when the butterfly is at rest (Fig. 5.1), were much less plastic than Cu1 hindwing eyespots, which are always exposed at rest. In addition, the size of the white center in forewing eyespots was not plastic at all (Monteiro et al. 2015). Subsequent work (Bhardwaj et al. 2017), examining plasticity in dorsal eyespots, similarly concluded that the wandering stage is the most temperature-sensitive stage for male eyespots (female eyespots are not plastic). In summary, eyespot size is primarily sensitive to temperature during the wandering stages of development, but size of Cu1 serial homologous eyespots on ventral forewings and hindwings does not respond to temperature in the same way.

Most examples of phenotypic plasticity known from insects seem to rely on a hormonal signal to translate variable environments into variable phenotypes (Nijhout 1999, Beldade et al. 2011). This prompted the search for the hormones responsible for the variation in wing pattern across *B. anynana* seasonal forms. Previous work on two different butterflies, the map butterfly *Araschnia levana* and the buckeye *Junonia coenia*, had discovered that differences in the presence and absence of a peak of the molting hormone, 20-hydroxyecdysone (20E), during the early pupal stage explained the different seasonal forms (spring and summer forms) of these butterflies, displaying different wing colors in response to day length (an important environmental cue used for regulating plasticity in these systems) (Koch and Buckmann 1987; Nijhout 1980; Rountree and Nijhout 1995). 20E became, thus, a candidate hormone to be investigated in connection with eyespot size plasticity in *B. anynana*.

Surprisingly, early work surrounding investigations into the physiological basis of eyespot size plasticity decided not to investigate physiological differences between the seasonal forms but instead focus on physiological differences observed between lines reared at the intermediate temperature of 20 °C, whose eyespots had been artificially selected to mimic the dry and wet season forms (Brakefield et al. 1998; Koch et al. 1996). In addition, titers of 20E were measured in individuals of these WS and DS form "genetic mimics" at different stages of development focusing primarily in the early pupal stages, as no differences were observed between these mimics during the wandering stages (Koch et al. 1996). Titers of 20E measured in the early pupal stage showed small differences between the seasonal form genetic mimics, and 20E injections into the dry season form mimic, which had a natural slower increase of 20E during the pupal stage, showed small (albeit significant) increases in eyespot size toward the phenotype of wet season forms (Koch et al. 1996). Later work, however, showed that these 20E titer differences observed between WS and DS form genetic mimics could more readily explain variation in pupal stage duration than eyespot size differences (Oostra et al. 2011).

Recent work finally measured 20E hemolymph titers in late larvae of temperature-induced WS and DS forms and discovered that levels of 20E differed significantly between the seasonal forms during the wandering stage of development (Monteiro et al. 2015). This is important because this stage of larval development is contained within the 5th and final larval stage, previously identified as the temperature-sensitive period for induction of eyespot size plasticity (Kooi and Brakefield 1999; Monteiro et al. 2015). Levels were higher in WS forms relative to DS forms, indicating a positive correlation between 20E and eyespot size.

To test whether these different levels in 20E were causing the variable wing phenotypes, hormone injections and hormone receptor manipulations were both done. These two types of manipulations, however, are not equivalent, but this has remained unrecognized by many researchers in this field (but see Zera 2007). To test whether the presence of a hormone at a given level is leading to the development of a phenotype, removal of the hormone or its producing cells/organs, or interfering with its specific receptor, are the best type of manipulations to test causation. If this cannot be done, adding hormone to the form with the lower natural levels to mimic the form with the highest levels is also possible. This latter type of manipulation, however, is more challenging to do because levels of the added hormone need to mimic rather than exceed the highest natural levels found in any of the plastic forms. If levels exceed the natural levels, this may lead to abnormal phenotypes that play no role in normal trait development. One way these abnormal phenotypes may emerge is if raising the levels of hormone A beyond some critical level stimulates the production of hormone B, which then impacts the trait of interest directly. In this situation, manipulations of hormone A would lead researchers to conclude incorrectly that it regulates the trait, when in fact it does so only via its effects on hormone B, which was induced due to high abnormal levels of A. Cross talk between hormonal systems is common, and special attention needs to be paid to this (Zera 2007; Orme and Leevers n.d.).

An example of the type of asymmetry in the response that can be observed with the two types of manipulation experiments described above was observed with 20E signal manipulations in the wandering larval stages of *B. anynana*. As mentioned above, WS wanderers have higher levels of 20E relative to DS wanderers. In order to test whether 20E levels at this stage of development were regulating adult eyespot size, injections of cucurbitacin B (CurcB), a EcR receptor antagonist (Dinan et al. 1997), and a control vehicle, were performed in WS wanderers to test whether they led to reduced adult eyespot size (Monteiro et al. 2015). CurcB is a small molecule that binds with high affinity to the ecdysone receptor (EcR), preventing 20E from binding it and preventing downstream signaling from taking place (Dinan et al. 1997). Injecting CurcB into WS forms led to adult butterflies exhibiting small eyespots resembling DS forms (Monteiro et al. 2015). However, Cu1 ventral forewing eyespots, which are less plastic than their Cu1 ventral hindwing counterparts, did not change in size. The asymmetry in the response of the two Cu1 eyespots to CurcB injections can be explained because the EcR receptor is present in Cu1 forewings eyespot centers but is absent in Cu1 hindwing eyespot centers (Monteiro et al. 2015). Absence of the receptor in forewing eyespot centers essentially makes them insensitive to the CurcB manipulation. What is important to note, however, is that these forewing eyespots, despite expressing no EcR, responded to injections of 20E and increased in size, just like their hindwing counterparts that expressed EcR. One possibility is that if 20E levels attained in DS forms via injections were beyond those observed in WS forms, they may have stimulated the production of a second hormone, which also contributed to the regulation of hindwing eyespot size via its own receptor.

To understand how temperature (and hormones) affected eyespot development, Brakefield et al. (Brakefield et al. 1996) looked at an early marker of eyespot development, the transcription factor *Distal-less (Dll)*, in late larvae and in early pupae. Dll showed comparable expression domains in 5th instar larval wings but had a broader domain of expression in the eyespot centers of WS forms in the early pupal stage. In addition, this gene also had a second domain of expression that corresponded to the much broader black disc of scales in an eyespot, which became visible later, around 12 h after pupation (Brunetti et al. 2001a; Monteiro et al. 2006). The larger group of cells expressing Dll clustered in the eyespot center, however, suggested that some time in between the late larvae and early pupal stages, the eyespot centers were becoming larger in response to temperature. A subsequent study looked at two other markers for eyespot development and found that Notch and Engrailed genes were expressed earlier in the eyespot centers of DS forms relative to their later expression in WS forms, suggesting that these genes could be downregulating eyespot size in DS forms (Oliver et al. 2013). The onset of Dll expression in the eyespot centers of WS and DS forms, however, was approximately the same (Oliver et al. 2013). A more recent study (Bhardwaj et al. 2017) showed that a fourth gene expressed in eyespot centers, the ecdysone receptor (EcR), showed an enlargement in its domain of expression during the second half of the wandering stage in WS forms. Cells in the center of dorsal forewing eyespots underwent cell division concurrently with the rise of 20E titers taking place at that stage of development. Other marker genes, such as Spalt, also increased their domains of expression at the same time, concurrently with local cell divisions. Cells in the dorsal eyespot centers of DS males, however, experiencing the lowest levels of 20E hormone, did not undergo cell division and produced a small eyespot center as well as an associated small eyespot. To test whether levels of 20E were directly responsible for the regulation of dorsal eyespot center size via a localized process of cell division, injections of 20E (into DS males) and CurcB (into WS forms) at 60% of wandering stage development were performed and confirmed an effect of 20E levels on the regulation of eyespot center sizes in WS individuals as well as in DS females, the odd sex with the abnormally large eyespots (Bhardwaj et al. 2017).

The experiments above pin the critical stage of regulation of eyespot center size, and eyespot size for both dorsal and ventral eyespots, to the second half of the wandering stage of development. At this stage, rearing temperature leads to variation in 20E titers, which in turn leads to localized patterns of cell divisions in cells that express the EcR receptor (Bhardwaj et al. 2017). These localized patterns of cell division determine the size of the eyespot centers, which are critical determinants of the size of the complete eyespot pattern (Monteiro and Brakefield 1994), and thus impact final eyespot size.

For many years, however, research into the physiological and genetic basis of eyespot size plasticity focused exclusively on the period of development following pupation, which is not as sensitive to temperature as the previous larval wandering stage (Kooi and Brakefield 1999; Monteiro et al. 2015). This period shows variation in timing of 20E titers in the seasonal form "genetic mimics" as well as in the actual seasonal forms (Mateus et al. 2014; Oostra et al. 2011). In particular, titers of 20E are low during the first 24 h (WS) (and 48 h in the DS) after pupation, which is the developmental window believed to be important for eyespot ring differentiation at high temperatures (French and Brakefield 1992; Brunetti et al. 2001b). This period of low hormone titers is followed by steadily rising 20E titers, where titers raise earlier in WS than in DS forms, relative to total development time. Furthermore, injections of large quantities of 20E (0.1 ug) into young pupae (0–6 h old) reared at 20 °C led to no changes in eyespot size (Koch et al. 1996). Eyespot size changed slightly only with injection of 20E doses larger than 0.25 ug at this early pupal stage (Koch et al. 1996). Note that injections of merely 0.006 ug of 20E (a dose that is 16 times smaller than 0.1 ug) into wanderers reared at 17 °C were sufficient to produce an almost complete seasonal form reversal in this butterfly species (Monteiro et al. 2015).

More recent experiments, focusing again on the early pupal stage, remeasured 20E titers in vehicle-injected and 20E-injected young pupae (3% of pupal development) reared at two different temperatures (19 and 27 °C) and documented small but significant differences in 20E hormone titers between vehicle-injected seasonal forms right after the injections (at 3.5% of pupal development) (Mateus et al. 2014). WS forms had slightly higher titers of 20E than DS forms. Differences in 20E titers in vehicle-injected seasonal forms, however, were no longer present at 8% of pupal development. While these titer measurements are not exactly "baseline" measurements for natural levels of 20E across these two rearing temperatures, they nevertheless show differences in 20E levels across the two seasonal forms (Mateus et al. 2014). In order to test the significance of these differences, injections of 20E were performed into both DS and WS seasonal forms at this early stage (3%) of pupal development, as well as at a later stage (16% pupal development), before the large raise in 20E titers. Special attention was paid to changes in the area of each of the color rings (white center, black, and gold ring) in a variety of different eyespots on dorsal, ventral, forewing, and hindwing surfaces, which are being determined at this stage of development (Brunetti et al. 2001b). One point of concern in these experiments, however, is that injections used 0.25 ug of 20E, a dose previously shown to produce effects on ventral wing patterns (Koch et al. 1996) but also shown to lead to unnaturally high levels of 20E titers in the hemolymph of pupae of both seasonal forms at both 3.5 and 8.5% development (Mateus et al. 2014).

These hormone manipulations showed that early (3%), but not later (16%), injections led to a variety of phenotypes. In particular, they affected the area of some of the color rings, of some of the eyespots, on some of the wing surfaces. When expression of EcR was examined across these different wing pattern traits, there was no clear correlation between the traits affected and the presence/absence of EcR expression in that trait (Mateus et al. 2014). It is possible that, as in the injection experiments performed during the wandering stage of development, these injections are stimulating a second hormonal system, which in turn is exerting its effects on the eyespot phenotypes via its own receptor. Alternatively, given that only those eyespots and eyespot traits that were shown to be especially plastic responded to the hormone injections, it is possible that 20E is regulating directly the expression of these traits, but the developmental stage examined only captures effects on individuals with extended periods of sensitivity or heightened sensitivity to the hormone. Alternatively, lower basal levels of EcR observed across the whole pupal wing epidermis are all that is required for 20E signaling to function at the period of development examined. None of the dorsal eyespots, however, responded to the injections (Mateus et al. 2014). This is likely because dorsal eyespot size plasticity, just as ventral eyespot size plasticity (Monteiro et al. 2015), is primarily controlled during the wandering stages of development (Bhardwaj et al. 2017), but perhaps these dorsal eyespots have fewer hormonal systems controlling their development, and cross talk between hormonal systems may have been minimized.

Going forward, future work on the physiological and genetic basis of wing pattern plasticity in any butterfly species should pay attention to a successive series of experiments that progressively narrows down the causative elements of trait plasticity. First, the critical period in development that is responsible for inducing trait plasticity should be identified using shifting experiments (see Monteiro et al. 2015). It is important to study each trait independently and not assume that the window of development controlling the features of a specific trait (say black ring of M1 eyespot on ventral forewing surface) will be the same as that controlling a similar but not identical trait (e.g., white center of the Cu1 eyespot on a different wing surface). Second, the physiological differences present at that stage (not later and not earlier) should be examined to pin down the physiological correlates that may underlie differences in trait development. Third, hormone depletion experiments (first) and hormone addition experiments (second) should be performed in order to mimic the physiological state of the two plastic forms, in a way that is independent of the environmental cue, to test causation. Here it is especially important to not raise hormone levels above those actually observed in the natural forms in order to avoid stimulating other hormone signaling systems in abnormal ways.

5.3 Evolution of Plasticity

Experiments on the evolution of plasticity in *B. anynana* have been of two types: microevolutionary population-level studies and macroevolutionary species-wide comparative studies. I will review these two types of experiments in turn.

The first type of study focused on testing whether genetic variation controlling the slope of a reaction norm, i.e., the sensitivity of ventral eyespot sizes to rearing temperature, was present in individuals of a single population. The initial rearing of different members of a family (representing similar genotypes) across different temperatures identified significant genetic variation for plasticity in a lab population of *B. anynana* (Windig 1994). In particular, variation in how each family responded to the same range of environments (temperatures) was captured via the presence of reaction norms with distinct slopes. However, further investigation concluded that this variation translated to minor changes to slopes when artificial selection was directly applied to the slope. These artificial selection experiments were of two types. The first type of experiment selected for steeper slopes by applying truncation selection for large eyespots at high temperature followed by truncation selection on small eyespots at low temperature, in the following generation (trying to increase the slope) (Wijngaarden and Brakefield 2001). Alternatively, truncation selection was applied for small eyespots at high temperature and large eyespots at low temperature in the following generation (trying to decrease the slope) (Wijngaarden and Brakefield 2001). The second type of experiment split many individual families into four different rearing temperatures, examined what the reaction norms for each family across the three highest temperatures looked like, and then selected those families that had either the steeper or the shallower slopes by breeding from their siblings that were developing at the slowest (and lowest) temperature (Wijngaarden et al. 2002). Both types of experiment indicated that there was little to no genetic variation for slope of the reaction norms.

A different type of experiment, where artificial selection was applied to the size of the eyespots at a constant temperature (28 °C), followed by a subsequent examination of how these populations diverged in eyespot size across a range of rearing temperatures showed, again, no effects on slope of the reaction norms. All eyespots, regardless of starting size, became smaller with decreasing rearing temperature (Holloway and Brakefield 1995).

Despite the microevolutionary experiments above indicating little to no available genetic variation for selection on plasticity in a single lab population of *B. anynana*, the reality is that plasticity did evolve in this species, and this called for a broader exploration regarding the presence of plasticity in different populations of *B. anynana* and different species of *Bicyclus*.

5.4 Plasticity Across Populations and Species

Field collections have concluded that different environmental cues must be used to regulate eyespot size plasticity in different species of *Bicyclus* across Africa. When eyespot measurements of field-collected specimens were correlated with records of environmental variables, it was clear that species from southern regions, where temperature and humidity are positively correlated (warm wet season, cool dry season), use temperature as a cue to regulate eyespot size plasticity, but species from northern regions, where temperature and humidity are negatively correlated (warm dry season, cool wet season), are likely using humidity as the environmental cue that regulates eyespot size plasticity (Roskam and Brakefield 1999).

These predictions were confirmed when five species of *Bicyclus* from Southern Africa (from savannah and savannah-rainforest ecotones) and two from Equatorial Africa (rainforest) were reared in the lab under a common range of temperatures. All the species responded to temperature in a broadly similar way – ventral "exposed" eyespots became larger with increasing rearing temperature (Roskam and Brakefield 1996; Oostra et al. 2014). However, the savannah-rainforest species had steeper reaction norms relative to savannah or seasonal rainforest species (Roskam and Brakefield 1996).

Similar results were obtained in lab experiments where two southern populations of *B. anynana* (although from geographically distant locations in Malawi and South Africa) both developed larger ventral eyespots when reared at warmer temperatures, despite having diverged in absolute eyespot size at each of the temperatures (de Jong et al. 2010).

While common garden rearing experiments have yet to be performed with northern African population/species of *Bicyclus* butterflies, the general consensus emerging is that phenotypic plasticity for eyespot size, where exposed eyespots increase in size with increasing temperature, is an ancestral property for the genus *Bicyclus*, as well as for other related sayrine genera (Roskam and Brakefield 1996; Brakefield and Frankino 2009). When species move to equatorial regions where there is almost no fluctuation in temperature across the year, they do not lose their plastic response, presumably because there are few costs associated with maintaining the genetic mechanisms of temperature sensitivity in wing patterns (Oostra et al. 2014).

Broader explorations of eyespot plasticity are now necessary, beyond the satyrids, for a more complete understanding of the evolution of eyespot size plasticity. Preliminary data (S. Bhardwaj, unpublished) indicates that many nymphalid butterflies outside the satyrids show the exact opposite pattern of plasticity in eyespot size in relation to rearing temperature. High rearing temperatures lead to smaller eyespots, instead of larger eyespots. The ecological significance of these patterns as well as their underlying physiological mechanisms needs to be examined in detail in the future for a more comprehensive examination of how plasticity in eyespots evolved.

5.5 Conclusions

The ecological significance of wing pattern plasticity in *Bicyclus anynana* is becoming increasingly well understood. In particular, exposed eyespots serve a cryptic function in the dry season, whereas they serve a deflection function in the wet season. Nonexposed eyespots serve a sexual signaling function and display their own patterns of plasticity, distinct from those of exposed eyespots. In addition, patterns of plasticity for each eyespot and for each of the color components within an eyespot are very eyespot-specific and need to be studied in isolation. The physiological basis of eyespot size plasticity in this species, unfortunately, focused for a very long time on a developmental period of low temperature sensitivity (the early pupal stage) instead of the more highly sensitive wandering larval stage of development. So, much of the early work in this system needs to be read and interpreted with caution. More recent experiments have clarified the developmental window and the physiological basis for size plasticity of both dorsal and ventral eyespots, and we have only begun to explore how different homologous wing pattern elements respond to the same environmental cue in different ways. Still, much work still remains to be done. For instance, as pointed out above, different species living in different environments are likely to use different cues to regulate homologous wing pattern elements. However, we still don't know which cues are used (besides temperature) and how they affect wing pattern development. We still don't understand how temperature regulates hormone titers in *B. anynana* and how 20E signaling regulates eyespot size, and we have no idea of the role of epigenetic processes, if any, on the regulation of this process. Finally, comparative work across species is necessary to understand when 20E hormone titers became regulated by rearing temperature at the wandering stage of development, when the ecdysone receptor became recruited to eyespot centers, making them sensitive to fluctuating 20E titers, and when genes from the eyespot gene regulatory network became sensitive to 20E signaling.

Acknowledgment I thank Patricia Beldade for the early input on the structure of this review and Shivam Bhardwaj for the lively discussion. Work in the lab regarding plasticity is funded by the Ministry of Education, Singapore, grant MOE2014-T2-1-146.

References

Bear A, Monteiro A (2013) Male courtship rate plasticity in the butterfly Bicyclus anynana is controlled by temperature experienced during the pupal and adult stages. PLoS One 8(5): e64061. doi:10.1371/journal.pone.0064061. PubMed PMID: WOS:000320362700095

Beldade P, Mateus ARA, Keller RA (2011) Evolution and molecular mechanisms of adaptive developmental plasticity. Mol Ecol 20(7):1347–1363. doi:10.1111/j.1365-294X.2011.05016.x. PubMed PMID: WOS:000288705300006

Bhardwaj S, Prudic KL, Bear A, Das Gupta M, Cheong WF, Wenk MR, et al. (2017) Sex differences in 20-hydroxyecdysone hormone levels control sexual dimorphism in butterfly wing patterns. bioRxiv, 124834.

Bradshaw AD (1956) Evolutionary significance of phenotypic plasticity in plants. In: Caspari EW (ed) Adv Genet. 13. Academic, New York, pp 115–155

Brakefield PM, Frankino WA (2009) Polyphenisms in Lepidoptera: multidisciplinary approaches to studies of evolution and development. In: Whitman DW, Ananthakrishnan TN (eds) Phenotypic plasticity in insects mechanisms and consequences. Science Publishers, Plymouth, pp 281–312

Brakefield PM, Larsen TB (1984) The evolutionary significance of dry and wet season forms in some tropical butterflies. Biol J Linn Soc 22:1–12

Brakefield PM, Mazzotta V (1995) Matching field and laboratory environments – effects of neglecting daily temperature-variation on insect reaction norms. J Evol Biol 8(5):559–573. doi:10.1046/j.1420-9101.1995.8050559.x. PubMed PMID: WOS:A1995RY61500002

Brakefield PM, Reitsma N (1991) Phenotypic plasticity, seasonal climate and the population biology of Bicyclus butterflies (Satyridae) in Malawi. Ecol Entomol 16:291–303. doi:10.1111/j.1365-2311.1991.tb00220.x

Brakefield P, Gates J, Keys D, Kesbeke F, Wijngaarden P, Monteiro A et al (1996) Development, plasticity and evolution of butterfly eyespot patterns. Nature 384(6606):236–242. doi:10.1038/384236a0. PubMed PMID: WOS:A1996VU38100042

Brakefield PM, Kesbeke F, Koch PB (1998) The regulation of phenotypic plasticity of eyespots in the butterfly *Bicyclus anynana*. Am Nat 152:853–860

Brunetti C, Selegue J, Monteiro A, French V, Brakefield P, Carroll S (2001a) The generation and diversification of butterfly eyespot color patterns. Curr Biol 11(20):1578–1585. doi:10.1016/S0960-9822(01)00502-4. PubMed PMID: WOS:000171651700015

Brunetti CR, Selegue JE, Monteiro A, French V, Brakefield PM, Carroll SB (2001b) The generation and diversification of butterfly eyespot color patterns. Curr Biol 11:1578–1585

Condamin M (1973) Monographie du genre *Bicyclus* (Lepidoptera, Satyridae). IFAN, Dakar

Costanzo K, Monteiro A (2007) The use of chemical and visual cues in female choice in the butterfly Bicyclus anynana. Proc R Soc Lond B Bio 274(1611):845–851. doi:10.1098/rspb.2006.3729. PubMed PMID: WOS:000243842100012

de Jong G (2005) Evolution of phenotypic plasticity: patterns of plasticity and the emergence of ecotypes. New Phytol 166(1):101–117. doi:10.1111/j.1469-8137.2005.01322.x. PubMed PMID: WOS:000227390500011

de Jong MA, Kesbeke F, Brakefield PM, Zwaan BJ (2010) Geographic variation in thermal plasticity of life history and wing pattern in Bicyclus anynana. Clim Res 43(1–2):91–102. doi:10.3354/cr00881. PubMed PMID: WOS:000280830000010

Dinan L, Whiting P, Girault JP, Lafont R, Dhadialla TS, Cress DE et al (1997) Cucurbitacins are insect steroid hormone antagonists acting at the ecdysteroid receptor. Biochem J 327:643–650. PubMed PMID: ISI:A1997YF86500003

Dion E, Monteiro A, Yew JY (2016) Phenotypic plasticity in sex pheromone production in *Bicyclus anynana* butterflies. Sci Report 6:39002. doi:10.1038/srep39002

Everett A, Tong XL, Briscoe AD, Monteiro A (2012) Phenotypic plasticity in opsin expression in a butterfly compound eye complements sex role reversal. BMC Evol Biol 12:232. doi:10.1186/1471-2148-12-232. PubMed PMID: WOS:000313877800001

Fischer K, Brakefield PM, Zwaan BJ (2003a) Plasticity in butterfly egg size: why larger offspring at lower temperatures? Ecology 84(12):3138–3147. doi:10.1890/02-0733

Fischer K, Eenhoorn E, Bot AN, Brakefield PM, Zwaan BJ (2003b) Cooler butterflies lay larger eggs: developmental plasticity versus acclimation. Proc R Soc B 270(1528):2051–2056. PubMed PMID: 14561294

French V, Brakefield PM (1992) The development of eyespot patterns on butterfly wings: morphogen sources or sinks? Development 116:103–109

Ho S, Schachat SR, Piel WH, Monteiro A (2016) Attack risk for butterflies changes with eyespot number and size. R Soc Open Sci 3:150614

Holloway GJ, Brakefield PM (1995) Artificial selection of reaction norms of wing pattern elements in Bicyclus-anynana. Heredity 74:91–99. PubMed PMID: ISI:A1995QA25300011

Koch PB, Buckmann D (1987) Hormonal-control of seasonal morphs by the timing of ecdysteroid release in Araschnia levana L (Nymphalidae, Lepidoptera). J Insect Physiol 33(11):823–829. PubMed PMID: ISI:A1987K176700006

Koch PB, Brakefield PM, Kesbeke F (1996) Ecdysteroids control eyespot size and wing color pattern in the polyphenic butterfly *Bicyclus anynana* (Lepidoptera, Satyridae). J Insect Physiol 42:223–230

Kooi RE (1995) The effect of food plant quality on wing pattern induction in the tropical butterfly Bicyclus anynana (Satyrinae). In: Sommeijer MJ, Francke PJ (eds), pp 107–112

Kooi RE, Brakefield PM. The critical period for wing pattern induction in the polyphenic tropical butterfly Bicyclus anynana (Satyrinae). J Insect Physiol 1999;45(3):201–212. PubMed

Lyytinen A, Brakefield PM, Mappes J (2003) Significance of butterfly eyespots as an anti-predator device in ground-based and aerial attacks. Oikos 100(2):373–379. PubMed PMID: ISI:000181854900018

Lyytinen A, Brakefield PM, Lindstrom L, Mappes J (2004) Does predation maintain eyespot plasticity in *Bicyclus anynana*? Proc R Soc Lond B Bio 271(1536):279–283. PubMed PMID: ISI:000188694400010

Macias-Munoz A, Smith G, Monteiro A, Briscoe AD (2016) Transcriptome-wide differential gene expression in Bicyclus anynana butterflies: female vision-related genes are more plastic. Mol

Biol Evol 33(1):79–92. doi:10.1093/molbev/msv197. PubMed PMID: WOS:000369992600006

Mateus ARA, Marques-Pita M, Oostra V, Lafuente E, Brakefield PM, Zwaan BJ et al (2014) Adaptive developmental plasticity: compartmentalized responses to environmental cues and to corresponding internal signals provide phenotypic flexibility. BMC Biol 12:97. doi:10.1186/s12915-014-0097-x. PubMed PMID: WOS:000348150100001

Monteiro A, Brakefield P (1994) French v. The evolutionary genetics and developmental basis of wing pattern variation in the butterfly bicyclus-anynana. Evol Int J Org Evol 48(4):1147–1157. doi:10.2307/2410374. PubMed PMID: WOS:A1994QE32300018

Monteiro A, Glaser G, Stockslager S, Glansdorp N, Ramos D (2006) Comparative insights into questions of lepidopteran wing pattern homology. BMC Dev Biol 6. doi:10.1186/1471-213X-6-52. PubMed PMID: WOS:000242189200001

Monteiro A, Tong XL, Bear A, Liew SF, Bhardwaj S, Wasik BR et al (2015) Differential expression of ecdysone receptor leads to variation in phenotypic plasticity across serial homologs. PLoS Genet 11(9). doi:10.1371/journal.pgen.1005529. PubMed PMID: WOS:000362269000043

Moran NA (1992) The evolutionary maintenance of alternative phenotypes. Am Nat 139 (5):971–989. PubMed PMID: ISI:A1992HU07100005

Nijhout HF (1980) Ontogeny of the color pattern on the wings of *Precis coenia* (Lepidoptera: Nymphalidae). Dev Biol 80:275–288

Nijhout HF (1999) Control mechanisms of polyphenic development in insects. Bioscience 49 (3):181–192. doi:10.2307/1313508

Nijhout HF (2003) Development and evolution of adaptive polyphenisms. Evol Dev 5(1):9–18. doi:10.1046/j.1525-142X.2003.03003.x. PubMed PMID: WOS:000179925400003

Oliver JC, Ramos D, Prudic KL, Monteiro A (2013) Temporal gene expression variation associated with eyespot size plasticity in Bicyclus anynana. PLoS One 8(6). doi:10.1371/journal.pone.0065830. PubMed PMID: WOS:000320440500052

Olofsson M, Jakobsson S, Wiklund C (2013) Bird attacks on a butterfly with marginal eyespots and the role of prey concealment against the background. Biol J Linn Soc 109(2):290–297. doi:10.1111/bij.12063. PubMed PMID: WOS:000318809500004

Oostra V, de Jong MA, Invergo BM, Kesbeke F, Wende F, Brakefield PM et al (2011) Translating environmental gradients into discontinuous reaction norms via hormone signalling in a polyphenic butterfly. Proc R Soc B Biol Sci 278(1706):789–797. doi:10.1098/rspb.2010.1560. PubMed PMID: ISI:000286507400021

Oostra V, Brakefield PM, Hiltemann Y, Zwaan BJ, Brattstrom O (2014) On the fate of seasonally plastic traits in a rainforest butterfly under relaxed selection. Ecol Evol 4(13):2654–2667. doi:10.1002/ece3.1114. PubMed PMID: WOS:000339494900004

Orme MH, Leevers SJ Flies on steroids: the interplay between ecdysone and insulin signaling. Cell Metab 2(5):277–278. doi:10.1016/j.cmet.2005.10.005

Prudic KL, Jeon C, Cao H, Monteiro A (2011) Developmental plasticity in sexual roles of butterfly species drives mutual sexual ornamentation. Science 331(6013):73–75. doi:10.1126/science.1197114. PubMed PMID: WOS:000285974000038

Prudic KL, Stoehr AM, Wasik BW, Monteiro A (2015) Invertebrate predators attack eyespots and promote the evolution of phenotypic plasticity. Proc R Soc Lond B Bio 282(1798):20141531

Robertson K, Monteiro A (2005) Female Bicyclus anynana butterflies choose males on the basis of their dorsal UV-reflective eyespot pupils. Proc R Soc Lond B Bio 272(1572):1541–1546. doi:10.1098/rspb.2005.3142. PubMed PMID: WOS:000231504300003

Roskam JC, Brakefield PM (1996) A comparison of temperature-induced polyphenism in African *Bicyclus* butterflies from a savannah-rainforest ecotone. Evol Int J Org Evol 50:2360–2372

Roskam JC, Brakefield PM (1999) Seasonal polyphenism in Bicyclus (Lepidoptera: Satyridae) butterflies: different climates need different cues. Biol J Linn Soc 66(3):345–356. doi:10.1111/j.1095-8312.1999.tb01895.x. PubMed PMID: WOS:000079376200005

Rountree DB, Nijhout HF (1995) Hormonal control of a seasonal polyphenism in precis coenia (Lepidoptera: Nymphalidae). J Insect Physiol 41(11):987–992

Stearns SC (1989) The evolutionary significance of phenotypic plasticity. Bioscience 39 (7):436–445. PubMed PMID: ISI:A1989AC84800005

Vlieger L, Brakefield PM (2007) The deflection hypothesis: eyespots on the margins of butterfly wings do not influence predation by lizards. Biol J Linn Soc 92(4):661–667. PubMed PMID: WOS:000251414500006

West-Eberhard MJ (2003) Developmental plasticity and evolution. Oxford University Press, New York, 794 p

Westerman E, Monteiro A (2016) Rearing temperature influences adult response to changes in mating status. PLoS One 11(2):e0146546. doi:10.1371/journal.pone.0146546. PubMed PMID: WOS:000370046600012

Wijngaarden PJ, Brakefield PM (2001) Lack of response to artificial selection on the slope of reaction norms for seasonal polyphenism in the butterfly Bicyclus anynana. Heredity 87:410–420. doi:10.1046/j.1365-2540.2001.00933.x. PubMed PMID: WOS:000172693900004

Wijngaarden PJ, Koch PB, Brakefield PM (2002) Artificial selection on the shape of reaction norms for eyespot size in the butterfly Bicyclus anynana: direct and correlated responses. J Evol Biol 15(2):290–300. doi:10.1046/j.1420-9101.2002.00380.x. PubMed PMID: WOS:000174709000012

Windig JJ (1994) Reaction norms and the genetic-basis of phenotypic plasticity in the wing pattern of the butterfly Bicyclus anynana. J Evol Biol 7(6):665–695. doi:10.1046/j.1420-9101.1994.7060665.x. PubMed PMID: ISI:A1994PV78800002

Windig JJ, Brakefield PM, Reitsma N, Wilson JGM (1994) Seasonal polyphenism in the wild: survey of wing patterns in five species of *Bicyclus* butterflies in Malawi. Ecol Entomol 19:285–298

Zera AJ (2007) Endocrine analysis in evolutionary-developmental studies of insect polymorphism: hormone manipulation versus direct measurement of hormonal regulators. Evol Dev 9 (5):499–513. PubMed PMID: ISI:000249321200009

6

Eyespot Focus Point Formation and Spatial Variation in Boundary Conditions

Toshio Sekimura and Chandrasekhar Venkataraman

Abstract Despite being the subject of widespread study, many aspects of the development of eyespot patterns in butterfly wings remain poorly understood. In this work, we examine, through numerical simulations, a mathematical model for eyespot focus point formation in which a reaction-diffusion system is assumed to play the role of the patterning mechanism. In the model, changes in the boundary conditions at the veins at the proximal boundary alone are capable of determining whether or not an eyespot focus forms in a given wing cell and the eventual position of focus points within the wing cell. Furthermore, an auxiliary surface reaction-diffusion system posed along the entire proximal boundary of the wing cells is proposed as the mechanism that generates the necessary changes in the proximal boundary profiles. In order to illustrate the robustness of the model, we perform simulations on a curved wing geometry that is somewhat closer to a biological realistic domain than the rectangular wing cells previously considered, and we also illustrate the ability of the model to reproduce experimental results on artificial selection of eyespots.

Keywords Butterfly patterning • Eyespot pattern • Focus point formation • Turing patterns • Reaction-diffusion system • Surface reaction-diffusion system • Surface finite element method

---

T. Sekimura
Department of Biological Chemistry, Graduate School of Bioscience and Biotechnology, Chubu University, Kasugai, Aichi 487-8501, Japan
e-mail: sekimura@isc.chubu.ac.jp

C. Venkataraman (✉)
School of Mathematics and Statistics, University of St Andrews, Fife KY16 9SS, UK
e-mail: cv28@st-andrews.ac.uk

6.1 Introduction

Eyespots, concentric bands of pigment patterning, constitute one of the most studied pattern elements on the wings of butterflies (c.f., Fig. 6.3 for an example). Each eyespot develops around a focus, a small group of cells that sends out a morphogenetic signal that determines the synthesis of circular patterns of pigments in their surroundings. In this work, we consider a model that provides a possible mechanism underlying the determination of the number and locations of eyespots on the wing surface. The model we consider, first described by Sekimura et al. (2015), provides a mechanism that places the foci around which eyespots form in various locations on the entire wing surface. We do not address here subsequent stages of eyespot formation that occurs after the development of the foci.

The model we consider is based on that of Nijhout (1990). The main novelty of the work in Sekimura et al. (2015) was to illustrate that simply changing the conditions assumed to hold at the proximal veins was sufficient to determine whether or not an eyespot formed in a given wing cell. In the present work, we extend the investigations of the models proposed in Sekimura et al. (2015). We show that it is possible to determine the location of eyespots within a wing cell simply by changing the conditions that are assumed to hold at the lateral wing veins that bound the wing cell. Furthermore, we illustrate that it is possible, using a two-stage model, to recapitulate the results of artificial selection experiments in terms of selection and location of eyespots in butterfly wings.

6.2 Modelling

In this section, we describe the mathematical model for focus point formation that we consider in the present work.

6.2.1 *Setting*

As butterfly wing patterns form in two layers that are thought to be separated completely by the middle tissue (e.g. Sekimura et al. 1998), we assume that the formation of eyespots takes place in a single layer of the wing disc. Hence, we model the domain in which eyespot formation occurs as a two-dimensional region. Furthermore, we assume that this two-dimensional region consists of several wing cells, regions bounded by the wing veins, and we consider a region of up to seven wing cells sufficient to represent the entire surface (front or back) of the wing disc. For the sake of simplicity, we assume that each of the wing cells is of the same shape and size.

The model we consider for the formation of focus points is based on that proposed by Nijhout (1990) and consists of a reaction-diffusion system of activator-inhibitor type (Gierer and Meinhardt 1972) posed in each wing cell with time-independent Dirichlet boundary conditions (i.e. a source of chemicals) on the wing veins and Neumann (zero flux) boundary conditions (i.e. no flux of chemicals) at the wing margin.

6.2.2 Mathematical Model

We denote by n_{seg} the number of wing cells. We denote by Ω_i the ith wing cell with boundaries $\Gamma_{m,i}$ (wing margin), $\Gamma_{v,i}, \Gamma_{v,i+1}$ (veins) and $\Gamma_{p,i}$ (proximal boundary). The boundary conditions for the activator (a_1) are Dirichlet (fixed) on the proximal boundary $\Gamma_{p,i}$ and the wing veins $\Gamma_{v,i}, \Gamma_{v,i+1}$ and Neumann (zero flux) on the wing margin $\Gamma_{m,i}$ (c.f., Fig. 6.1). The boundary conditions for the inhibitor (a_2) are zero

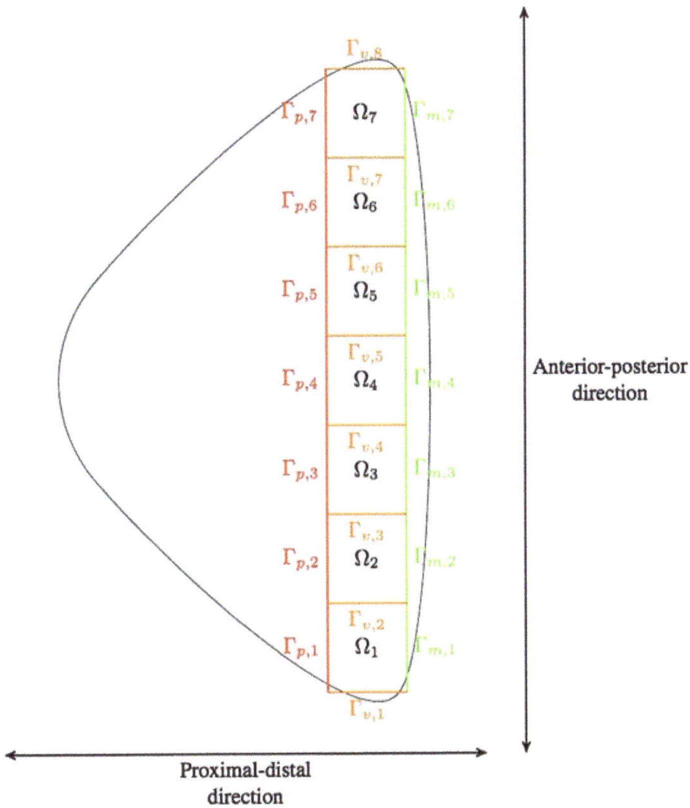

Fig. 6.1 A sketch of the domain on which we model the formation of eyespot focus points

flux on all four boundaries of each wing cell. The Dirichlet boundary condition on each vein $\Gamma_{v,i}$ is the same for each vein. We take the initial data for both activator and inhibitor to be the positive spatially homogeneous steady state of the Gierer-Meinhardt (GM) equation. Thus, our model for focus pattern formation consists of n_{seg}-independent GM equations. The model system equations may be stated as follows:

For $i = 1, \ldots, n_{\text{seg}}$, find $\vec{a}(\vec{x}, t)$, $(\vec{x}, t) \in \Omega \times (0, T)$, such that

$$\begin{aligned} &\partial_t \vec{a}(\vec{x}, t) - D\Delta\vec{a}(\vec{x}, t) = \vec{f}(\vec{a}(\vec{x}, t)) && (\vec{x}, t) \in \Omega_i \times (0, T) \\ &a_1(\vec{x}, t) = u(\vec{x}) && \vec{x} \in \partial\Omega_i / \Gamma_{m,i} \\ &\nabla a_1(\vec{x}, t) \cdot \vec{n}(\vec{x}, t) = 0 && (\vec{x}, t) \in \Gamma_{m,i} \times (0, T) \\ &\nabla a_2(\vec{x}, t) \cdot \vec{n}(\vec{x}, t) = 0 && (\vec{x}, t) \in \partial\Omega_i \times (0, T) \\ &\vec{a}(\vec{x}, t) = \vec{a}^{ss} && \vec{x} \in \Omega_i, \end{aligned} \tag{6.1}$$

where D is a diagonal matrix of positive diffusion coefficients and the reaction kinetic vector $\vec{f}(\vec{v})$ is given by $f_1(\vec{v}) = \alpha\left(\left(\kappa_1 v_1^2 / v_2\right) - \kappa_2 v_1\right)$ and $f_2(\vec{v}) = \alpha\left(\kappa_1 v_1^2 - \kappa_3 v_2\right)$, with $\kappa_1, \kappa_2, \kappa_3 > 0$. The choice of kinetics yields that the corresponding ODE system has a positive steady $\vec{a}^{ss} = (\kappa_2/\kappa_2, \kappa_1\kappa_3/\kappa_2)^T$.

Nijhout (1990, 1994) showed that the above model was capable of generating source profiles consistent with the formation of an eyespot focus within a wing cell. In Sekimura et al. (2015), we showed that changes in the Dirichlet boundary condition for a_1 at the proximal boundary $\Gamma_{p,i}$ alone were sufficient to determine whether or not an eyespot focus forms in a wing cell. For the proximal boundary profile, we consider two different cases firstly, prescribed boundary conditions, and secondly, in order to propose a full model, we consider that the boundary profiles are themselves generated by a patterning mechanism that is posed along the entire proximal boundary, i.e. the *curved surface* $\Gamma_p := \cup_i \Gamma_{p,i}$. For this one-dimensional patterning mechanism, for consistency with the two-dimensional model above, we consider a surface reaction-diffusion system which for illustrative purposes we choose to be the activator-depleted substrate model of Schnakenberg (1979), stated as follows:

Find $\vec{u}(\vec{x}, t)$ such that

$$\partial_t \vec{u}(\vec{x}, t) - D_u \Delta_\Gamma \vec{u}(\vec{x}, t) = \vec{h}(\vec{u}(\vec{x}, t)) \qquad \text{on } \Gamma_p, \tag{6.2}$$

where D_u is a diagonal matrix of positive diffusion coefficients, Δ_Γ is the Laplace-Beltrami operator (the analogue to the usual Cartesian Laplacian on the surface) and the function $\vec{h}(\vec{u})$ is given by $h_1(\vec{u}) = \gamma(\vec{x})\left(a - u_1 + u_1^2 u_2\right)$ and $h_2(\vec{u}) = \gamma(\vec{x})\left(b - u_1^2 u_2\right)$, with $a, b > 0$. u_1 and u_2 are the concentrations of two chemicals (the activator and substrate, respectively). The function γ can be thought of as a reaction rate and is typically taken to be constant in most studies that employ such systems to model biological pattern formation. However, if such an approach is adopted, patterns with a constant wavelength across Γ_p are to be expected. In the present context, this would be insufficient to explain butterfly wing patterning in which the distribution of eyespots occurs with differing frequency in different parts of the wing. For this reason, we allow the reaction rate to be a function of space, which appears to provide sufficient freedom to generate the necessary source profiles from this one-dimensional model that produces any arbitrary eyespot configuration observed on butterfly wings. The resulting model is a *two-stage* model for focus point formation in which the first stage corresponds to solving the Schnakenberg surface reaction-diffusion system Eq. (6.2) to steady state and in the second stage the solution u_2 to this model is used to determine the proximal boundary profiles for a_1 in the eyespot reaction-diffusion system model Eq. (6.1) within each of the wing cells.

6.3 Computational Approximation

For the approximation of the eyespot reaction-diffusion system models posed within each of the wing cells, we employ an implicit-explicit finite element method developed and analysed in Lakkis et al. (2013). An advantage of such an approach is that arbitrary, potentially evolving, geometries can be considered. In particular, one does not need to assume that the wing cells are rectangular, and indeed using open-source meshing software, it is even possible to solve the systems on geometries obtained from image data, which may be a worthwhile extension. For the approximation of the surface reaction-diffusion system, we employ the surface finite element method (Dziuk and Elliott 2013). We refer to the above two references for further details on the numerical approach.

6.4 Results

6.4.1 Gradients in Source Strength on the Wing Veins Can Determine Eyespot Location in the Wing Cell

We start by illustrating that in the eyespot focus point formation model of Sect. 6.2, it is possible to change the location of eyespots by allowing the Dirichlet boundary condition at the wing veins to vary in space. To this end, we suppose that the wing cells are trapezoidal with parallel sides corresponding to the proximal and marginal boundaries that are chosen to be of length 1.5 and 2.5, respectively and are such that the height (proximal-marginal) is 3. We set the proximal boundary condition to be a convex profile of the form $u(\vec{x}) = 2a_1^{ss}(1 - \sin^2(\pi d(\vec{x})/1.5))$ where $d(\vec{x})$ is the distance from the boundary points of the proximal boundary. The boundary

Table 6.1 Parameter values for simulations of Sect. 6.4.1

D_1	D_2	α	κ_1	κ_2	κ_3
0.0031	0.03	20	0.03	0.03	0.0125

condition thus takes the value $2a_1^{ss}$ at the boundary points of $\Gamma_{p,i}$ and decays to 0 at the centre of the proximal boundary. For the wing veins, we consider a gradient in the Dirichlet boundary condition by considering a linear boundary condition of the form $u(\vec{x}) = 2a_1^{ss}(1 - s_1x_2/3)$, where x_2 denotes the distance in the proximal-distal direction from the wing margin and $s_1 > 0$ is a parameter that governs the magnitude of the gradient. Thus the boundary condition takes the value $2a_1^{ss}$ at the point where the vein meets the marginal boundary and decays towards the proximal boundary with slope given by $s_1 > 0$. The remaining parameter values we select are given in Table 6.1. For the discretisation we used linear finite elements on a grid with 2145 degrees of freedom (DOFs) and a time step of 0.01. The system was solved until the discrete solution was (approximately) at steady state.

Figure 6.2a–d shows snapshots of the activator a_1 concentration at different times for different values of s_1. In each of the subfigures, the value of $s_1 = 0, 0.15, 0.25, 0.35, 0.45, 0.5$ reading from left to right. We see that in the case of constant boundary conditions or if the gradient is small ($s_1 = 0, 0.15, 0.25, 0.35$), the centreline peak, characteristic of the Nijhout model, does not extend very far from the margin. The focus point forms near the middle of the wing cell and migrates towards the wing margin with the steady state corresponding to a single focus near the margin. For larger values of the gradient ($s_1 = 0.45, 0.5$), the centreline peak extends much further, almost reaching the proximal boundary, and the resulting focus point forms close to the proximal boundary. The focus point migrates downwards only until around the centre of the wing cell, and the resulting steady state is a single focus point around the centre of the wing cell.

6.4.2 *A Surface Reaction-Diffusion System Model with Piecewise Constant Reaction Rate Generates Boundary Profiles and Resulting Eyespot Foci Recapitulate Those Observed in Artificial Selection*

We now report on simulations in which we illustrate that the two-stage model proposed in Sect. 6.2 (see also, Sekimura et al. 2015) is capable of reproducing the differing selection of dorsal forewing eyespots observed in artificial selection experiments on *Bicyclus anynana*. Beldade et al. (2002) showed that, through artificial selection, it is possible to generate different phenotypes of *B. anynana* with either zero, one (anterior or posterior) or two forewing eyespots (anterior and posterior) (c.f., Fig. 6.3). To investigate whether our two-stage model is capable of

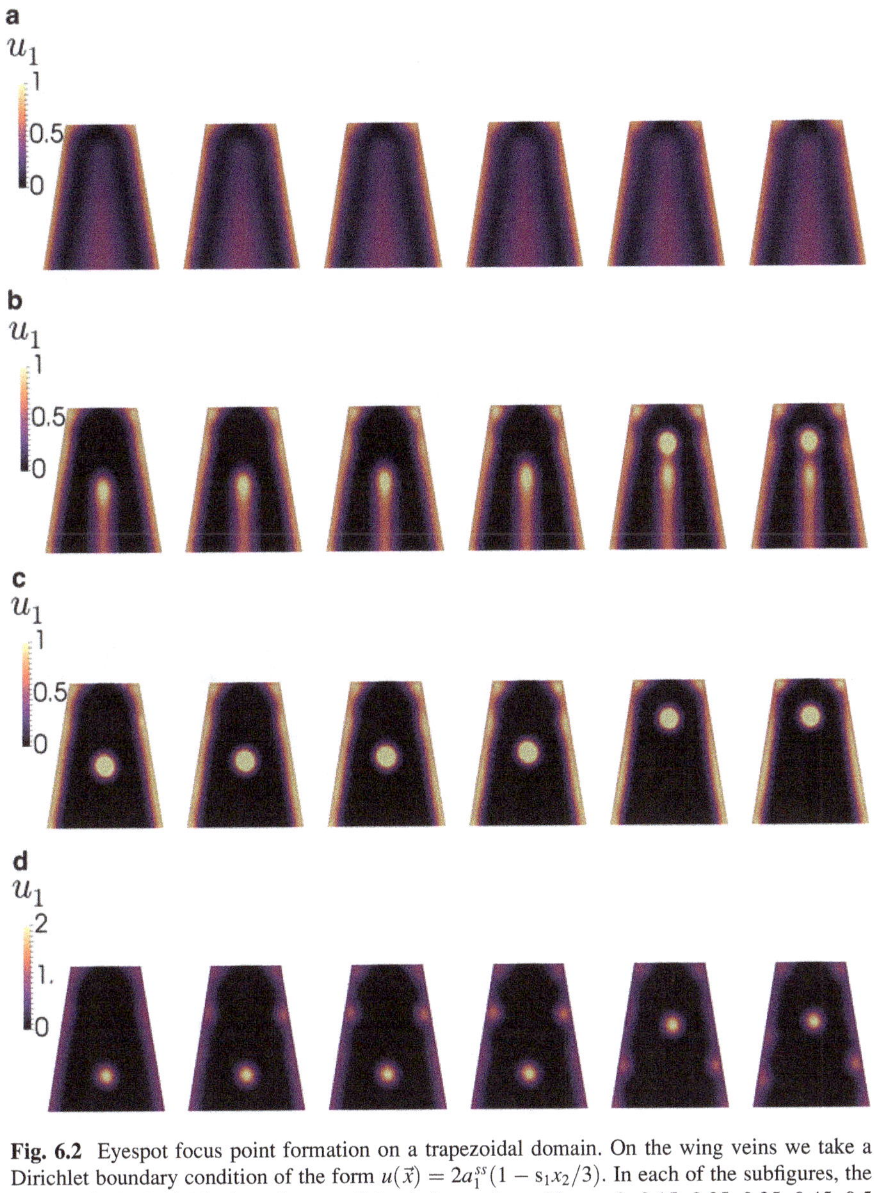

Fig. 6.2 Eyespot focus point formation on a trapezoidal domain. On the wing veins we take a Dirichlet boundary condition of the form $u(\vec{x}) = 2a_1^{ss}(1 - s_1x_2/3)$. In each of the subfigures, the gradient in the Dirichlet boundary condition is increasing with $s_1 = 0, 0.15, 0.25, 0.35, 0.45, 0.5$ reading from *left* to *right*. Thus the leftmost snapshot in each subfigure corresponds to constant Dirichlet boundary conditions on the wing veins, whilst the rightmost snapshot in each subfigure corresponds to the steepest linear gradient with $u(\vec{x}) = 2a_1^{ss}$ at the point where the wing veins meet the margin and $u(\vec{x}) = a_1^{ss}$ at the point where the wing veins meet the proximal boundary. In all the subfigures, we only display snapshots of the activator a_1 concentration; the inhibitor concentrations are in phase with those of the activator and are thus omitted. For remaining parameter values, see text. (**a**) t=0.1. (**b**) t= 0.2. (**c**) t=0.5. (d) Steady state

Fig. 6.3 Eyespot phenotypes of *B. anynana* produced in artificial selection experiments (Beldade et al. 2002) (Figure reproduced with permission of the publisher)

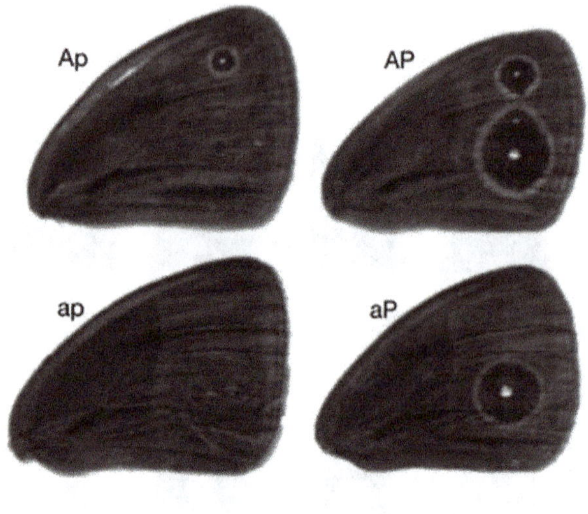

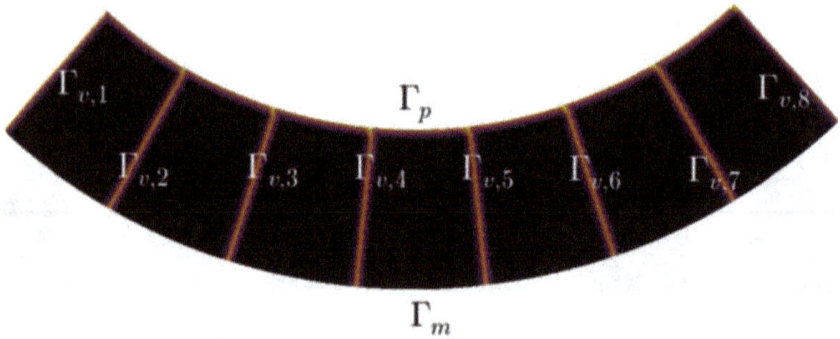

Fig. 6.4 Sketch of the geometry used to model the entire region of the wing disc on which eyespot formation occurs for the experiments of Sect. 6.4.2

reproducing these observations, we consider a wing as shown in Fig. 6.4. The proximal (Γ_p) and marginal (Γ_m) boundaries are curves corresponding to a portion of the circumference of two concentric circles of radius 9 and 12, respectively. The wing veins ($\Gamma_{v,i}$) are assumed to be radial and of length 3, whilst the proximal and marginal boundaries of each of the wing cells are approximately of length 1.88 and 3.35, respectively. We consider the two-stage model described in Sect. 6.2. In the first stage, we solve the surface reaction-diffusion system with the Schnakenberg kinetics to steady state. We select Dirichlet (prescribed) boundary conditions for u_1 with $u_1 = {u_1}^{ss}$ on one boundary and $u_1 = 2{u_1}^{ss}$ at the other boundary point. For u_2 we set zero-flux boundary conditions. The initial data is taken to be the steady state value for both u_1 and u_2. We consider the case that the function γ is piecewise

Table 6.2 Parameter values for simulations of Sect. 6.4.2

D_{u1}	D_{u2}	a	b	D_1	D_2	α	κ_1	κ_2	κ_3
1	15	0.1	0.9	0.005	0.03	20	0.03	0.03	0.0125

constant (e.g. McMillan et al. 2002); in particular, we allow it to take two distinct values on either side of the midpoint (anterior-posterior) of the proximal boundary curve. The remaining parameter values we employed are shown in Table 6.2. After solving the Schnakenberg system to steady state, we assume the Dirichlet boundary condition at the proximal boundary for the reaction-diffusion system posed in each wing cell is of the form

$$a_1(\vec{x},t) = 1.9\bar{u}_2(\vec{x})a_1^{SS}\vec{x}\in\Gamma_{p,i},$$

where $\bar{u}_2(\vec{x})$ is the spatially inhomogeneous steady state of the substrate in the Schnakenberg equation. At the veins, we set Dirichlet boundary conditions for the activator equal to twice the steady state value. The remaining parameter values are given in Table 6.2. We note that each wing cell in this simulation is slightly larger in area than those considered in Sect. 6.4.1, and it is due to this fact that we require a slightly larger activator diffusivity, D_1, than that which was used in Sect. 6.4.1.

For the numerical parameters, we used a mesh with 3927 DOFs to represent the entire wing disc. The surface reaction-diffusion system was solved on the trace mesh corresponding to the boundary edges of the bulk mesh; the corresponding one-dimensional mesh had 1793 DOFs. We used a piecewise linear finite element method for both the surface and bulk reaction-diffusion systems with a time step of 0.05, and we solved the system until the concentration profiles were (approximately) at steady state. Figure 6.5 shows the steady state values obtained for simulations in which we vary the value of the piecewise constant reaction rate γ. We see that when γ is zero in both the anterior and posterior, as expected the substrate concentration (that satisfies zero-flux boundary conditions) in the one-dimensional system simply converges to a constant. Using this profile in the proximal boundary conditions for the model posed in each wing cell, we generate a wing with no foci similar to the *ap* case of Fig. 6.3. If we allow γ to be large on one half of the proximal boundary and small on the other half, then we generate boundary profiles from the one-dimensional system that results in a single eyespot in the half of the wing in which γ is large, similar to the *Ap* and *aP* phenotypes of Fig. 6.3. Finally, if γ is large and constant across the entire proximal boundary, we generate a profile that leads to both the anterior and posterior foci forming as in the *AP* phenotype of Fig. 6.3. The choice of Dirichlet boundary conditions for u_1 leads the substrate troughs to form in the correct locations for the eventual eyespots dependent on whether they are anterior or posterior; as for zero-flux or symmetric Dirichlet boundary conditions, we would expect solutions that are symmetric along the midpoint of the proximal boundary. We note that this asymmetry need not be

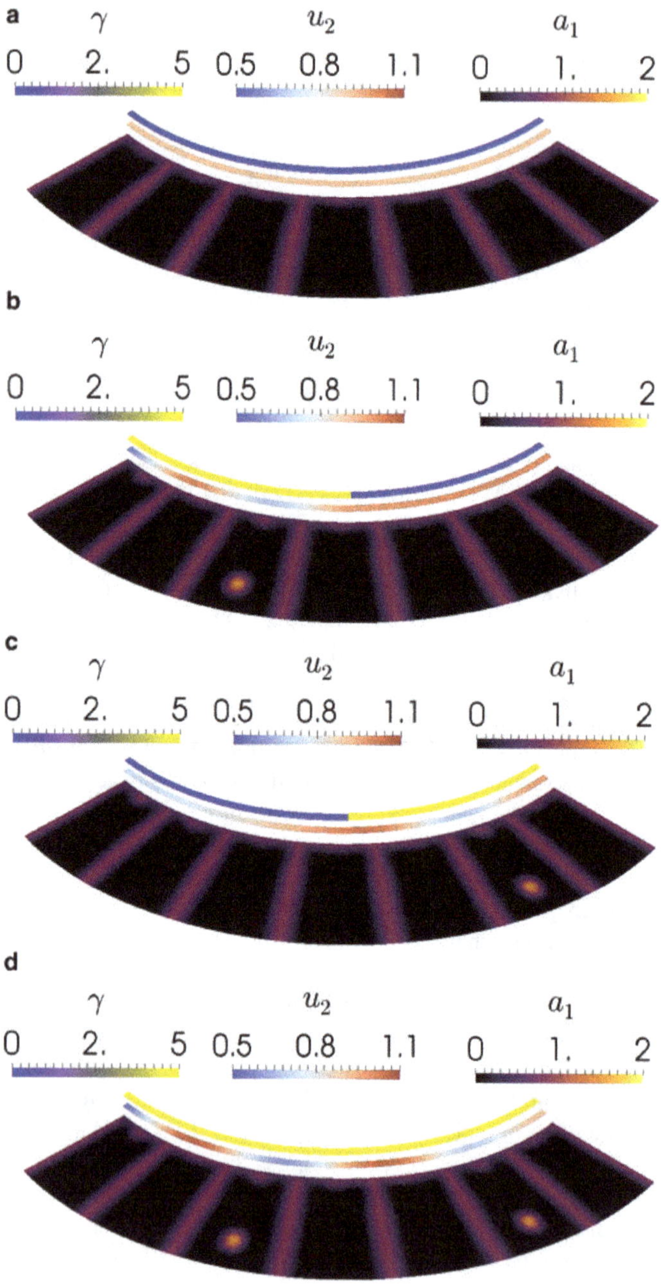

Fig. 6.5 Simulations of eyespot focus point formation using a two-stage model. Initially a reaction-diffusion system with the Schnakenberg kinetics is solved to steady state on the curved

through Dirichlet boundary conditions and could be the result of differences between individual wing cells or some other aspect which is thus far neglected in the modelling.

6.5 Discussion

In this study, we reported on further investigations of a model for the selection and distribution of eyespot foci, originally presented in the paper (Sekimura et al. 2015). The basic idea of the model is that whether an eyespot focus forms in a given wing cell and its eventual position in the wing cell can be determined through changing only the boundary conditions that are assumed to hold at the veins. Furthermore, we considered a two-stage model consisting of two related pattern-forming mechanisms, one posed along the proximal vein and the other posed in each wing cell. The two-stage model appears capable of reproducing the results of artificial selection experiments in terms of eyespot selection. A hypothesis within the two-stage model is that patterning in the first stage could be governed by a reaction-diffusion mechanism in which the reaction rate is dependent on the spatial position. Such an assumption is consistent with assuming different levels of gene activation in different regions of the wing (e.g. McMillan et al. 2002). We note however that the present model is still sensitive to changes in the parameter values and crucially, changes in the geometry. In particular, the naturally observed variations in wing cell size across butterflies appear too large for the present model to be applicable. Hence a potentially attractive avenue for future studies is to investigate Turing systems with a degree of scale invariance as has been attempted in other contexts (e.g. Othmer and Pate 1980).

Fig. 6.5 (continued) proximal boundary using a piecewise constant value for the parameter γ, Dirichlet boundary conditions for u_1 and zero-flux boundary conditions for u_2 (see text for further details). The Dirichlet boundary condition on the proximal boundary is taken to be proportional to the substrate concentration u_2 of the Schnakenberg equation. The remaining boundary conditions and parameter values are given in the text. (**a**) Steady state values of u_2 and a_1 for constant $\gamma = 0$, corresponding to no eyespot foci. (**b**) Steady state values of u_2 and a_1 for piecewise constant $\gamma = 500$ on one half of the wing and $\gamma = 10$ on the other half, corresponding to one eyespot focus on the half of the wing with increased γ. (**c**) Steady state values of u_2 and a_1 for piecewise constant $\gamma = 10$ on one half of the wing and $\gamma = 500$ on the other half, corresponding to one eyespot focus on the half of the wing with increased γ. (**d**) Steady state values of u_2 and a_1 for constant $\gamma = 500$, corresponding to two eyespot foci

References

Beldade P, Koops K, Brakefield PM (2002) Developmental constraints versus flexibility in morphological evolution. Nature, 416(6883):844–847. PMID: 11976682

Dziuk G, Elliott CM (2013) Finite element methods for surface PDEs. Acta Numerica 22:289–396

Gierer A, Meinhardt H (1972) A theory of biological pattern formation. Biol Cyber 12(1):30–39

Lakkis O, Madzvamuse A, Venkataraman C (2013) Implicit–explicit timestepping with finite element approximation of reaction–diffusion systems on evolving domains. SIAM J Num Anal 51(4):2309–2330

McMillan WO, Monteiro A, Kapan DD (2002) Development and evolution on the wing. Trends Ecol Evol 17(3):125–133

Nijhout HF (1990) A comprehensive model for colour pattern formation in butterflies. Proc R Soc Lond B 239(1294):81–113

Nijhout HF (1994) Genes on the wing. Science, 265(5168):44–45. PMID: 7912450

Othmer HG, Pate E (1980) Scale-invariance in reaction-diffusion models of spatial pattern formation. Proc Nat Acad Sci 77(7):4180–4184

Sekimura T, Maini PK, Nardi JB, Zhu M, Murray JD (1998) Pattern formation in lepidopteran wings. Comments in Theoretical Biology 5(2–4):69–87

Sekimura, T., Venkataraman, C. and Madzvamuse, A. (2015). A model for selection of eyespots on butterfly wings. PLoS ONE. http://dx.doi.org/10.1371/journal.pone.0141434

Schnakenberg J (1979). Simple chemical reaction systems with limit cycle behaviour. J Theor Biol, 81 (3):389–400. PMID: 537379

7

Butterfly Wings: Color Pattern Rules and Formal Models

Joji M. Otaki

> Anyhow, exploring the *consequences* of self-similarity was proving full of extraordinary surprises, helping me to understand the fabric of nature.—Benoit B. Mandelbrot (1983). *The Fractal Geometry of Nature*. Revised edition, Page 423

Abstract The morphology of multicellular organisms can be viewed as structures of three-dimensional bulges and dents of an otherwise nearly two-dimensional epithelial sheet. Morphogenesis is thus a process to stably form those physical distortions over time through differential cellular adhesion, contraction, and aggregation and through cellular changes in size, shape, and number. Such physical distortions may be hierarchically repeated with modifications, which is suggested by self-similar structures in organisms. Butterfly wings are nearly two-dimensional but contain three-dimensional bulges and dents that correspond to organizing centers for color pattern elements. Importantly, an eyespot and its corresponding parafocal element on a wing, constituting the border symmetry system, are self-similar. From this perspective, I review here the color pattern rules and several formal models that have been proposed, clarifying their relationships with the induction model for positional information. To reinforce the induction model, I propose the distortion hypothesis, in which dynamic epithelial distortion forces at organizing centers, such as the center of a presumptive eyespot, that are produced through changes in cell size spread to surrounding immature cells over distances as morphogenic signals in developing butterfly wings. The physical distortion forces open stretch-activated calcium channels that cause calcium signals in the cell and activate the expression of regulatory genes. These regulatory gene products initiate

J.M. Otaki (✉)
The BCPH Unit of Molecular Physiology, Department of Chemistry, Biology and Marine Science, Faculty of Science, University of the Ryukyus, Senbaru, Nishihara, Okinawa 903-0213, Japan
e-mail: otaki@sci.u-ryukyu.ac.jp

a cascade of structural genes that eventually produce eyespot black rings. Calcium waves also activate a process of genome duplication, resulting in an increase in cell size, as the ploidy hypothesis states. A new distortion of epithelial cells is induced at the center of a presumptive parafocal element through an increase in cell size, producing self-similarity of the eyespot and the parafocal element. The self-similar configuration of the border symmetry system further suggests the essence of morphogenesis as the DCG cycle: repeated sequential events of epithelial distortions (D), calcium waves (C), and gene expression changes (G). Future studies should examine these hypotheses and speculations that constitute the induction model in butterfly wings and the generality of the DCG cycle in other organisms.

Keywords Butterfly wing • Color pattern rule • Distortion hypothesis • Eyespot • Induction model • Morphogen • Parafocal element • Pattern formation • Ploidy hypothesis • Self-similarity

7.1 Introduction

One of the important goals of developmental biology is to understand how morphological structures are produced during development. Morphological structures are usually three-dimensional, but they are initiated as physical changes in a two-dimensional epithelial sheet to create three-dimensional bulges and dents. Developmentally speaking, the origin of morphology in amphibian embryogenesis can be traced back to the blastula stage, which is the stage when a sheet of cells emerges for the first time after fertilization. Subsequently, the plain cellular sheet undergoes dynamic cellular movement for gastrulation and eventually forms an embryo and, later, a complete adult individual. These processes are understood as mechanical changes of the epithelial cells. In this sense, a center of physical distortion forces could correspond to an organizing center. In insects, early embryogenesis is executed in the syncytial blastoderm, which may not be similar to this concept of mechanical changes, but a process of adult tissue formation from imaginal disks in the prepupal and pupal stages involves dynamic physical distortions of the epithelial cells.

In this view, morphogenesis can be considered to be a process of forming physical distortions over time through differential cellular contraction, adhesion between cells, and aggregation among cells and through cellular changes in size, shape, and number. Furthermore, the whole biological structure of a given organism can be viewed as a series of repetitions of epithelial distortions, despite their superficial dissimilarity. This kind of repetition unit may be called the "**morphogenesis unit**." Therefore, the mechanism employed to produce the morphogenesis unit is the essence of morphogenesis.

This view of morphogenesis has been derived from observations of diverse butterfly wing color patterns and from interpretations of physiologically induced color pattern changes (Otaki 2008a). Butterfly wings are mostly two-dimensional, but careful examinations reveal that they are indeed three-dimensional (Taira and Otaki 2016), as in other tissues and organs in animals, and therefore likely involve mechanical forces that are generated by cellular changes in size, shape, and number. Butterfly wing disks at the larval and pupal stages are sheets of epithelial cells (more specifically, epidermal cells) that may be ready to accept mechanical changes. Butterfly wings additionally produce three-dimensional microstructures of scales and bristles, the processes of which are interesting but beyond the scope of this paper. In this paper, I endeavor to extract "the essence of morphogenesis" from the color pattern development of the border symmetry system. The border symmetry system is one of the symmetry systems in nymphalid color patterns and consists of border ocelli (eyespots) and parafocal elements (PFEs), which will be explained shortly below.

The repetition unit in biological entities may be identified by seeking homologous structures. **Serial homology** or **modularity** is a popular concept in the field of animal development. A good example of serial homology is serial eyespots on a single wing surface of nymphalid butterflies (Nijhout 1991; Beldade et al. 2002, 2008; Monteiro et al. 2003; Monteiro 2008, 2014). However, in this paper, I focus on **self-similarity**, a concept that is different from serial homology and modularity. Eyespots on a single wing surface are homologous but not self-similar; self-similarity is hierarchical repetition but not parallel repetition.

In the following sections, I first introduce the concept of self-similarity in biological entities using plants as examples. I use plants because they often manifest self-similar structures that are relatively easy to pinpoint, and many of them have been analyzed well mathematically (Mandelbrot 1983; Barnsley et al. 1986; Ball 1999, 2016).

7.2 Self-Similarity in Plants and Animals

In self-similar structures, a large structure contains its own smaller structures, wherein the small ones are nested within the larger one; they are hierarchically produced. In other words, the whole and its partial structures are similar to each other, but they are not necessarily morphologically identical in actual biological systems because of the extreme modifications of the essential process for their morphogenesis. These modifications often make identification of self-similarity difficult in actual biological systems.

One of the most famous self-similar structures in biological entities may be a fern or leaf structure that is produced by fractal branching patterns (Barnsley et al. 1986). Many leaves exhibit clear self-similarity, but the way it manifests is greatly dependent on the plants. A similar leaf branching pattern is also seen in bacterial growth (Ben-Jacob et al. 1994), blood vessels (Family et al. 1989), seaweeds, sponges and corals (Kaandorp and Kübler 2001), and other systems (Ball 1999, 2016), suggesting the universality of the branching fractal structures in biological systems.

Fig. 7.1 Examples of self-similarity in plants. (**a**) Buds of cauliflower romanesco. An *inset* shows the whole structure. (**b**) Buds of a common cauliflower. An *inset* shows the whole structure. (**c–f**) Flowers of the crown of thorns, *Euphorbia milii*

The spiral floret arrangement of cauliflower romanesco (*Brassica oleracea var. botrytis*) is another famous example of self-similarity (Fig. 7.1a). A common cauliflower also exhibits self-similarity, but it is less clear (Fig. 7.1b). A similar spiral arrangement can be found in shells (Meinhardt 2009) and other systems (Ball 1999, 2016), suggesting that animals, too, have an ability to produce spiral fractal structures.

A more important and illuminating example salient to a discussion of butterfly color patterns can be found in the flowering pattern of *Euphorbia milii* (Fig. 7.1c-f). A single flower can produce a few smaller flowers from its own flower. This is a nested or hierarchical configuration, and these flowers are self-similar. It appears that this type of self-similarity in a complex biological entity (i.e., a flower in this example) that is not either simple branching or spiral patterns is relatively rare. A potential explanation for this finding is that the original self-similar structures are extensively modified to a degree unnoticeable by human eyes in most biological systems.

These examples in plants, animals, and other organisms demonstrate that organisms have an ability to form self-similar structures. I turn to butterfly wing color patterns from a viewpoint of self-similarity below, but before discussing self-similarity, I first discuss the symmetry in butterfly wing color patterns. Also in the following sections, I propose possible rules for color pattern formation in butterfly wings, which contain my own speculations. I then propose models and hypotheses that incorporate my speculations. For the readers' convenience, I summarize the color pattern rules at the elemental and sub-elemental levels that are discussed below in Table 7.1 and the additional color pattern rules at the scale and cellular levels in Table 7.2. I also summarize the models and hypotheses that are discussed in this paper in Table 7.3.

Table 7.1 Color pattern rules at the elemental and sub-elemental levels

1. Symmetry rule (color symmetry rule)	Pigment distribution is symmetric in a given system or element
2. Core-paracore rule	A unit of a symmetry system is composed of a single core element and a pair of paracore element
3. Self-similarity rule (nesting rule)	An eyespot and its accompanying parafocal element are self-similar
4. Binary rule (binary color rule)	Eyespot (and other elements) is depicted in dark color against light background color
5. Imaginary ring rule	An eyespot has a vanishingly weak light ring outside the outer-most dark ring
6. Inside-wide rule	In a full eyespot, the inner dark core ring (disk) is larger in width than the outer dark ring
7. Uncoupling rule	Sub-elements of an eyespot can be uncoupled from the rest of the eyespot
8. Midline rule	Center of a natural eyespot is placed at the midline of a wing compartment

Table 7.2 Color pattern rules at the scale and cellular levels

1. One-cell one-scale rule	A single scale cell produces a single scale throughout a butterfly wing
2. Color-size correlation rule for scales	Scales of elements (dark-colored scales) are larger than scales of background nearby (light-colored scales)
3. Central maxima rule for elemental scale size	Scales at the center of an element have the largest size in that element
4. Size-ploidy correlation rule for scales and cells	Scale size is correlated with the degree of ploidy of scale cells
5. Distortion rule for organizing centers	Organizing centers are physically distorted as bulges and dents that are reflected in pupal cuticle spots

7.3 Part I: Color Pattern Rules

7.3.1 Symmetry in Butterfly Wing Color Patterns

Highly diverse butterfly wing color patterns are thought to have been derived from a basic overall wing color pattern called the **nymphalid ground plan**. The nymphalid ground plan is a sketch of a general color pattern that was obtained by inductive reasoning from observations of many actual butterflies. This pattern was independently proposed by Schwanwitsch (1924) and Süffert (1927). Based on these two original schemes, a modern version was proposed by Nijhout (1991, 2001), and a few minor revisions were introduced by Otaki (2012a).

The nymphalid ground plan is composed of color pattern "**elements**," which are placed on a "**background**" (Fig. 7.2). The important point is that the elements are symmetrically arranged regarding pigment composition (i.e., coloration) (Nijhout

Table 7.3 Models and hypothesis for color pattern formation

1. Concentration gradient model (gradient model)	The classical model based on diffusive morphogen gradient that is released from organizing center. Thresholds are inherently set in signal-receiving cells
2. Transient models (collective)	Models that have been proposed transiently and withdrawn readily to investigate the simplest models for color pattern determination, including the two sub-step model and the multiple morphogen model
3. Adopted models (collective)	Models that have been proposed fragmentally but adopted to be synthesized as the induction model. Adopted models are the wave model, the two-morphogen model, and the heterochronic uncoupling model
4. Threshold change model	The most popular model that could explain color pattern modifications induced by physical damage and by pharmacological or temperature treatment
5. Induction model	An alternative model that proposes a sequential release of wavelike morphogenic signals from organizing center and dynamic interactions between signals
6. Rolling-ball model	A way of signal dispersion in the induction model, mainly based on the results of pharmacological modifications of parafocal elements and eyespots
7. Signal settlement mechanisms	A ways of signal settlement in the induction model. Three mechanisms are proposed: time-out mechanism, spontaneous velocity-loss mechanism, and repulsive velocity-loss mechanism. The latter has two sub-mechanisms: self-repulsive and nonself-repulsive velocity-loss mechanisms
8. Ploidy hypothesis	The hypothesis that morphogenic signal for color patterns is a ploidy signal. Scale color is determined as a result of cell size and the degree of ploidy
9. Physical distortion hypothesis	The hypothesis that morphogenic signal for color patterns is physical distortions of epithelial sheet
10. DCG cycle	The essence of morphogenesis in the revised version of the induction model, producing self-similar structures. D, distortion waves; C, calcium waves; G, gene expression

1994); this principle may be called the **symmetry rule** or, more accurately, **color symmetry rule**. In contrast, elemental shape is often very asymmetric. It has been believed that elemental symmetry in coloration comes from circular arrangements of morphogenic signals (signals that function as morphogens) from the organizing center located at the center of a prospective element.

There are three major symmetry systems (the basal, central, and border symmetry systems) and two peripheral systems (the wing root and marginal band systems) on the wings of nymphalid butterflies (Nijhout 1991, 2001; Otaki 2009, 2012a; Taira et al. 2015), although Martin and Reed (2014) stated reasonably that the basal

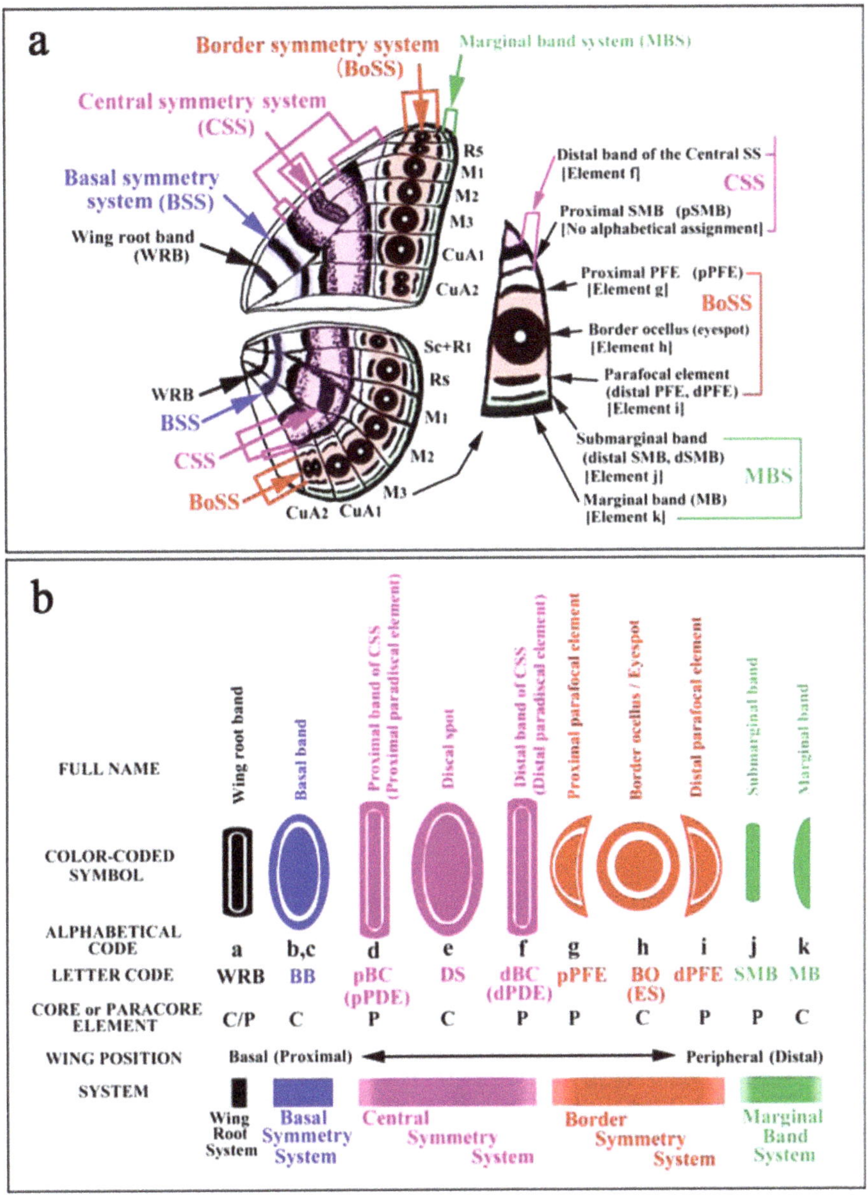

Fig. 7.2 The nymphalid ground plan. Reproduced and modified from Otaki (2012a) and Taira et al. (2015). (**a**) A standard scheme. In this scheme, dSMB is a part of the distal band of the central symmetry system (dBC) and thus may be omitted from the ground plan. (**b**) A simplified scheme. Elements are aligned from the basal (*left*) to the peripheral (*right*)

symmetry system may be associated with the central symmetry system. The two peripheral systems are also likely symmetric, but simply because they are placed at the wing margins, only a portion of them are expressed on a wing. It is likely that all five systems share the same developmental mechanism. In other words, they can be considered to have been derived from modifications of the basic "**ground pattern**" for a single symmetry system. In this sense, they are homologous. Importantly, it is reasonable to assume that each unit of a symmetry system is primarily organized by a single organizing center during development.

7.3.2 The Core-Paracore Rule and Self-Similarity Rule

Because eyespots and PFEs belong to the border symmetry system (Otaki 2009; Dhungel and Otaki 2009), it is likely that the unit of color pattern (or the basic "ground pattern") in any symmetry system of the nymphalid ground plan is composed of the **core element** and a pair of **paracore elements** (Otaki 2012a), which may be dubbed the **core-paracore rule**. The single elemental system containing the core and paracore elements is symmetric, and a single core element is symmetric regarding pigment composition. Likewise, a single paracore element is symmetric. Importantly, the pigment composition of a paracore element is often similar to that of a corresponding core element. Thus, the core-paracore rule may be elaborated as the **self-similarity rule (the nesting rule)**. Based on the core-paracore rule and the self-similarity rule, the diversity of the symmetry system can be understood as various modifications of the basic process of elemental formation (Fig. 7.3).

7.3.3 The Border Symmetry System and Its Self-Similarity

To understand the core-paracore relationship, I hereafter mainly focus on the border symmetry system in nymphalid butterfly wings. The core and paracore elements in this system are border ocelli (BOs or eyespots) and PFEs, respectively. PFEs are often found on the distal side of eyespots (dPFEs), and those on the proximal sides (pPFEs) are less frequent (Otaki 2009). When it is simply known as a parafocal element, dPFE is meant.

Examples of the border symmetry system are shown here. In *Argyreus hyperbius*, BOs and PFEs are both beige in color, although they have different shapes (Fig. 7.4a, left). In contrast, the submarginal bands are differently colored. This coloration pattern probably arises because BOs and PFEs belong to the same system, and they are different from submarginal bands, which belong to the marginal band system (Taira and Otaki 2016). In *Vanessa indica*, BOs and PFEs are similar both in coloration and shape (Fig. 7.4a, middle). In *Araschnia burejana*, PFEs are elongated oval rings with or without blue filling inside (Fig. 7.4a, right).

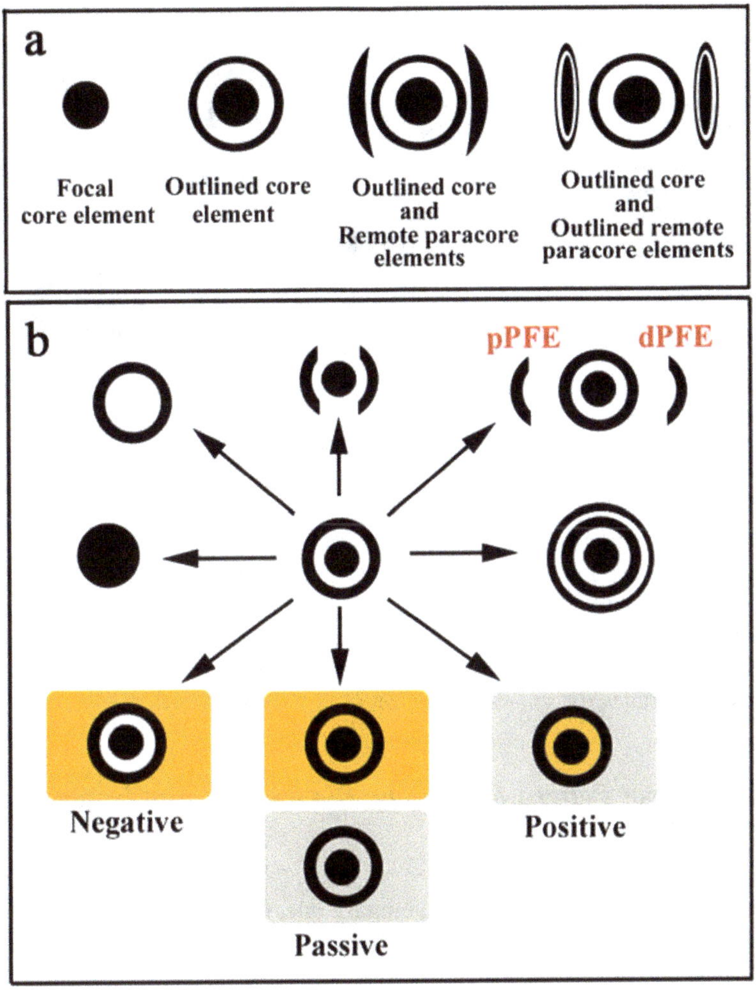

Fig. 7.3 Morphological transformation of color patterns of the border symmetry system Reproduced from Otaki (2012a). (**a**) Stepwise changes from the simplest black dot (*left*) to the complicated self-similar pattern (*right*). (**b**) Diverse transformation of a standard eyespot to various eyespots. Coloration of the inner light ring in a negative, passive, or positive fashion also contributes to eyespot diversity

This configuration of the border symmetry system appears to be typical in nymphalid butterflies.

Self-similarity between BOs and PFEs is not always clear in the cases above, but in *Tarattia lysanias*, the outer ring of a BO is isolated from the inner black disk, which is similar in shape to a PFE (Fig. 7.4b, left). The inner black disk of a BO is also divided into two rods in *Symbrenthia leoparda*, making a distinction in shape between BO and PFE difficult (Fig. 7.4b, right). Rod-shaped BOs and eyespot-

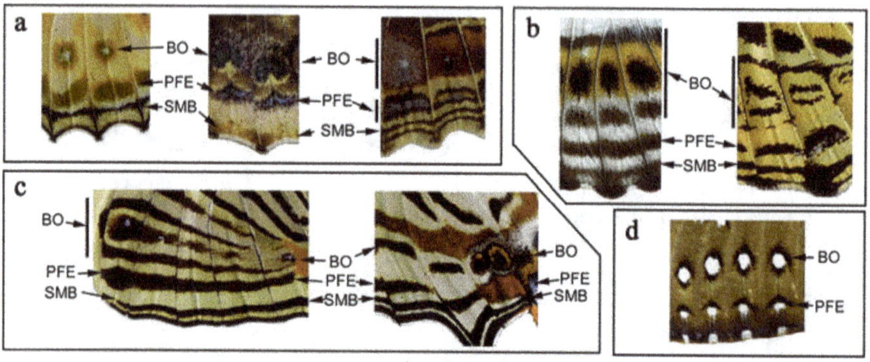

Fig. 7.4 Examples of the border symmetry system in nymphalid butterflies. BO, border ocellus; PFE, parafocal element; SMB, submarginal band. (**a**) *Argyreus hyperbius* (*left*), *Vanessa indica* (*middle*), and *Araschnia burejana* (*right*). (**b**) *Tarattia lysanias* (*left*) and *Symbrenthia leopard* (*right*). (**c**) *Colobura dirce* (*left*) and *Cyrestis camillus* (*right*). (**d**) *Hamanumida daedalus*

shaped BOs coexist on the identical wing surface in *Colobura dirce* and *Cyrestis camillus* (Fig. 7.4c). I believe, therefore, that a PFE is equivalent to an eyespot ring (Dhungel and Otaki 2009; Otaki 2009).

Another intriguing case is found in *Hamanumida daedalus*, where both BOs and PFEs are circular (not rod-shaped) and are similar to each other (Fig. 7.4d). This case strongly argues for self-similarity between BOs and PFEs.

7.3.4 Eyespot Pattern Rules: The Binary Rule and Inside-Wide Rule

To developmentally understand the symmetry rule, the core-paracore rule, and the self-similarity rule discussed above, additional rules regarding nymphalid butterfly color patterns will be discussed here.

An eyespot (BO) is composed of its parts, which may be called sub-elements. Typically, from the center to the peripheral regions, an eyespot is composed of a white dot, a dark (usually black) inner ring (disk), a light ring, and the outermost dark ring. Often, the light ring is variously colored, and the white dot may be absent. Additional rings may exist. The overall shape also varies from a near-true circle to extreme elongation such as rods and lines. Despite these diverse cases, the simplest eyespot is composed of two dark rings (inner and outer dark rings) and one light ring between them. Importantly, the light ring is similar or even identical to the background in coloration. That is, an eyespot is depicted in a dark color against a light background. This is called the **binary rule (binary color rule)** (Otaki 2011a). The binary rule can be revealed when BOs are expressed as rods or lines. *Symbrenthia leopard* (Fig. 7.4b, right) and *Colobura dirce* (Fig. 7.4c, left) illustrate this point: the light rings are continuous with the background, and they are colored without any distinction from the background. The binary rule also implies that the outer dark ring (including PFE) is remotely located from the inner dark ring (disk). This means that morphogenic signals for the outer ring and PFE can travel long distances from the center of the symmetry system.

However, it is also true that in many eyespots, a light ring is not completely identical to the background but may be variously colored. I consider the light ring coloration to be evolutionary modifications. Because it is sandwiched by two dark regions, the light region has to have means to inhibit the invasion of black pigmentation during development. That is, I believe that the inhibitory signal is upregulated in the light ring. This inhibitory signal might have linked to pigment synthesis pathways later in evolution. The inhibitory signal also exists in the background region in contact with the outermost dark ring. This region often shows a vanishingly weak "light ring," which is called the imaginary ring (Otaki 2011b). This pattern may be dubbed the **imaginary ring rule**.

In nymphalid eyespots, the dark inner ring is almost always larger than the outer rings in width. This is called the **inside-wide rule** (Otaki 2012b). A "typical" eyespot without distortion that illustrates the inside-wide rule well can be found frequently in Satyrinae (Fig. 7.5a). Non-Satyrinae eyespots are likely more diverse but still largely follow the inside-wide rule (Fig. 7.5b). However, an exception to this inside-wide rule is small "immature" eyespots (Otaki 2011b), which were probably still developing when the signaling and reception steps were terminated (see below for the four-step process). Alternatively, inhibitory signals were upregulated earlier in the immature eyespot than in the mature eyespot (see

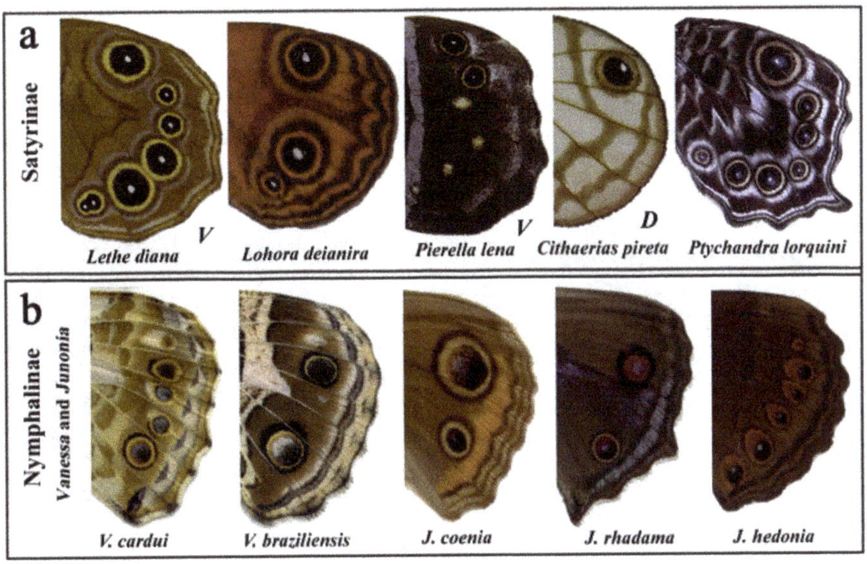

Fig. 7.5 Examples of nymphalid butterfly eyespots. (**a**) Satyrinae. (**b**) Nymphalinae

below for the induction model). These immature eyespots can be found among consecutive eyespots on a wing in many species (Fig. 7.5a, b).

The behavior of PFEs is worth mentioning. A PFE becomes larger when it is displaced toward the corresponding eyespot by pharmacological treatment (Otaki 2008a, 2012b), which follows the inside-wide rule. However, the PFE is sometimes larger than the entire BO, contrary to the inside-wide rule (if it is considered to be a part of an eyespot system as discussed above), as seen in *Argyreus hyperbius* (Fig. 7.4a, left). This exception probably occurs because, once moved away from the core element, the PFE behaves independently as a source of morphogenic signal, as the self-similarity rule suggests.

7.3.5 Eyespot Pattern Rules: The Uncoupling Rule and Midline Rule

An analysis of diverse eyespots indicates that the dark inner ring and the outer ring are not always placed on a single symmetry axis (Otaki 2011b). They appear to be independent of each other to some extent. This conclusion is also supported by physical damage experiments in which the outer ring enlarges and the inner ring diminishes in size in a single eyespot in response to damage (Otaki 2011c). Similarly, an eyespot white spot ("focus") behaves independently from the rest of the eyespot (eyespot body) (Iwata and Otaki 2016a). The uncoupling of the white spot is probably somewhat surprising for those who are not familiar with the genus *Calisto*, which has a white spot not at the center but outside an eyespot (Fig. 7.6). This type of semi-independent behavior of sub-elements is dubbed the **uncoupling rule**. The uncoupling behavior of sub-elements has been suggested in Nijhout (1990), Monteiro (2008), and Iwata and Otaki (2016a, b).

Despite the uncoupling, elemental centers are primarily located on the midline of a compartment (one of the Nijhout's design principles for formal models described in Nijhout (1990)); this may be called the **midline rule**. In contrast, damage-

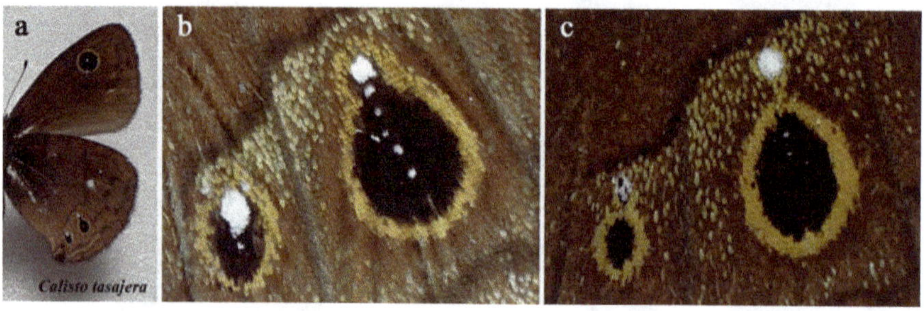

Fig. 7.6 Eyespots of *Calisto tasajera*. Reproduced and modified from Iwata and Otaki (2016a). (**a**) Ventral side of the whole wings. (**b, c**) Ventral hindwing eyespots of two different individuals. White spots are often located outside of the main eyespots in this species

induced elements can emerge at the non-midline (Otaki 2011c). Because the midline is defined by the wing veins, there is no doubt that the wing veins and compartments play critical roles in determining the location of a given element, as elaborated in Nijhout (1978, 1990, 1991).

7.4 Part II: Formal Models toward the Induction Model

7.4.1 Four Steps for Color Pattern Formation as a Starting Frame

It is first important to recognize as a starting frame that there are four sequential steps of color pattern formation: signaling, reception, interpretation, and expression (Otaki 2008a, 2012b). The signaling step was executed by organizing cells, whereas the other three steps were executed by immature scale cells that receive positional information. Most models, including the induction model below, focus on the signaling step and do not pay much attention to the latter three steps. However, the diversity of actual butterfly color patterns may be realized by changes in any single step, at least theoretically.

7.4.2 Gradient Model for Positional Information

The **concentration gradient model** for positional information is probably still the most popular model to explain butterfly eyespot formation (Nijhout 1978, 1980a, 1981, 1990, 1991; French and Brakefield 1992, 1995; Brakefield and French 1995; Monteiro et al. 2001). However, the gradient model cannot easily explain the pattern rules discussed above. Furthermore, it is difficult to explain the additional features of diverse color patterns in actual butterfly wings such as multiple dark rings and differences between small and large eyespots that have drastically different morphology in adjacent compartments using this model (Otaki 2011a, b). Additionally, this model cannot explain dynamic signal interactions (Otaki 2011a, b, c). Time series of color deposition in pupal wings have revealed that red color for an eyespot light ring that develops earlier is "overwritten" by black color that develops later and that a given black area develops as a fusion of patchy black islands (Iwata et al. 2014). These ontogenic observations are not compatible with the gradient model.

However, these facts do not completely deny the usefulness of the gradient model. I believe that a concentration gradient of signaling molecules may play an important role in finalizing the expression of genes for pigment synthesis in a relatively short range (e.g., within a given eyespot ring) (see below). In this sense, gene expression changes may be a result (not a cause) of upstream long-range signals from the center of a prospective eyespot.

7.4.3 Transient Models for TS-Type Modifications and Parafocal Elements

Although I mentioned that the conventional gradient model was not satisfactory, I did not immediately reach this conclusion; I devised a few models before the induction model. I here collectively call them **transient models** because they were transiently proposed and readily discarded. Nonetheless, these models are important to determining the simplest (most parsimonious) model that reasonably explains experimentally induced and naturally occurring eyespots and PFEs. The inclusion of PFEs in the process of making a formal model is critical because both eyespots and PFEs belong to the same symmetry system.

To explain the PFE formation in eyespot-forming and eyespot-less compartments based on the gradient model, the two sub-step model for eyespots and PFEs has been proposed (Otaki 2008a). In this model, a diffusive gradient is first formed to determine the location of PFEs in both eyespot-forming and eyespot-less compartments. After the determination of the PFE location by the periphery of the gradient, the gradient entirely disappears quickly and does not form an eyespot in an eyespot-less compartment. Note that the presence and absence of an eyespot cannot be attributed to threshold differences between the two compartments because they have the same threshold levels if the thresholds exist at all, as shown in an eyespot that occupies two or more compartments (Otaki 2011b). This two sub-step model should also mean that the reception step first takes a snapshot of the PFE, and after the disappearance of the eyespot signal, another snapshot should be taken. This model is too awkward to be accurate, but it hints at the importance of uncoupling the behavior of the PFE from the eyespot proper.

Multiple morphogens (and multiple receptors) for PFEs and eyespots may also save the gradient model. In this multiple morphogen model, there are a few different chemicals that act as morphogens. This model explains a difference between the eyespot-forming and eyespot-less compartments. That is, a morphogen for a PFE is secreted in both compartments, but a morphogen for an eyespot is not secreted in an eyespot-less compartment. However, considering that the PFE is equivalent to the outer eyespot ring belonging to the border symmetry system, multiple morphogen factors are not likely. The introduction of multiple factors in a model can produce all-around models but violates the parsimony of model construction.

Despite these efforts, it is better to abandon the idea of the gradient model, considering its difficulty in explaining the color pattern rules and other points discussed in the previous section. An alternative model is the wave model, in which the signal is transmitted as a series of waves (Otaki 2008a). In this context, the two sub-step model discussed above may be modified to support the wave model, in which the first morphogen for a PFE is released as the first wave and the second morphogen is released as the second wave for the eyespot (more precisely, as the second wave for the eyespot outer ring and then as the third wave for the eyespot inner ring) (Otaki 2008a; Dhungel and Otaki 2009). In this two-morphogen model (wave model), two (or three) morphogens are identical in chemical (or physical) qualities (and therefore different from multiple chemical factors) but are released heterochronically as a train of pulses, being consistent with the heterochronic uncoupling model for TS-type changes (see below). These two models (the wave model and the two-morphogen model) have not been discarded. Rather, they have been adopted, together with the heterochronic model below, and synthesized as the induction model. They may collectively be called the **adopted models**.

There is a weakness in this wave model (Otaki 2008a). Focal damage produces a smaller-than-usual eyespot, indicating the source dependence of the signal. In general, wave signals are not source dependent, theoretically. However, the results of damage experiments can be explained well by the revised version of the induction model (see below).

7.4.4 *Heterochronic Uncoupling Model for TS-Type Changes*

I have examined the color pattern modifications induced by temperature shock or pharmacological treatments (collectively called the TS-type modifications) (Hiyama et al. 2012; Otaki and Yamamoto 2004a, b; Otaki et al. 2005b, 2006, 2010; Otaki 2007, 2008b, c; Mahdi et al. 2010, 2011) (Fig. 7.7a). It is worth noting that temperature treatments (Nijhout 1984) and pharmacological treatments (Otaki 1998, 2008a; Serfas and Carroll 2005) are the only means that can efficiently create

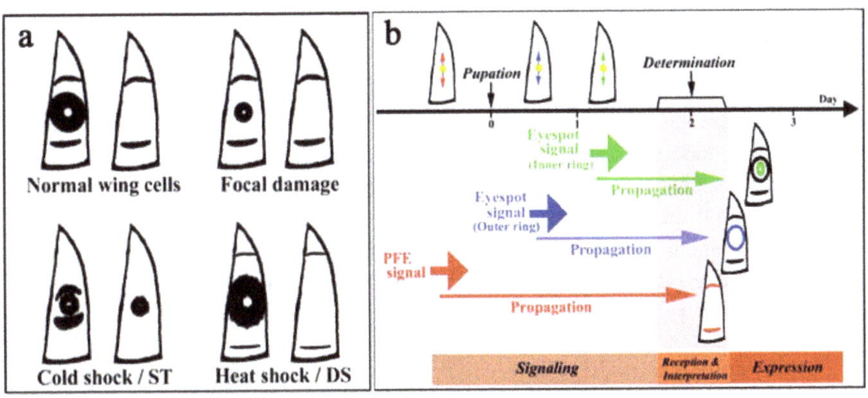

Fig. 7.7 Effects of physiological treatments on eyespot and parafocal element. Reproduced and modified from (Otaki 2011a). (**a**) Modification patterns of various treatments. Two wing compartments (one with an eyespot and a parafocal element and the other with a parafocal element only). ST and DS indicate treatment with sodium tungstate and dextran sulfate, respectively. (**b**) Interpretation of the modifications. Signals are released in the order parafocal element, eyespot outer ring, and eyespot inner ring

this "artificial rearrangement of elements" or "elemental transformation," which is reminiscent of evolutionary trial and error to invent new color patterns based on the nymphalid ground plan. These color pattern modifications are evolutionarily and physiologically relevant (Hiyama et al. 2012; Otaki and Yamamoto 2004a, b; Otaki et al. 2005b, 2006, 2010; Otaki 2007, 2008b, c; Mahdi et al. 2010, 2011), justifying their use as an important method to construct a formal model. The **threshold change model** is the most popular interpretation of the TS-type modifications (Otaki 1998, 2008a; Serfas and Carroll 2005) as well as of physically induced modifications (Nijhout 1980a, 1985; French and Brakefield 1992, 1995; Brakefield and French 1995). However, the TS-type modifications cannot be reproduced by simple threshold changes, as not only relative locations but also the size and colors of the elements are changed. For example, modifications of PFE in an eyespot-less compartment often produce eyespot-like spots (Otaki 2008a).

Because the TS-type modifications are interpreted as a series of possible color pattern snapshots during development, the modifications are likely consequences of a delay of the signaling step (slow signal propagation) or an acceleration of the reception step (Otaki 2008a). That is, temperature shock and pharmacological treatments introduce a time difference between the signaling and reception steps, leading to the heterochronic uncoupling model for TS-type changes. This model simply notes that the TS-type modifications are products of snapshots of propagating signals, which is part of the basis of the induction model (Fig. 7.7b).

7.5 Part III: Induction Model

7.5.1 An Overview

To be consistent with the color pattern rules discussed in Part I above and to reflect a few relevant models discussed in Part II, an integrated model is required. To this end, I have proposed the **induction model** (Otaki 2011a, 2012b). This model is largely based on the "movement" of PFEs and eyespots by tungstate injection and other physiological treatments (Fig. 7.7a). In other words, the induction model is not based on the putative diffusive molecule, which is in contrast to the gradient model.

The physiological modifications can be interpreted as follows, which is indeed a simplified version of the induction model to explain a determination process of the border symmetry system (Fig. 7.7b). Signals for PFEs, the outer ring, and the inner ring are released independently in this order with defined intervals, and each signal propagates independently. These signals are simultaneously received by immature scale cells at the reception step.

7.5.2 *Early and Late Stages*

The induction model can be separated into many steps but roughly into two stages: the early and late stages (Fig. 7.8a). The early stage is the primary signal expansion and settlement. The late stage is the induction of activating signals (and their self-enhancement) and inhibitory signals and their stabilizing interactions. The late stage of the induction model employs the concept of "the short-range activation and long-range lateral inhibition" (Fig. 7.8b), which is the core concept of the reaction-diffusion model (Gierer and Meinhardt 1972; Meinhardt and Gierer 1974, 2000; Meinhardt 1982). In the induction model, the dark and light areas in an eyespot correspond to the areas of activator and inhibitor signals, respectively.

In contrast, the early stage does not follow the reaction-diffusion mechanism because the method of signal propagation is different; the signal is thought to be propagated according to the **rolling-ball model** (Otaki 2012b). The signal behaves like numerous minute balls rolling on a board of even friction (constant deceleration) (Fig. 7.9a). This behavior is described by classical mechanics. The propagation is thus determined by the initial velocity of each minute unit signal. In addition, the interval of signal release determines the overall shape of an eyespot. The signals propagate slowly and gradually slow down. These properties of signals satisfy the binary rule and the inside-wide rule and produce natural and experimentally induced eyespots and PFEs. These properties also satisfy the uncoupling and heterochronic nature of the signal. It is also possible to simulate morphological differences between small and large eyespots (Fig. 7.9b).

7.5.3 *Settlement Mechanisms*

In the induction model, there are different modes of **signal settlement** that are proposed (Otaki 2012b). First, a snapshot of propagating signals may be taken by the transition from the signaling to reception steps (the time-out mechanism). Second, propagating signals stop when velocity is lost spontaneously because of low initial velocity (spontaneous velocity-loss mechanism) and when the propagation is blocked by an inhibitory signal nearby (repulsive velocity-loss mechanism). The repulsion comes not only from a nearby element (non-self-repulsive velocity-loss mechanism) but also from the signals for the imaginary ring (the outermost inhibitory ring that is not well expressed) that are induced by the outermost dark ring (self-repulsive velocity-loss mechanism). In this sense, the speed and level of the inhibitory signal induction primarily determine the final size of an eyespot. The self-repulsive mechanism thus ensures autonomous determination of an eyespot.

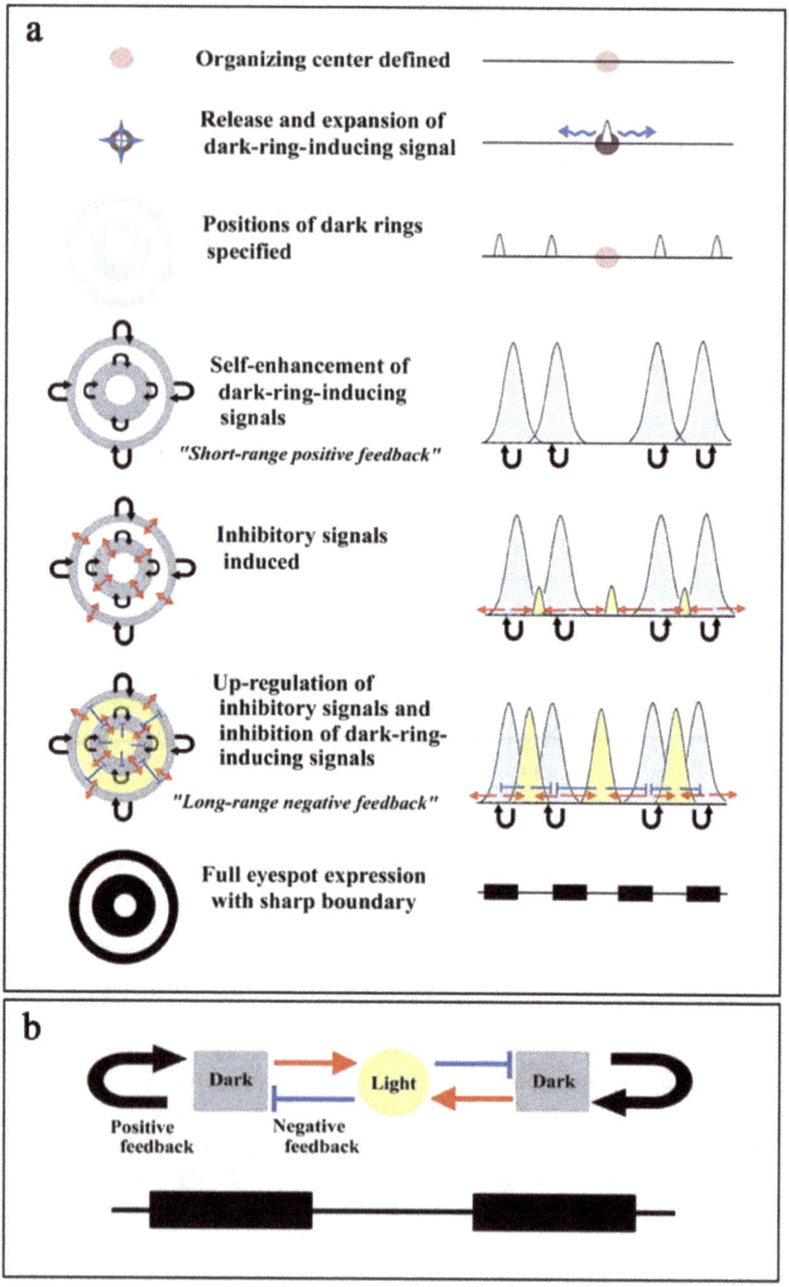

Fig. 7.8 Induction model for positional information. Reproduced from Otaki (2011a). (**a**) Sequential steps of eyespot formation. (**b**) Short-range activation and long-range inhibition in the late stage of the induction model

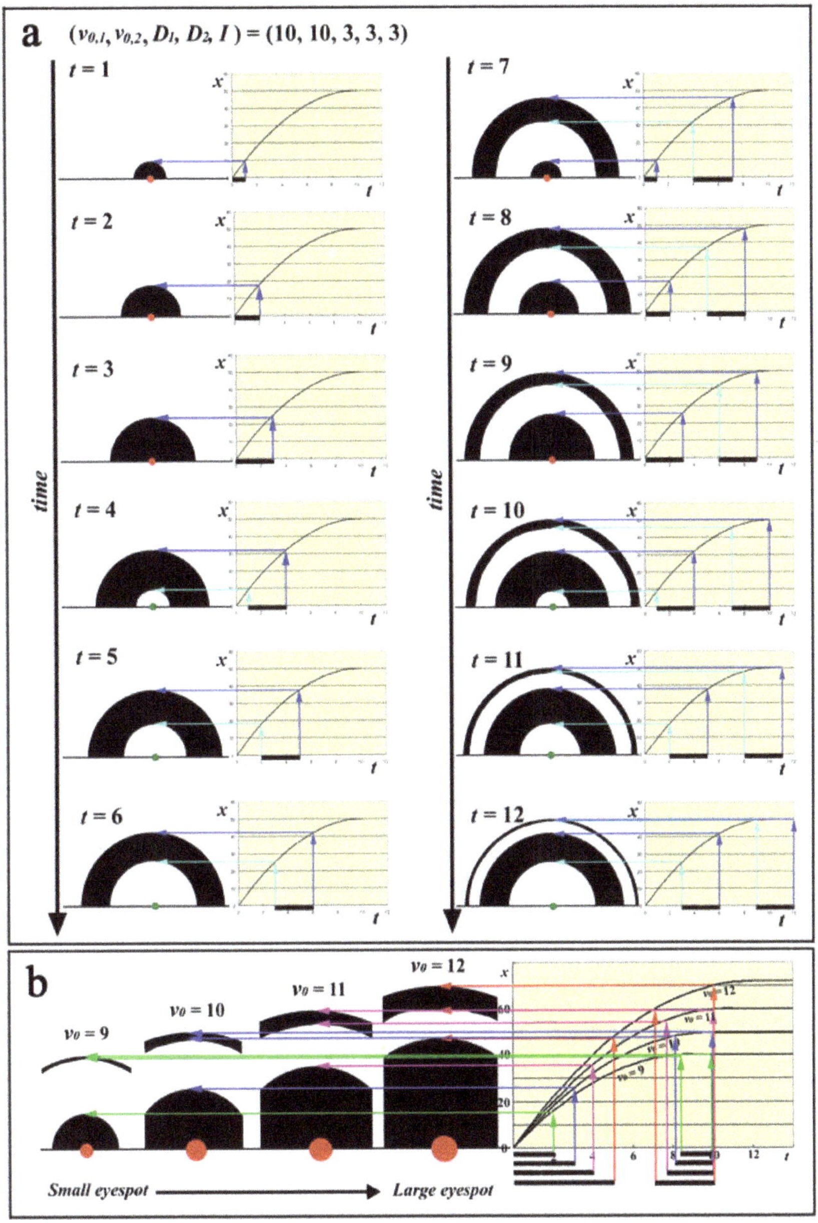

Fig. 7.9 Simulation of eyespot formation based on the rolling-ball model. Reproduced from Otaki (2012b). (**a**) Time course of developmental signals for a typical eyespot. The signals follow the curve shown on the right side of each time point. Initial velocity (v_0) and signal duration (D) are set for two black rings together with their signal interval (I). (**b**) Effect of various initial velocities (v_0). Various eyespots are produced

7.5.4 *Mechanisms for Self-Similarity*

There should be a mechanism that produces self-similar structures, which is based on the following mechanism: highly enhanced activating (black-inducing) signals in the late stage would signify a new organizing center. This mechanism can be explained by the ploidy hypothesis (Iwata and Otaki 2016b), which states that the morphogenic signal for color patterns is indeed a ploidy signal that induces polyploidization and cellular size increase (see below), together with the physical distortion hypothesis, which states that cellular and epithelial distortions act as morphogenic signals (see below). The origin of distortions can be considered as organizing centers. Importantly, the self-similarity of eyespots and PFEs argues for the repulsive velocity-loss mechanism and against the time-out mechanism because the signal dynamics should still persist after the possible time-out for the primary organizing centers for eyespots, when the secondary organizing centers for parafocal elements are determined and become activated. That is, the time-out mechanism cannot explain the heterochronic behaviors of the primary and secondary signal dynamics.

7.5.5 *Reality Check*

Is there any signal that can follow the rolling-ball model in biological systems? In the mesoscopic world (not microscopic world explained by quantum physics nor macroscopic world explained by classical mechanics) of cells and molecules in water, Brownian motion and non-covalent molecular interactions prohibit the rolling-ball-like behavior of a molecule. In contrast, mechanical force can be transmitted easily via an epithelial sheet if epithelial cells are connected firmly but flexibly. That is, epithelial distortions may show rolling-ball-like behavior and act as morphogenic signals from organizing centers. In Part IV below, I review evidence for the ploidy hypothesis and the distortion hypothesis.

7.6 Part IV: Ploidy, Calcium Waves, and Physical Distortions

7.6.1 *Scale Size of Elements*

At the cellular level, one cell builds one scale (Nijhout 1991), which may be dubbed the **one-cell one-scale rule.** Therefore, any morphological features of scales directly indicate the developmental status of the scale-building cells (or simply scale cells). Scale size distribution is graded from the basal to peripheral areas of a wing in butterflies and moths (Kristensen and Simonsen 2003; Simonsen and

Kristensen 2003). Similar size gradation has been found in the background scales in *Junonia orithya*, *J. almana*, *Vanessa indica*, and *V. cardui* (Kusaba and Otaki 2009; Dhungel and Otaki 2013; Iwata and Otaki 2016b).

What about the size of scales that constitute elements? In *J. orithya* and *J. almana*, the scale size of an element is larger than that of its surrounding background (Kusaba and Otaki 2009; Iwata and Otaki 2016b) (Fig. 7.10). In this sense, scale color and size are reasonably correlated, which can be called the **color-size correlation rule for scales**. This rule may sound trivial but is indeed important as a clue to understanding the possible nature of morphogenic signals for color patterns (see below). Furthermore, the largest scales in an element are found roughly at the center of an element (Kusaba and Otaki 2009; Iwata and Otaki 2016b). This may be called the **central maxima rule for elemental scale size**. It is important to recognize that scale size changes suddenly at the boundary between the inner black ring of an eyespot and a yellow ring. There are similar abrupt changes at the outer ring boundary and the PFE boundary. These abrupt size changes may reflect the independence of black areas (the binary rule and the uncoupling rule) rather than gradual changes of positional information.

Additionally, scales of different colors differ in their structure, such as overall scale shape and scale ultrastructure (Gilbert et al. 1988; Nijhout 1991; Janssen et al. 2001). Our laboratory also obtained similar results using *Junonia* and other butterflies (Kusaba and Otaki 2009; Iwata and Otaki unpublished data; Kazama et al. 2017).

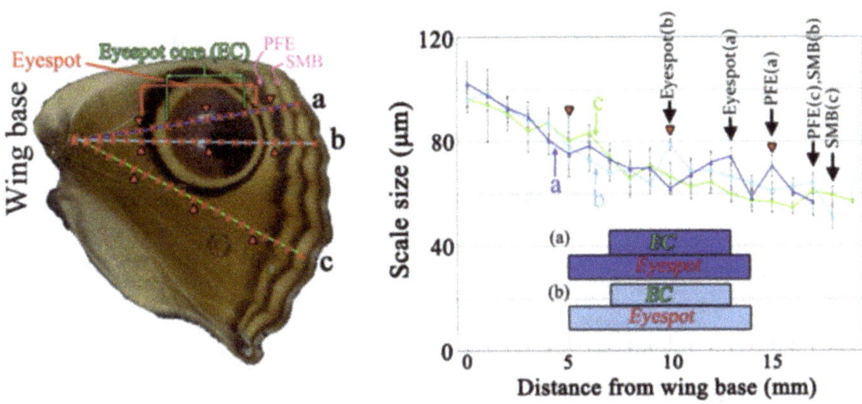

Fig. 7.10 Scale size distribution on a wing of *Junonia almana*. Reproduced from Iwata and Otaki (2016b). A dorsal forewing was examined along lines *a*, *b*, and *c* in 1.0 mm intervals (*left*). Results are shown in the graph (*right*). Along line *b*, scale size peaked at the center of the eyespot. Along line *a*, the peak was located at the distal edge of the eyespot core and also at the center of the parafocal element (PFE). Line *c* did not show conspicuous peaks. All lines showed the size decrease from the basal to the peripheral areas except at the elemental positions

7.6.2 Ploidy Hypothesis

According to Henke (1946) and Henke and Pohley (1952), scale size reflects the degrees of ploidy of the cell in moths (Sonhdi 1963; Cho and Nijhout 2013). This size-ploidy relationship, or the **size-ploidy correlation rule for scales and cells**, is probably applicable to butterflies. This leads us to propose the **ploidy hypothesis** (Fig. 7.11a) (Dhungel and Otaki 2013; Iwata and Otaki 2016a, b). This hypothesis states that morphogenic signals induce polyploidization of signal-receiving cells. The higher the ploidy level, the larger the cell. The larger the cell, the larger the scale it can produce. Simply because a high ploidy level means high numbers of genes for pigment synthesis enzymes, the concentration of pigment in the scales can

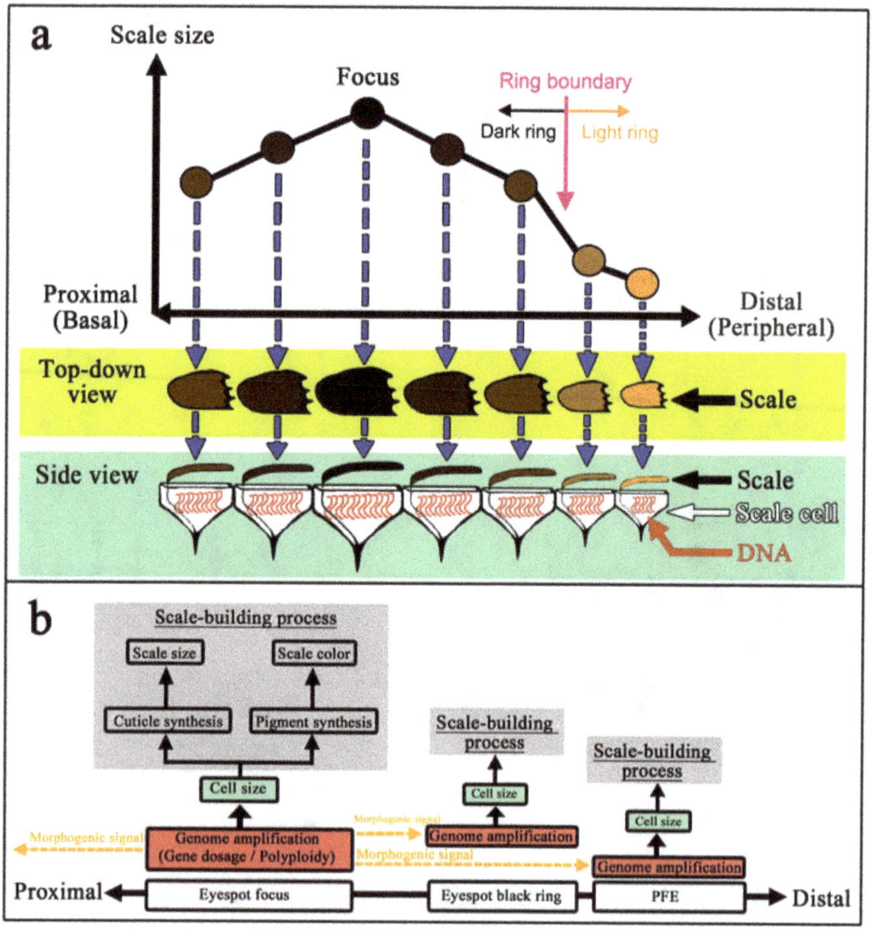

Fig. 7.11 Ploidy hypothesis. Reproduced from Iwata and Otaki (2016b). (**a**) Scale size distribution and its relationship with cell size. (**b**) A hypothetical determination process for scale color and size based on the induction model

change their coloration. Alternatively, gene dosage determines which pigment to be synthesized. In this way, the level of morphogenic signals indirectly determines the levels of pigment chemicals in a scale through the regulation of polyploidization or gene dosage. The ploidy hypothesis is an important component of the induction model (Fig. 7.11b).

The recent discovery that a cell cycle regulator, *cortex*, plays a role in the darkening of the wings in butterflies and moths (Nadeau et al. 2016; van't Hof et al. 2016) may be a surprise for many biologists, but this discovery fits well with the ploidy hypothesis, although it is not discussed in these papers. This cell cycle regulator may control a process of polyploidization of immature scale cells, which determines the final coloration of scales according to the ploidy hypothesis.

7.6.3 Calcium Waves

Recently, spontaneous long-range calcium waves have been discovered in the developing pupal wings *in vivo* (Ohno and Otaki 2015b). Calcium waves have been found to be released from the prospective eyespot centers and from damage sites (Fig. 7.12), although wave origins are not restricted to known elemental centers. At least four different types of waves are observed: expanding ring or traveling line, wandering line or point, oscillating area, and traveling oscillating area. Color patterns are disrupted by the injection of thapsigargin, a well-characterized inhibitor of Ca^{2+}-ATPase in the endoplasmic reticulum. For example, fuzzy boundaries of pattern elements have been reported in thapsigargin-treated individuals (Otaki et al. 2005b; Ohno and Otaki 2015b). I speculate that the calcium waves act as the activator in the late stage of the induction model, but calcium waves are not morphogenic signals themselves. Morphogenic signals are likely to be physical distortions (see below), and calcium waves may be released from these distortion waves.

7.6.4 Physical Distortion Hypothesis

What are the morphogenic signals? Despite the prediction of the rolling-ball model, it is difficult to imagine numerous minute "balls" rolling out from the center of a prospective eyespot. A hint comes from a study on pupal cuticle spots and their associated structures. Remarkably, organizing centers are often marked inherently as pupal cuticle focal spots in butterflies (Nijhout 1980a, b, 1990, 1991; Otaki et al. 2005a; Taira and Otaki 2016) (Fig. 7.13). This feature is especially notable in *Junonia* butterflies, but it is widely seen in many nymphalid butterflies that have eyespots or black spots (Otaki et al. 2005a). In addition, some cuticle focal spots are accompanied by cuticle marks. These spots and marks are likely produced by organizing cells for adult eyespots. The epithelial distortion structures of the

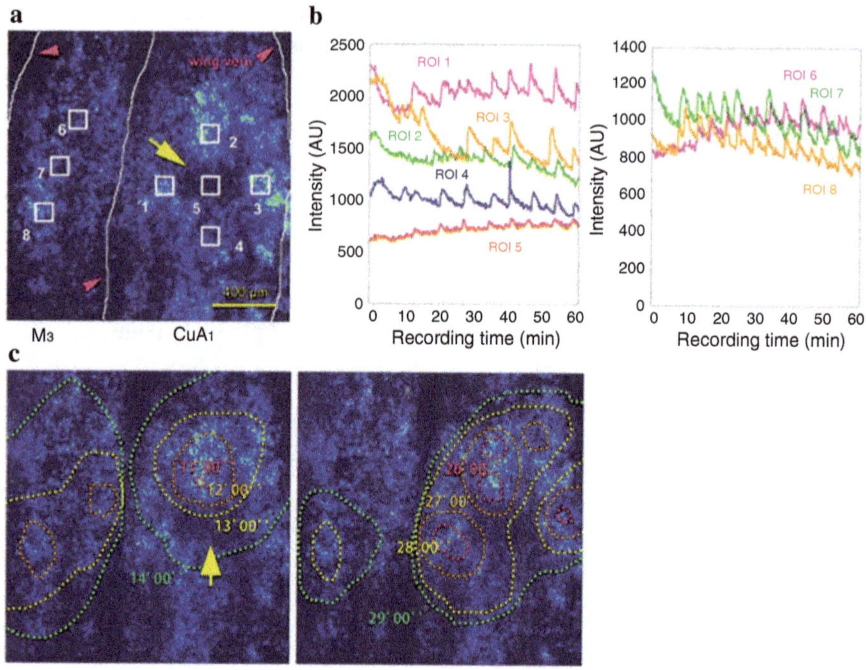

Fig. 7.12 Spontaneous calcium waves from the prospective organizing center for eyespot. Reproduced and modified from Ohno and Otaki (2015b). (**a**) Calcium signals (*blue*) in the M_3 and CuA_1 compartments. ROIs 1–8 were examined for intensity changes in the following panels. The *yellow arrow* indicates the prospective eyespot (also in **c**). The *red arrowheads* indicate the wing veins. (**b**) Fluorescence intensity changes of Fluo-4 in ROIs. (**c**) Propagating calcium signals around the prospective eyespot area. Panels in **a** and **c** show a single identical visual field at different time points. The shape of a wave at a given time point (in min) is depicted by a *dotted circle*

prospective elements have also been confirmed by *in vivo* imaging of the living tissue (Ohno and Otaki 2015a; Iwasaki et al. 2017). The association of the organizing centers with distortion structures may be called the **distortion rule for organizing centers**.

It is likely that the cellular volume increase or change in shape at the particular position results in the formation of the pupal cuticle spot as a by-product. The cellular changes would cause epithelial distortions, which could expand as a series of waves. The slow contraction of the wing tissue during the early pupal stage revealed by time-lapse movies (Iwata et al. 2014) probably helps to expand the distortion waves. That is, the **physical distortion hypothesis** states that morphogenic signals are physical distortions of an epithelial sheet. The distortion hypothesis thus states that morphogenic signals cannot be reduced to a substance. Rather, these signals are a wave, i.e., a physical phase change of a medium (the epithelial sheet). To realize this signaling system, the epithelial sheet has to have a tension or at least cellular connections in some way, which is likely the case (Ohno and Otaki

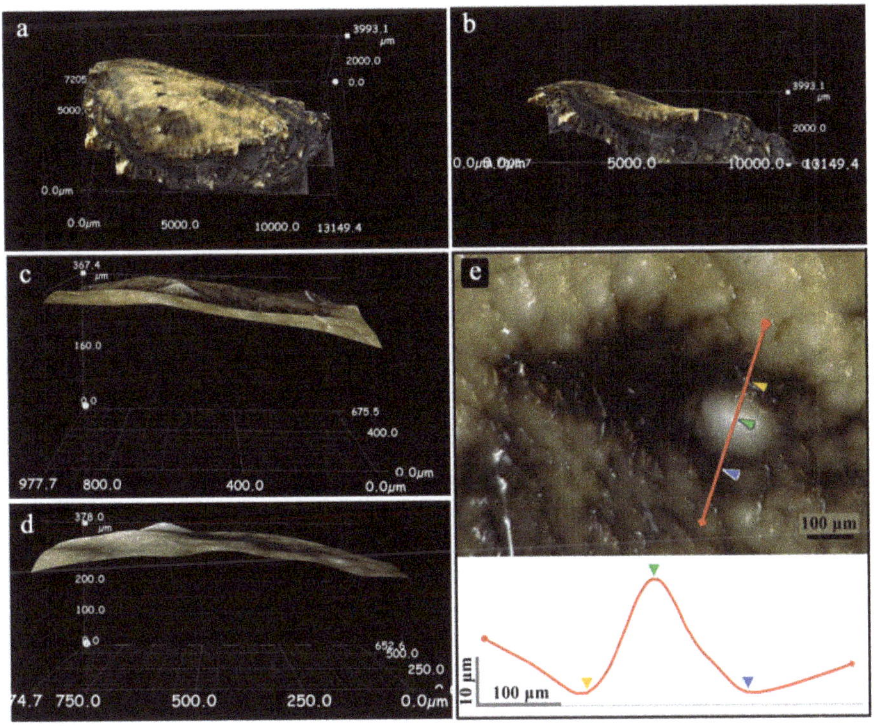

Fig. 7.13 Pupal focal cuticle spot of *Junonia orithya* based on three-dimensional reconstruction. Reproduced from Taira and Otaki (2016). (**a**) Top-down view of the entire left forewing surface. (**b**) Side view. (**c, d**) A pupal cuticle spot. (**e**) High-magnification image of a pupal cuticle spot and its cross-sectional height. *Colored arrowheads* in the image indicate the site of measurement in the graph

2015a). A physical distortion could open stretch-activated calcium channels, as in other systems (Lee et al. 1999; Tracey et al. 2003; O'Neil and Heller 2005; Hillyard et al. 2010). Epithelial cells that have received enough calcium ions inside could duplicate their genome and differentiate into scale cells that harbor specified cell size and specified scale size. In this time series, gene expression changes are downstream (not upstream) events; in other words, these changes are not a cause but a result of morphogenic signal propagation.

The distortion hypothesis states that mechanical disturbance of an epithelial sheet functions as morphogenic signals. This idea may sound unfamiliar to biologists, but this should not be a reason to reject this model as long as the model is consistent with experimental and observational results. Fortunately, mechanobiology is an expanding interdisciplinary field between biology and physics (Iskratsch et al. 2014). Changes in the mechanical property of a cellular sheet may be caused by physical damage and subsequent wound-healing processes (Antunes et al. 2013) and by cell death (Teng and Toyama 2011; Toyama et al. 2008) in addition to cellular size and shape changes.

7.6.5 Damage-Induced Ectopic Elements

Physical damage at the prospective eyespot center immediately after pupation has been shown to reduce or eliminate eyespots, but damage at the prospective background induces ectopic elements in butterfly wings (Nijhout 1985; Brakefield and French 1995; French and Brakefield 1992, 1995; Otaki et al. 2005a, b; Otaki 2011c). Ectopic eyespots are most likely by-products of a wound-healing process. I believe that physical damage elicits physical distortions of the epithelial sheet. Interestingly, the genes expressed are similar in normal development and in the healing process (Monteiro et al. 2006). Likewise, physical damage elicits calcium waves in normal development and in the healing process (Ohno and Otaki 2015b). Thus, the wound-healing process and the normal process of color pattern development would share similar mechanisms not only at the phenotypic level but also at the molecular level.

If the putative morphogen from a natural organizing center is a specific substance, it is difficult to imagine that physical damage confers an ability in immature epithelial cells to synthesize that specific substance. Probably partly for this reason, it is often interpreted that physical damage (and also pharmacological treatments) increases or decreases the "preset" threshold levels of signal-receiving immature scale cells in the conventional gradient model (Nijhout 1985; Brakefield and French 1995; French and Brakefield 1992, 1995; Otaki et al. 2005a, b; Otaki 2011c). Although it is entirely possible that this interpretation explains many damage-induced effects, dynamic interactions between two adjacent eyespots during development, shown in *J. almana*, suggest that a simple change in threshold levels is not realistic; when one eyespot becomes smaller as a result of damage, the other eyespot becomes larger (Otaki 2011c). It should also be noted that a possible mechanism of how damage lowers threshold, if this is the case, has never been well explained.

7.6.6 Focal Damage

What will occur when physical damage at the eyespot focal site is elicited? At the early stage of pupae, a smaller-than-normal eyespot is produced. Interestingly, late damage produces a larger-than-normal eyespot. The late damage result is explained by the addition of a new signal because this is similar to the fact that background damage produces a new signal for an eyespot or a black spot. The early damage result is explained as the damage of the signal-producing cells, resulting in the low-level signal. However, this result may indicate the source dependence of the signal, whereas wave signals are supposed to be source independent.

Considering the physical distortion hypothesis, the focal damage during the signal release may simply relax the distortion of the epithelial sheet. As a result, a distortion wave cannot go away. It may even go back to the original state. In contrast, at the later stage, epithelial distortions may have already been relaxed and the signal is ready to settle. Thus, the late focal damage may recreate the distortion, such as the background damage, resulting in a larger-than-normal eyespot.

7.7 Part V: Generalization and Essence

7.7.1 *Reinforced Version of the Induction Model*

To summarize, the reinforced version of the induction model for eyespot development is explained below (Fig. 7.14). This scheme includes many speculations to bridge the fragmented knowledge of the butterfly wing system. For simplicity, the development of a simple black disk (i.e., black spot) is delineated first below (Fig. 7.14a).

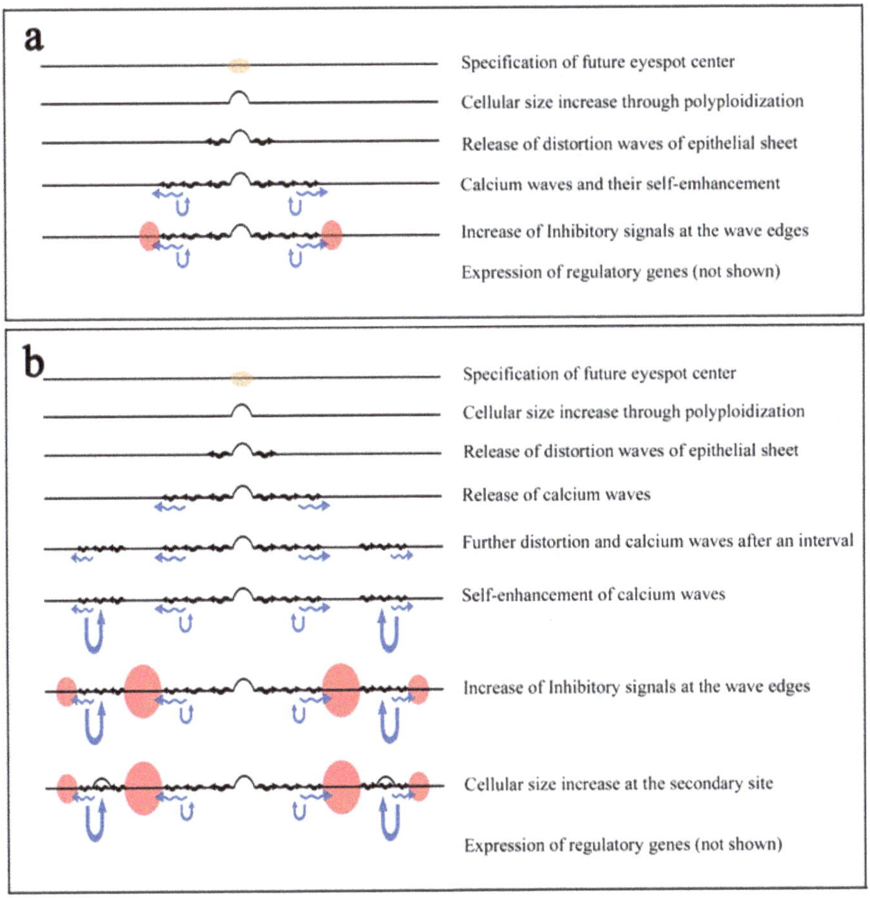

Fig. 7.14 Reinforced induction model. Time series of events from the top to the bottom. (**a**) Black spot formation. (**b**) Eyespot formation

In the beginning, a future eyespot center (organizing center) is first specified. Physical distortion of the epithelial sheet is formed due to cellular size changes and deformations. These cellular changes would cause distortion waves that propagate radially to surrounding cells, according to the rolling-ball model. The propagating waves are "translated" into chemical signals, i.e., calcium waves, possibly through a stretch-activated calcium ion channel on the membrane, acting as an activator in a reaction-diffusion model, as traveling calcium waves have been detected (Ohno and Otaki 2015b). As the physical distortions and their associated calcium signals propagate, calcium signals may be enhanced by themselves as oscillations, as oscillating calcium waves have also been detected (Ohno and Otaki 2015b). Calcium oscillations induce unknown inhibitory signals in cells located in the periphery of the oscillations. The induced inhibitory signals inhibit further propagation of the original calcium signals, finalizing the position and shape of the black spot. Calcium oscillations stimulate cells to undergo genome amplification and to express a set of regulatory genes such as *Wnt*-family genes (Monteiro et al. 2006; Martin and Reed 2014), *spalt* and *Distal-less* (Monteiro et al. 2013; Adhikari and Otaki 2016; Dhungel et al. 2016; Zhang and Reed 2016). Alternatively, calcium oscillations may be stabilized by the Wnt/Ca^{2+} transduction pathway that involves intracellular calcium release (Kühl et al. 2000; Kohn and Moon 2005). Cellular size increases in the prospective black ring according to the genome size or ploidy level. This process may be regulated by the *cortex* gene, which has been identified recently (Nadeau et al. 2016; van't Hof et al. 2016). The final cellular size or the degrees of polyploidy then determine a repertoire of pigment synthesis genes to be expressed.

When an eyespot is produced, the scheme is more complicated (Fig. 7.14b). A released distortion wave does not readily induce calcium waves, but it progresses for some time. In the meantime, the distortion wave for the outer black ring is terminated, but after an interval, a new distortion wave for the inner black ring is released. At this point, calcium wave induction and its self-enhancement occur, and inhibitory signals are produced at the wave edges, which finalize the position of the black rings. Genome amplification and the expression of regulatory genes follow. Cellular size increases at the prospective black rings according to the number of genomes in a cell. Where the calcium oscillations by self-enhancement are highly active, the high degree of cellular size increase occurs, resulting in the formation of a secondary organizing center, which is often seen in PFEs. This second round of color pattern determination ensures self-similarity between the eyespot and PFE.

In this series of events, the three most important events are distortion waves (D), calcium waves (C), and gene expression changes (G), which may be called the **DCG cycle**. This series of events repeats twice to create the self-similarity between the eyespot and PFE.

7.7.2 *Generalization to Other Systems*

Thus far, I have discussed the nymphalid wing color pattern system. The applicability of the information above to other butterfly systems has not been examined, but the lycaenid system is probably similar because the symmetry rule and the core-paracore rule hold true, at least in the lycaenid central symmetry system (Iwata et al. 2013, 2015). The fish skin system is different from the butterfly wing system in that epidermal cells in fish can move in response to surrounding cells, whereas butterfly cells cannot move. Nonetheless, the inductive nature of different colors based on short-range activation and long-range inhibition is likely shared in fish and butterflies; both systems can produce ectopic patterns associated with calcium waves after physical damage (Ohno and Otaki 2012).

Morphogenesis is three-dimensionally dynamic in any developmental system, but a good example of three-dimensional dynamism of the epithelial sheet is the morphogenetic furrow in the *Drosophila* retina (Greenwood and Struhl 1999; Schlichting and Dahmann 2008; Sato et al. 2013). The furrow is a physical distortion of the imaginal eye disk. This epithelial fold moves, and its movement coincides with cellular differentiation. The furrow may physically elicit the expression of morphogenetic genes such as *hedgehog* and *decapentaplegic* if the furrow is not a physical by-product of cellular differentiation.

7.7.3 *DCG Cycle for Self-Similarity and Its Implications*

Nearly two-dimensional butterfly wing color patterns can be viewed, somewhat ironically, as a developmental and evolutionary application of three-dimensional bulges and dents that are used in general morphogenesis. To achieve self-similar structures, organisms evolve to transmit a signal from the primary to secondary organizing centers through distortion waves of the epithelial sheet. This mechanical lateral signaling mechanism can cover a long distance with simplicity. Thus, it may be a very early evolutionary innovation. Evolution of the signal translator, mechanosensory calcium channels, might have followed, together with several genes that stabilize calcium oscillations and inhibition. In conclusion, the DCG cycle for self-similar structures has deep implications for biological evolution and development.

Acknowledgments The author is grateful to the organizers of the meeting, Professor T. Sekimura (Chubu University, Japan) and Professor H. F. Nijhout (Duke University, USA), for giving me an opportunity to present my research at the meeting and to write this meeting report. I also thank the members of the BCPH Unit of Molecular Physiology for discussion.

References

Adhikari K, Otaki JM (2016) A single-wing removal methods to assess correspondence between gene expression and phenotype in butterflies: a case of *Distal-less*. Zool Sci 33:13–20

Antunes M, Pereira T, Cordeiro JV, Almeida L, Jacinto A (2013) Coordinated waves of actomyosin flow and apical cell constriction immediately after wounding. J Cell Biol 202:365–379

Ball P (1999) The self-made tapestry: pattern formation in nature. Oxford University Press, Oxford

Ball P (2016) Patterns in nature: why the natural world looks the way it does. The University of Chicago Press, Chicago

Barnsley MF, Ervin V, Hardin D, Lancaster J (1986) Solution of an inverse problem for fractals and other sets. Proc Natl Acad Sci U S A 83:1975–1977

Beldade P, Koops K, Brakefield PM (2002) Modularity, individuality, and evo-devo in butterfly wings. Proc Natl Acad Sci U S A 99:14262–14267

Beldade P, French V, Brakefield PM (2008) Developmental and genetic mechanisms for evolutionary diversification of serial repeats: eyespot size in *Bicyclus anynana* butterflies. J Exp Zool Mol Dev Evol 310B:191–201

Ben-Jacob E, Shochet O, Tenenbaum A, Cohen I, Czirok A, Vicsek T (1994) Generic modelling of cooperative growth patterns in bacterial colonies. Nature 368:46–49

Brakefield PM, French V (1995) Eyespot development on butterfly wings: the epidermal response to damage. Dev Biol 168:98–111

Cho EH, Nijhout HF (2013) Development of polyploidy of scale-building cells in the wings of *Manduca sexta*. Arthropod Struct Dev 42:37–46

Dhungel B, Otaki JM (2009) Local pharmacological effects of tungstate in the color-pattern determination of butterfly wings: a possible relationship between the eyespot and parafocal element. Zool Sci 26:758–764

Dhungel B, Otaki JM (2013) Morphometric analysis of nymphalid butterfly wings: number, size and arrangement of scales, and their implications for tissue-size determination. Entomol Sci 17:207–218

Dhungel B, Ohno Y, Matayoshi R, Iwasaki M, Taira W, Adhikari K, Gurung R, Otaki JM (2016) Distal-less induces elemental color patterns in *Junonia* butterfly wings. Zool Lett 2:4

Family F, Masters BR, Platt DE (1989) Fractal pattern formation in human retinal vessels. Physica D: Nonlinear Phenomena 38:98–103

French V, Brakefield PM (1992) The development of eyespot patterns on butterfly wings: morphogen sources or sinks? Development 116:103–109

French V, Brakefield PM (1995) Eyespot development on butterfly wings: the focal signal. Dev Biol 168:112–123

Gierer A, Meinhardt H (1972) A theory of biological pattern formation. Kybernetik 12:30–39

Gilbert LE, Forrest HS, Schultz TD, Harvey DJ (1988) Correlations of ultrastructure and pigmentation suggest how genes control development of wing scales of *Heliconius* butterflies. J Res Lepidoptera 26:141–160

Greenwood S, Struhl G (1999) Progression of the morphogenetic furrow in the *Drosophila* eye: the roles of Hedgehog, Decapentaplegic and the Raf pathway. Development 126:5795–5808

Henke K (1946) Ueber die verschiendenen Zellteilungsvorgänge in der Entwicklung des beschuppten Flügelepithelis der Mehlmotte *Ephestina kühniella* Z. Biol Zentralbl 65:120–135

Henke K, Pohley H-J (1952) Differentielle Zellteilungen und Polyploidie bei der Schuppenbildung der Mehlmotte *Ephestia kühniella* Z. Z Naturforsch B 7:65–79

Hillyard SD, Willumsen NJ, Marrero MB (2010) Stretch-activated cation channel from larval bullfrog skin. J Exp Biol 213:1782–1787

Hiyama A, Taira W, Otaki JM (2012) Color-pattern evolution in response to environmental stress in butterflies. Front Genet 3:15

Iskratsch T, Wolfenson H, Sheetz MP (2014) Appreciating force and shape – the rise of mechanotransduction in cell biology. Nat Rev Mol Cell Biol 15:825–833

Iwasaki M, Ohno Y, Otaki JM (2017) Butterfly eyespot organizer: *in vivo* imaging of the prospective focal cells in pupal wing tissues. Sci Rep 7:40705

Iwata M, Otaki JM (2016a) Focusing on butterfly eyespot focus: uncoupling of white spots from eyespot bodies in nymphalid butterflies. SpringerPlus 5:1287

Iwata M, Otaki JM (2016b) Spatial patterns of correlated scale size and scale color in relation to color pattern elements in butterfly wings. J Insect Physiol 85:32–45

Iwata M, Hiyama A, Otaki JM (2013) System-dependent regulations of colour-pattern development: a mutagenesis study of the pale grass blue butterfly. Sci Rep 3:2379

Iwata M, Ohno Y, Otaki JM (2014) Real-time *in vivo* imaging of butterfly wing development: revealing the cellular dynamics of the pupal wing tissue. PLoS One 9:e89500

Iwata M, Taira W, Hiyama A, Otaki JM (2015) The lycaenid central symmetry system: color pattern analysis of the pale grass blue butterfly *Zizeeria maha*. Zool Sci 32:233–239

Janssen JM, Monteiro A, Brakefield PM (2001) Correlations between scale structure and pigmentation in butterfly wings. Evol Dev 3:415–423

Kaandorp JA, Kübler JE (2001) The algorithmic beauty of seaweeds, sponges, and corals. Springer, Berlin

Kazama M, Ichinei M, Endo S, Iwata M, Hino A, Otaki JM (2017) Species-dependent microarchitectural traits of iridescent scales in the triad taxa of Ornithoptera birdwing butterflies. Entomol Sci 20:255–269

Kohn AD, Moon RT (2005) Wnt and calcium signaling: β-Catenin-independent pathways. Cell Calcium 38:439–446

Kristensen NP, Simonsen TJ (2003) Hairs and scales. In: Kristensen PN (Ed). Lepidoptera, moths and butterflies: morphology, physiology, and development. Handbook of zoology, arthropoda: insecta, vol IV. Walter de Gruyter, Berlin, pp 9–22

Kühl M, Sheldahl LC, Park M, Miller JR, Moon RT (2000) The Wnt/Ca^{2+} pathway: a new vertebrate Wnt signaling pathway takes shape. Trends Genet 16:279–283

Kusaba K, Otaki JM (2009) Positional dependence of scale size and shape in butterfly wings: wing-wide phenotypic coordination of color-pattern elements and background. J Insect Physiol 55:174–182

Lee J, Ishihara A, Oxford G, Johnson B, Jacobson K (1999) Regulation of cell movement is mediated by stretch-activated calcium channels. Nature 400:382–386

Mahdi SHA, Gima S, Tomita Y, Yamasaki H, Otaki JM (2010) Physiological characterization of the cold-shock-induced humoral factor for wing color-pattern changes in butterflies. J Insect Physiol 56:1022–1031

Mahdi SHA, Yamasaki H, Otaki JM (2011) Heat-shock-induced color-pattern changes of the blue pansy butterfly *Junonia orithya*: physiological and evolutionary implications. J Therm Biol 36:312–321

Mandelbroit BB (1983) The fractal geometry of nature, Revised edn. W. H. Freeman, New York

Martin A, Reed RD (2014) *Wnt* signaling underlies evolution and development of the butterfly wing pattern symmetry systems. Dev Biol 395:367–378

Meinhardt H (1982) Models of biological pattern formation. Academic Press, London

Meinhardt H (2009) The algorithmic beauty of sea shells, Fourth edn. Springer, New York

Meinhardt H, Gierer A (1974) Applications of a theory of biological pattern formation based on lateral inhibition. J Cell Sci 15:321–346

Meinhardt H, Gierer A (2000) Pattern formation by local self-activation and lateral inhibition. BioEssays 22:753–760

Monteiro A (2008) Alternative models for the evolution of eyespots and of serial homology on lepidopteran wings. BioEssays 30:358–366

Monteiro A (2014) Origin, development, and evolution of butterfly eyespots. Annu Rev Entomol 60:253–271

Monteiro A, French V, Smit G, Brakefield PM, Metz JA (2001) Butterfly eyespot patterns: evidence for specification by a morphogen diffusion gradient. Acta Biotheor 49:77–88

Monteiro A, Prijs J, Bax M, Hakkaart T, Brakefield PM (2003) Mutants highlight the modular control of butterfly eyespot patterns. Evol Dev 5:180–187

Monteiro A, Glaser G, Stockslager S, Glansdorp N, Ramos D (2006) Comparative insights into questions of lepidopteran wing pattern homology. BMC Dev Biol 6:52

Monteiro A, Chen B, Ramos D, Oliver JC, Tong X, Guo M, Wang W-K, Fazzino L, Kamal F (2013) *Distal-less* regulates eyespot patterns and melanization in *Bicyclus* butterflies. J Exp Zool B Mol Dev Evol 320:321–331

Nadeau N, Pardo-Diaz C, Whibley A, Supple MA, Saenko SV, Wallbank RWR, Wu GC, Maroja L, Ferguson L, Hanly JJ, Hines H, Salazar C, Merrill RM, Dowling AJ, French-Constant RH, Laurens V, Joron M, WO MM, Jiggins CD (2016) The gene *cortex* controls mimicry and crypsis in butterflies and moths. Nature 534:106–110

Nijhout HF (1978) Wing pattern formation in lepidoptera: a model. J Exp Zool 206:119–136

Nijhout HF (1980a) Pattern formation on lepidopteran wings: determination of an eyespot. Dev Biol 80:267–274

Nijhout HF (1980b) Ontogeny of the color pattern on the wings of *Precis coenia* (Lepidoptera: Nymphalidae). Dev Biol 80:275–288

Nijhout HF (1981) The color patterns of butterflies and moths. Sci Am 254:145–151

Nijhout HF (1984) Colour pattern modification by coldshock in Lepidoptera. J Embryol Exp Morphol 81:287–305

Nijhout HF (1985) Cautery-induced colour patterns in *Precis coenia* (Lepidoptera: Nymphalidae). J Embryol Exp Morpholog 86:191–203

Nijhout HF (1990) A comprehensive model for color pattern formation in butterflies. Proc R Soc Lond B 239:81–113

Nijhout HF (1991) The development and evolution of butterfly wing patterns. Smithsonian Institution Press, Washington, DC

Nijhout HF (1994) Symmetry systems and compartments in Lepidopteran wings: the evolution of a patterning mechanism. Development 1994(Suppl):225–233

Nijhout HF (2001) Elements of butterfly wing patterns. J Exp Zool 291:213–225

Ohno Y, Otaki JM (2012) Eyespot colour pattern determination by serial induction in fish: mechanistic convergence with butterfly eyespots. Sci Rep 2:290

Ohno Y, Otaki JM (2015a) Live cell imaging of butterfly pupal and larval wings *in vivo*. PLoS One 10:e0128332

Ohno Y, Otaki JM (2015b) Spontaneous long-range calcium waves in developing butterfly wings. BMC Dev Biol 15:17

O'Neil RG, Heller S (2005) The mechanosensitive nature of TRPV channels. Pflugers Arch 451:193–203

Otaki JM (1998) Color-pattern modifications of butterfly wings induced by transfusion and oxyanions. J Insect Physiol 44:1181–1190

Otaki JM (2007) Reversed type of color-pattern modifications of butterfly wings: a physiological mechanism of wing-wide color-pattern determination. J Insect Physiol 53:526–537

Otaki JM (2008a) Physiologically induced color-pattern changes in butterfly wings: mechanistic and evolutionary implications. J Insect Physiol 54:1099–1112

Otaki JM (2008b) Phenotypic plasticity of wing color patterns revealed by temperature and chemical applications in a nymphalid butterfly *Vanessa indica*. J Therm Biol 33:128–139

Otaki JM (2008c) Physiological side-effect model for diversification of non-functional or neutral traits: a possible evolutionary history of *Vanessa* butterflies (Lepidoptera, Nymphalidae). Trans Lepid Soc Jpn 59:87–102

Otaki JM (2009) Color-pattern analysis of parafocal elements in butterfly wings. Entomol Sci 12:74–83

Otaki JM (2011a) Generation of butterfly wing eyespot patterns: a model for morphological determination of eyespot and parafocal element. Zool Sci 28:817–827

Otaki JM (2011b) Color-pattern analysis of eyespots in butterfly wings: a critical examination of morphogen gradient models. Zool Sci 28:403–413

Otaki JM (2011c) Artificially induced changes of butterfly wing colour patterns: dynamic signal interactions in eyespot development. Sci Rep 1:111

Otaki JM (2012a) Colour pattern analysis of nymphalid butterfly wings: revision of the nymphalid groundplan. Zool Sci 29:568–576

Otaki JM (2012b) Structural analysis of eyespots: dynamics of morphogenic signals that govern elemental positions in butterfly wings. BMC Syst Biol 6:17

Otaki JM, Yamamoto H (2004a) Species-specific color-pattern modifications of butterfly wings. Develop Growth Differ 46:1–14

Otaki JM, Yamamoto H (2004b) Color-pattern modifications and speciation in butterflies of the genus *Vanessa* and its related genera *Cynthia* and *Bassaris*. Zool Sci 21:967–976

Otaki JM, Ogasawara T, Yamamoto H (2005a) Morphological comparison of pupal wing cuticle patterns in butterflies. Zool Sci 22:21–34

Otaki JM, Ogasawara T, Yamamoto H (2005b) Tungstate-induced color-pattern modifications of butterfly wings are independent of stress response and ecdysteroid effect. Zool Sci 22:635–644

Otaki JM, Kimura Y, Yamamoto H (2006) Molecular phylogeny and color-pattern evolution of *Vanessa* butterflies (Lepidoptera, Nymphalidae). Trans Lepid Soc Jpn 57:359–370

Otaki JM, Hiyama A, Iwata M, Kudo T (2010) Phenotypic plasticity in the range-margin population of the lycaenid butterfly *Zizeeria maha*. BMC Evol Biol 10:252

Sato M, Suzuki T, Nakai Y (2013) Waves of differentiation in the fly visual system. Dev Biol 380:1–11

Schlichting K, Dahmann C (2008) Hedgehog and Dpp signaling induce cadherin Cad86C expression in the morphogenetic furrow during *Drosophila* eye development. Mech Dev 125:712–728

Schwanwitsch BN (1924) On the ground plan of wing-pattern in nymphalid and certain other families of rhopalocerous Lepidoptera. Proc Zool Soc London 34:509–528

Serfas MS, Carroll SB (2005) Pharmacologic approaches to butterfly wing patterning: sulfated polysaccharides mimic or antagonize cold shock and alter the interpretation of gradients of positional information. Dev Biol 287:416–424

Simonsen TJ, Kristensen NP (2003) Scale length/wing length correlation in Lepidoptera (Insecta). J Nat Hist 37:673–679

Sondhi KH (1963) The biological foundations of animal patterns. Q Rev Biol 38:289–327

Süffert F (1927) Zur vergleichende Analyse der Schmetterlingsaeinchnung. Biol Zentralbl 47:385–413

Taira W, Otaki JM (2016) Butterfly wings are three-dimensional: pupal cuticle focal spots and their associated structures in *Junonia* butterflies. PLoS One 11:e0146348

Taira W, Kinjo S, Otaki JM (2015) The marginal band system in the nymphalid butterfly wings. Zool Sci 32:38–46

Teng X, Toyama Y (2011) Apoptotic force: active mechanical function of cell death during morphogenesis. Dev Growth Differ 53:269–276

Toyama Y, Peralta XG, Wells AR, Kiehart DP, Edwards GS (2008) Apoptotic force and tissue dynamics during *Drosophila* embryogenesis. Science 321:1683–1686

Tracey WD Jr, Wilson RI, Laurent G, Benzer S (2003) *painless*, a *Drosophila* gene essential for nociception. Cell 113:261–273 van't Hof AE, Campagne P, Rigden DJ, Yung CJ, Lingley J, Quail MA, Hall N, Darby AC, Saccheri IJ (2016) The industrial melanism mutation in British peppered moths is a transposable element. Nature 534:102–105

Zhang L, Reed RD (2016) Genome editing in butterflies reveals that *spalt* promotes and *Distal-less* represses eyespot colour patterns. Nat Commun 7:11760

Part III
Butterflies and Developmental Genetics

8

Genome Editing in Lepidoptera: Application of CRISPR/Cas9 Methods

Linlin Zhang and Robert D. Reed

Abstract CRISPR/Cas9 genome editing has revolutionized functional genetic work in many organisms and is having an especially strong impact in emerging model systems. Here we summarize recent advances in applying CRISPR/Cas9 methods in Lepidoptera, with a focus on providing practical advice on the entire process of genome editing from experimental design through to genotyping. We also describe successful targeted GFP knockins that we have achieved in butterflies. Finally, we provide a complete, detailed protocol for producing targeted long deletions in butterflies.

Keywords Genome editing • Knockin • Butterfly • Transgenic • Transformation • Evo-devo

8.1 Introduction

The order Lepidoptera represents a tenth of the world's described species and includes many taxa of economic and scientific importance. Despite strong interest in this group, however, there has been a frustrating lack of progress in developing routine approaches for manipulative genetic work. While the last two decades have seen examples of transgenesis and targeted knockouts using methods like transposon insertion (Tamura et al. 2000), zinc-finger nucleases (Takasu et al. 2010; Merlin et al. 2013), and TALENs (Takasu et al. 2013; Markert et al. 2016), especially in the silk moth *Bombyx mori*, these approaches have resisted widespread application due to their laborious nature. We see two other main reasons manipulative genetics has failed to become routine in Lepidoptera. The first is that many lepidopterans are sensitive to inbreeding, and in some species it can be difficult to maintain experimental lines without special effort. The second is that lepidopterans appear to have an unusual resistance to RNAi (Terenius et al. 2011; Kolliopoulou and Swevers 2014), a method that has dramatically accelerated work in other groups of insects.

L. Zhang • R.D. Reed (✉)
Department of Ecology and Evolutionary Biology, Cornell University, 215 Tower Rd., Ithaca, NY 14853-7202, USA
e-mail: robertreed@cornell.edu

Given this history of challenges in Lepidoptera, it is with great excitement that over the last few years we have seen an increasing number of studies that demonstrate the high efficiency of CRISPR/Cas9-mediated genome editing in this group. Our own lab began experimenting with genome editing in butterflies in 2014, and we and our collaborators have now successfully edited over 15 loci across six species, generating both targeted deletions and insertions. The purpose of this review is to briefly summarize the current state of this fast-moving field and to provide practical advice for those who would like to use this technology in their own work.

8.2 Published Examples of Cas9-Mediated Genome Editing in Lepidoptera

Between 2013 and early 2017, we identify 22 published studies applying CRISPR/Cas9 methods in Lepidoptera (Table 8.1). The earliest published reports of Cas9-mediated genome editing in Lepidoptera, from 2013 and 2014, all describe work done in *B. mori* (Wang et al. 2013; Ma et al. 2014; Wei et al. 2014) – an experimental system that benefits from a large research community that had already developed efficient methods for injection, rearing, and genotyping. To our knowledge, Wang et al. (2013) represent the first published report of Cas9-mediated genome editing in Lepidoptera and set three important precedents. First, they established the protocol that has been more or less emulated by most following studies, where single-guide RNAs (sgRNAs) are co-injected with Cas9 mRNA into early-stage embryos. Second, they demonstrated that it is possible to co-inject dual sgRNAs to produce long deletions. In this respect, the 3.5 kb deletion they produced was an important early benchmark for demonstrating the possibility of generating long deletions in Lepidoptera. Third, they showed that deletions could occur in the germ line at a high enough frequency to generate stable lines.

After Wang et al. (2013), one of the next most important technical advancements came from Ma et al. (2014), who showed that knockins could be achieved using a donor plasmid to insert a DsRed expression cassette using ~1 kb homology arms. Following this, Zhu et al. showed successful epitope tagging of *BmTUDOR-SN* gene by CRISPR/Cas9-mediated knockin in *Bombyx* cells (Zhu et al. 2015). Unfortunately, to our knowledge, these remain the only two examples of lepidopteran knockins outside of the new data we present below. The first example of Cas9 genome editing in a species besides *B. mori* was described by Li et al. (2015a), who produced deletions in three genes in the swallowtail butterfly *Papilio xuthus*. This was an important case study because it showed that the general approach used by Wang et al. (2013) in *B. mori* could be transferred to other species and still retain the same level of high efficiency. Two more notable technical advancements include the production of an 18 kb deletion in *B. mori* by Zhang et al. (2015) – the longest deletion we know of in Lepidoptera, and much longer than anything in

Table 8.1 Comparison of CRISPR genome editing in Lepidoptera

Species	References	Knockout/ knockin	Knockout strategy	Delivery	Genes targeted	Genotyping	Mosaic (%)	Germ line
Bombyx mori	Wang Y. et al. (2013)	Knockout	Small indels and long deletions	Cas9 mRNA and sgRNA	*BLOS2*	PCR	94–100%	Yes
	Ma S. et al. (2014)	Knockout/ knockin	Small indels	Plasmid of Cas9 and gRNA	*ku70*	T7E1	16.8–30.3%	Yes
	Wei W. et al. (2014)	Knockout	Small indels and long deletions	Cas9 mRNA and sgRNA	*ok/KMO/TH/tan*	Sequencing	16.7–35.0%	Yes
	Liu Y. et al. (2014)	Knockout	Small indels	Plasmid of Cas9 and gRNA	*Th/re/fl/yel-e/ kynu/e*	T7E1	5.7–18.9%	No
	Zhu L. et al. (2015)	Knockout/ knockin	Small indels	All-in-one vector	*Ku70/Ku80/Lig IV/XRCC4/XLF*	T7E1	20–80%	No
	Ling L. et al. (2015)	Knockout	Long deletions	Cas9 mRNA and sgRNA	*awd/fng*	PCR	40–61%	No
	Li Z, et al. (2015b)	Knockout	Long deletions	Cas9 mRNA and sgRNA	*EO*	PCR	60.60%	No
	Xin H. et al. (2015)	Knockout	Small indels	Cas9 mRNA and sgRNA	*sage*	Sequencing	46.67%	Yes
	Zhang Z. et al. (2015)	Knockout	Long deletions	Cas9 mRNA and sgRNA	*wnt1*	T7E1	42.5–90.6%	No
	Zeng B. et al. (2016)	Knockout	Small indels	Cas9 mRNA and sgRNA	*GFP/BLOS2*	T7E1	NA	Yes
Helicoverpa armigera	Khan S. et al. (2017)	Knockout	Small indels	Cas9 mRNA and sgRNA	*w/st/bw/ok*	T7E1	NA	Yes

(continued)

Table 8.1 (continued)

Species	References	Knockout/knockin	Knockout strategy	Delivery	Genes targeted	Genotyping	Mosaic (%)	Germ line
Papilio xuthus	Li X, et al. (2015a)	Knockout	Small indels and long deletions	Cas9 mRNA and sgRNA	*Abd-B/e/fz*	T7E1	18.33–90.85%	No
	Zhang L. et al. (2017)	Knockout	Long deletions	Cas9 protein and sgRNA	*y*	NA	83.33%	No
	Perry M. et al. (2016)	Knockout	Long deletions	Cas9 protein and sgRNA	*y/ss*	Sequencing	NA	No
Vanessa cardui	Zhang L. and Reed R. (2016)	Knockout	Long deletions	Cas9 protein and sgRNA	*Ddc/Dll/spalt*	PCR and sequencing	51.7–56%	No
	Zhang L. et al. (2017)	Knockout	Long deletions	Cas9 protein and sgRNA	*y/b/e/yel-d*	PCR	42–80%	No
	Perry M. et al. (2016)	Knockout	Long deletions		*y/ss*	Sequencing	NA	No
Junonia coenia	Zhang L. and Reed R. (2016)	Knockout	Long deletions	Cas9 protein and sgRNA	*Dll/spalt*	PCR and sequencing	33–41%	No
	Zhang L. et al. (2017)	Knockout	Long deletions	Cas9 protein and sgRNA	*Ple/e*	NA	27–42%	No
Bicyclus anynana	Zhang L. et al. (2017)	Knockout	Long deletions	Cas9 mRNA/protein and sgRNA	*y/e*	T7E1	31–72%	No
	Beldade P. and Peralta C. (2017)	Knockout	NA	NA	*Ddc/e*	NA	NA	No
Danaus plexippus	Markert M. et al. (2016)	Knockout	Small and long deletions	Cas9 mRNA and sgRNA	*cry2/clk*	Cas9 in vitro cleavage	57.9–61.5%	Yes

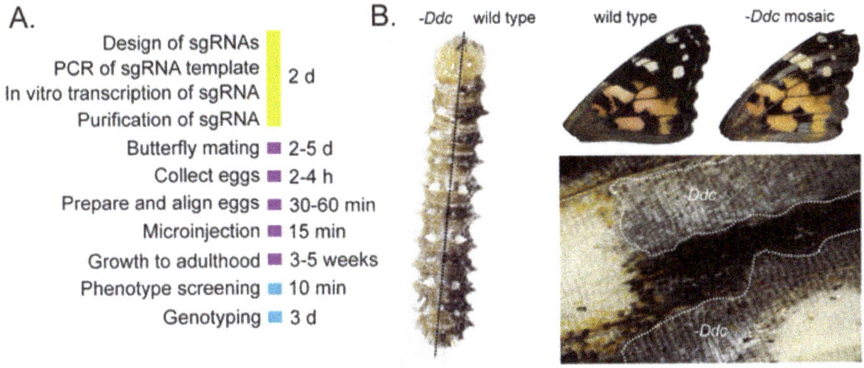

Fig. 8.1 Timeline and example outcome of G_0 CRISPR/Cas9 mosaic knockout experiments in butterflies. (**a**) Overview and timeline of mutant generation by CRISPR/Cas9 injection to butterfly embryos. (**b**) Example of larval and adult wing somatic mosaic phenotypes resulting from knockout of the melanin pigmentation gene *Ddc*

Drosophila reports we have seen – and the direct injection of recombinant Cas9 protein instead of Cas9 mRNA (Zhang and Reed 2016; Perry et al. 2016), which was an important improvement to the protocol that significantly simplifies the genome editing workflow.

Through our lab's research on butterfly wing pattern development, we have tried most of the methods described in the studies cited above, and we have gained significant experience in porting these protocols across species. We now perform targeted long deletions routinely and with a fairly high throughput. As of the end of 2016, we and our colleagues have successfully applied this general approach in six butterfly and two moth species (*Vanessa cardui*, *Junonia coenia*, *Bicyclus anynana*, *Papilio xuthus*, *Heliconius erato*, *Agraulis vanillae*, *B. mori*, and *Plodia interpunctella*), with each species requiring only minor modifications to physical aspects of egg injection protocol. As we describe below, we have also successfully achieved protein coding knockins similar to Zhu et al. (2015), although our efficiency levels remain similarly low. Below we outline the approach that we have found to be the most time- and cost-efficient and transferable between species (Fig. 8.1a).

8.3 Experimental Design

Deletions Loss-of-function deletion mutations can be generated by nonhomologous end joining (NHEJ) following double-strand breaks (DSBs). Both small indel (single cleavage) and long deletion knockout strategies (co-injection of two sgRNAs) have been employed in Lepidoptera (Table 8.1). Our lab currently favors long deletions using dual sgRNAs because it facilitates rapid screening and genotyping of mutants using PCR and regular agarose gel electrophoresis. Small indels produced by single cleavages are too small to detect easily using normal agarose gels. Dual sgRNA deletions, however, can be tens, hundreds, or thousands of base pairs long and are easy to identify in gels. sgRNA target sites can be easily identified simply by scanning the target region for $GGN_{18}NGG$ or $N_{20}NGG$ motifs on either strand using the CasBLASTR web tool (http://www.casblastr.org/). In our experience, the relative strandedness of sgRNAs does not appear to have a significant effect on the efficiency of double sgRNA long deletion experiments. If a reference genome is available, candidate sgRNA sequences should be used for a blast search to confirm there are not multiple binding sites that may produce off-target effects. The injection mix we typically use is 200 ng/μl Cas9 and 100 ng/μl of each sgRNA – this will tend to give larger effects and is suitable for less potentially lethal loci. For targets that may result in more deleterious effects, we recommend decreasing the amount of Cas9/sgRNA mix and injecting later in embryonic development to induce fewer and smaller clones. We have been able to induce mosaic mutants (e.g., Fig. 8.1b) using as low as 20 ng/μl Cas9 and 50 ng/μl of each sgRNA in different butterfly species.

Insertions CRISPR/Cas9-induced-site-specific DSBs can be precisely repaired by homology-directed recombination repair (HDR). The HDR pathway can replace an endogenous genome segment with a homologous donor sequence and can thus be used for knockin of foreign DNA into a selected genomic locus. To our knowledge, there are only two published examples of this approach in Lepidoptera, both of which

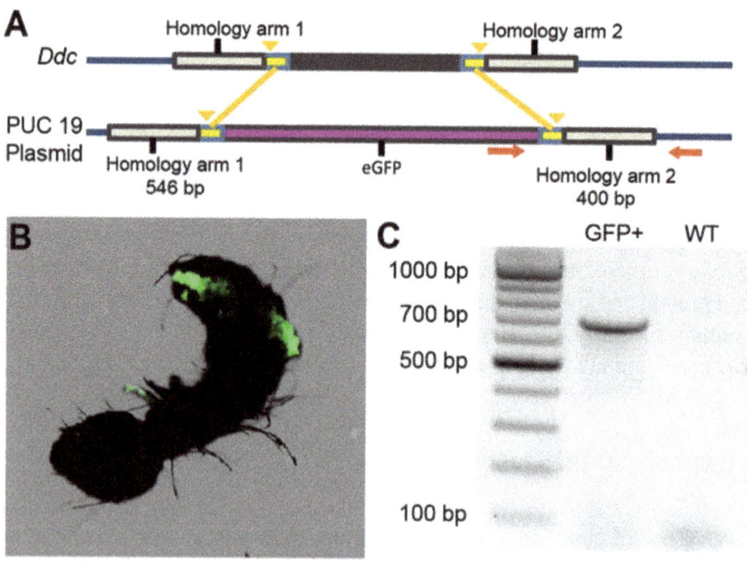

Fig. 8.2 Knockin tagging of the *Ddc* gene in *V. cardui*. (**a**) Schematic overview of the *Ddc* locus and donor construct consisting of homology arms, EGFP coding region, and genotyping primers. PAM regions are marked by *yellow*, cut sites are marked by *yellow arrowhead*, and genotyping primers are marked by *red arrows*. (**b**) Strong mosaic EGFP expression in knockin caterpillars visualized by fluorescent microscopy. (**c**) PCR analysis demonstrates using the primers in (**a**) showing the insertion of EGFP into the *Ddc* coding region

are in *B. mori* (Ma et al. 2014; Zhu et al. 2015). To test the feasibility of this approach in butterflies, we sought to insert an in-frame EGFP coding sequence into to the *V. cardui dopa decarboxylase* (*Ddc*) locus using a donor plasmid containing the EGFP coding sequences and homologous arms matching endogenous sequences flanking the Cas9 cut sites (Fig. 8.2a). As shown in Fig. 8.2b, EGFP fluorescence was detected in clones in the mutant caterpillars. In addition, PCR analysis with primers flanking the 5′ and 3′ junctions of the integration shows a clear band in mutants, but not in wild type (Fig. 8.2c). Our results show that donor DNA with ~500 bp homology arms is sufficient for precise in-frame knockins. Compared to NHEJ-mediated high efficiency knockouts (69% in the case of *V. cardui Ddc* deletion knockouts (Zhang and Reed 2016)), the rate of HDR-mediated targeted integration is low, at ~3% in our most recent trials. It has been shown that knocking out factors in the NHEJ pathway can enhance the HDR pathway and increase gene targeting efficiency in *Bombyx* (Ma et al. 2014; Zhu et al. 2015). Some Cas9-mediated homology-independent knockin approaches have shown higher efficiency rates in zebrafish (Auer et al. 2014) and human cell lines (He et al. 2016), suggesting NHEJ repair may provide an alternate strategy to improve incorporation of donor DNA in Lepidoptera.

8.4 Embryo Injection

When adapting CRISPR/Cas-9 genome editing to a new species, the greatest technical challenges we face typically lie in optimizing the injection protocol. The main reason for this is that the eggs of different species can be quite different in terms of how difficult they are to puncture with a glass needle and how they react to mechanical injection, especially in terms of internal pressure and postinjection backflow.

Injection Needles Proper needle shape is critical for achieving successful egg injections in Lepidoptera. In our experience some taxa like *Heliconius* spp. have very soft, easy-to-inject eggs that present very few problems and are relatively robust to variation in needle shape. Many lepidopterans, however, have difficult-to-puncture eggs with high internal pressure. The key challenge for these eggs is to use needles that are strong enough to penetrate tough eggshells but are not so wide as to weaken pressure balance or destroy embryos. For instance, needles that are too long and narrow can break easily when used on tough eggs and will clog at a high frequency. Conversely, needles that have a very wide diameter will tend to have problems with pressure loss and backflow. We recommend the needle shape shown in Fig. 8.3a which is characterized by a short rapid taper to a fine point. We have found that this shape provides enough strength to puncture fairly tough eggs, yet is relatively resistant to clogging and pressurization problems. Our initial attempts at pulling needles like this with a traditional gravity needle puller failed. We now pull our needles using a velocity-sensitive Sutter P-97 programmable needle puller, which works very well for crafting nuanced needle shapes. We currently prefer to

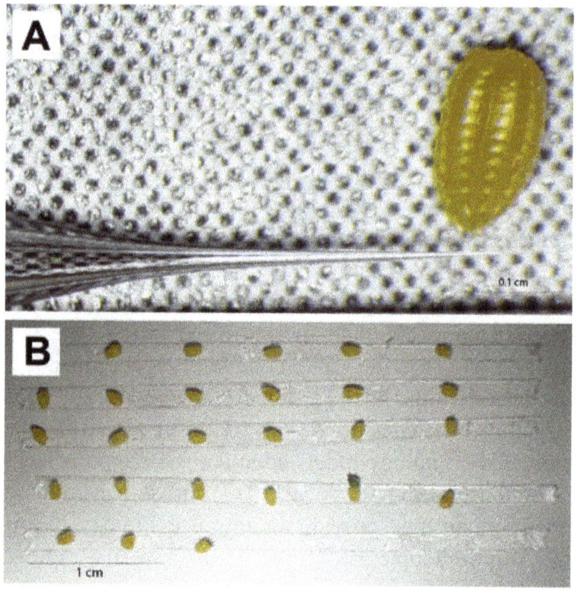

Fig. 8.3 Needle shape and egg arrangement for butterfly embryo injections. (**a**) The injection needle shape we prefer has a steep taper and a relatively large orifice. Here a preferred needle is shown next to a *Heliconius* egg. (**b**) An example of arranging *Heliconius* eggs on double-sided tape on a microscope slide just before injection

use Sutter Instrument 0.5 mm fire-polished glass capillary needles (Sutter BF-100-50-10) and 3 mm square box heating filaments (Sutter FB330B). Although settings will vary by instrument and filament, we use a single-cycle program on our puller with parameters HEAT 537, PULL strength 77, VELOCITY (trip point) 16, and TIME mode (cooling) 60. Among these parameters, the HEAT value has to be adjusted relative to the RAMP value, which is specific to certain instruments – different pullers can produce slightly different needle shapes even with the same parameter setting. We provide our settings as a starting point for other users to work toward optimizing production of needles with a steep taper and a large orifice as shown in Fig. 8.3a.

Egg Treatment Egg treatment is different for eggs from different taxa. For species with soft eggs like *Heliconius*, *Agraulis*, and *Danaus*, freshly collected eggs can be immediately arranged on double-sided adhesive tape on a microscope slide (Fig. 8.3b) and injected. For those eggs with relatively soft chorion but high pressure, like *V. cardui*, collected eggs should be arranged on a slide and then kept in a desiccation chamber for 15 min before injection. We use a simple sealed petri dish filled with desiccant for this purpose. For species with thick-shelled eggs like *J. coenia*, we recommend that eggs be dipped in 5% benzalkonium chloride (Sigma-Aldrich, St Louis, MO, USA) for 90s to soften the chorion and then washed in water for 2 min before mounting on microscope slide. We also tried treatment with 50% bleach solution to soften eggs; however, this significantly reduced the hatch rate. Softened eggs can then be dried in a desiccation chamber for 15 min and injected.

Injection Timing In all published cases we are aware of, injections of sgRNA and Cas9 (either mRNA or recombinant protein) were completed between 20 m and 4 h after oviposition, when embryos are presumed to be in an acellular syncytial state. Most of our injecting experience has been in eggs 1–3 h old. Although we have not rigorously quantified this effect, after extensive work with pigmentation genes in *V. cardui*, we found that injecting earlier (e.g., at 1 h) typically produces more and larger mutant clones compared to injection performed later (e.g., at 4 h). This is consistent with previous studies that have found a higher deletion frequency when embryos are injected at earlier versus later stages (Li et al. 2015a). Thus, for most of our deletion experiments, we aim to inject ~1–2 h after oviposition. If we expect that deletion of the locus will have a strongly deleterious or embryonic lethal effect, we will begin by injecting at 2–4 h to decrease the magnitude of somatic deletions.

Egg Injection The key concern during injection is to minimize damage as much as possible. An optimum angle for needle insertion is about 30°–40° in our experience. We prefer to use a Narishige MM-3 micromanipulator for full three-dimensional control of the needle during injection. In the butterfly species we have worked with, the location of injection does not seem to have a major impact on editing efficiency, although in *V. cardui* we get a slightly higher survival rate by injecting into the side near the base of the egg. Proper positive balance pressure is critical for successful injection. Users should adjust balance pressure to a point where the needle is just able to retain the solution. Prior to any egg injection, adjust the injection pressure and time to ensure the flowing droplet is visible when pressing the injector's footswitch. We have worked extensively with two different injectors: a Harvard apparatus PLI-100 Pico-Injector and a Narishige IM 300 Microinjector. In our experience, PLI-100 Pico-Injector has better sensitivity in terms of balance pressure, which is very important for species with high-pressure eggs like *V. cardui* and *J. coenia*. The IM 300 does not perform as well with these eggs. The other two injectors we know of that also work well for butterfly eggs are Eppendorf FemtoJet microinjector and Drummond Nanoject III. We use 10 psi injection pressure and 0.5 psi balance pressure for soft-shelled eggs with the Narishige IM 300 injector and 20 psi injection pressure and 0.8 psi balance pressure for *V. cardui* and *J. coenia* eggs with the PLI-100 Pico-Injector. After injection we maintain slides with the injected eggs in a petri dish and move larvae to their rearing containers immediately upon emergence.

8.5 Interpreting Somatic Mosaics

While several studies have been published that describe the germ line transmission of edited alleles in *B. mori*, thus far most studies in Lepidoptera have focused on interpreting deletion phenotypes in G_0 somatic mosaics. Maintaining edited genetic lines is necessary for looking at the homozygous effects of specific edited alleles and will also be essential for a future generation of more sophisticated knockin studies. Maintaining edited lines presents a few challenges in Lepidoptera, however. First, the deletion phenotypes of many interesting genes would likely be embryonic lethal. For example, our lab has thus far been unsuccessful in efforts to produce living larvae with *wingless* or *Notch* coding region deletions, which is unsurprising because these genes are known to be essential for early embryonic development in insects. For loci like these, we can confirm deletions by PCR and sequencing, but all embryos with deletions die before or shortly after hatching. Second, based on our experience with inbreeding attempts in *Heliconius* spp., *V. cardui*, and *J. coenia*, and through discussions with colleagues working in other systems, it is clear that many lepidopterans are sensitive to inbreeding, and lines will die out quickly unless fairly large stocks are kept. Large stocks then make it more difficult to identify individuals with specific genotypes. So while maintaining lines is possible in many lab-adapted species, it is not always a trivial endeavor.

Because of the challenges posed by the embryonic lethality of many target genes, along with the difficulty of maintaining and genotyping edited lines, most of our attention has focused on analysis of mosaic G_0 phenotypes. One obvious advantage of focusing on somatic mosaics is that data can be collected in a single generation. Another advantage is that the phenotypic effects of lesions are limited to the subset of cell lineages (clones) hosting deletion alleles, thus reducing the deleterious effects of many deletions. Because of their clear phenotypes, knockout work on melanin pigmentation genes has allowed a very useful visual demonstration of the nature of somatic mosaicism in injected animals. Our work on eight pigmentation genes across several butterfly species (Zhang et al. 2017) has allowed us some general insights into work with mosaics. First, as described above, we found a loose association between the number and size of clones and the timing of injection, where earlier injections with higher concentrations of Cas9/sgRNA complexes tend to produce larger clones. We have not attempted to quantify this effect, but across replicated experiments, our tentative conclusion is that this is a real and consistent phenomenon. This is important because it gives rough control over the strength of a phenotype and can thus be important for trying to get small non-deleterious clones for an otherwise lethal gene. Conversely, by injecting at very early stages to knock out minimally pleiotropic genes, we can often produce animals with very large clones, such as entire wings.

One challenge of working with somatic mosaics lies in detecting and interpreting more subtle phenotypes. Most of the phenotypes published to date, such as loss of wing pattern features like eyespots or production of discolored patches, are fairly obvious and/or far outside the range of natural variation. Without having a dramatic phenotype or independent clone boundary marker, however, minor or highly localized effects can be difficult to differentiate from natural variation. It is possible that quantitative image analysis approaches could address this issue, although we are unaware of published examples of this. In our own work, we have relied on two main criteria to validate putative deletion phenotypes: (1) replicates, which, of course, are useful for increasing confidence (we typically aim for a minimum of three, although the number of replicates required to make a particular inference is somewhat arbitrary and there is no standard), and (2) asymmetry, which is perhaps the most powerful criterion for inferring deletion phenotypes. Because natural variation is ordinarily symmetrical, strongly asymmetric phenotypes are best explained by left/right variation in clonal mosaics.

8.6 Genotyping

To validate that genome editing is occurring as expected at the appropriate locus, it is necessary to perform genotyping on experimental animals. We have found that the simplest and most robust genotyping approach is to design PCR primers flanking the deletion sites and then to compare PCR product sizes between wild-type and experimental animals. We recommend that genotyping amplicons cover less than 1.5 kb and be at least 100 bp outside of the closest sgRNA site to allow proper band size resolution and detection of large deletions. This approach works best for long deletions produced by double sgRNAs – indels induced by repair of a single Cas9 cut site will usually be too small to detect by PCR alone. For this reason, our lab always uses double sgRNAs to produce deletions. These PCR products can also be cloned and sequenced for further validation, as well as to better characterize the diversity and nature of deletion alleles. If a single sgRNA is used, it is likely that deletion alleles will need to be sequenced to confirm lesions. To genotype insertions, PCR primers flanking the insertion site may be used similarly, or one may also use a primer inside the transgene (e.g., Fig. 8.2c).

A current challenge in genotyping edited animals is the lack of tools to rigorously confirm specific deletion alleles in specific cell populations. First, there is the physical problem of isolating a population of cells representing a single pure clone. To our knowledge, this has not been done in insects outside of using transgenic cell sorting methods (Böttcher et al. 2014). Even carefully dissected presumptive clones cannot be assumed to be pure clonal cell populations. Indeed, to our knowledge there is not yet a practical method developed to firmly associate specific alleles with specific phenotypes. This challenge also makes it difficult to decisively confirm whether a clone is monoallelic (i.e., has a single edited allele) or biallelic (i.e., has two edited alleles), thus making it difficult to infer dosage effects without additional information. Therefore, even though some previous studies present DNA sequences of edited alleles isolated from tissues including cells with deletion phenotypes (e.g., whole embryos), none of these studies rigorously associate individual alleles with specific clones because it cannot be ruled out that the tissue samples maybe have contained multiple monoallelic or bialleleic clones. A second challenge for genotyping specific clones is that some tissues of special interest, such as adult cuticle structures, including wing scales, do not have genomic DNA of sufficient quality to permit straightforward PCR genotyping, especially for longer amplicons. Thus, even if methods become available for isolating specific clone populations, there will still be limitations when dealing with some tissue types. Given the challenges outlined above, readers should understand that most genotyping to date should be seen as a validation of the experimental approach (editing accuracy and efficiency) and not necessarily as decisive confirmation that specific alleles underlie a certain clone phenotype.

8.7 Future Prospects

CRISPR/Cas9 genome editing is rapidly revolutionizing genetic work in Lepidoptera, as it is across all of biology. It is now fairly straightforward to quickly and cheaply induce long, targeted deletions in virtually any species that can be reared in captivity. Published reports to date have focused on producing deletions in gene coding regions; however, we anticipate there will be significant interest in also applying long deletion approaches to test the function of noncoding regulatory regions, especially now that *cis*-regulatory elements can be functionally annotated with high resolution, thanks to methods like ChIP-seq (Lewis et al. 2016). Pilot work shown here and elsewhere also demonstrates that targeted insertions are possible as well, thus promising even further developments on the near horizon such as protein tagging, reporter constructs, and tissue-specific expression constructs. Right now the main challenge with knockin strategies is the relatively low efficiency rate, although newer technologies such as NHEJ mediated knockin (Auer et al. 2014) promise to dramatically improve this. Perhaps the most exciting thing about CRISPR-associated genome editing approaches, though, is the straightforward portability of the technology between species. This is truly an exciting time to be a comparative biologist.

Acknowledgments We thank Arnaud Martin for extensive discussions regarding genome editing methods in butterflies, Joseph Fetcho for early assistance with needle pulling and injection procedures, and Katie Rondem for helpful comments on the manuscript. This work was funded by the US National Science Foundation grant DEB-1354318.

Appendix: A Detailed Example of CRISPR/Cas9 Genome Editing in the Painted Lady Butterfly *V. cardui*

The following procedure provides guidelines to generate genomic deletions in the butterfly *V. cardui* using the CRISPR/Cas9 nuclease system. This protocol includes a specific example of the Reed Lab's work deleting the melanin pigmentation pathway gene *Ddc* as previously reported (Zhang and Reed 2016).

Target Design

No genome reference was available for *V. cardui* when we first began our experiment, so we used a transcriptome assembly (Zhang et al. 2017) to identify sequences of the *Ddc* coding region. Primers GCCAGATGATAAGAGGAGGTT AAG and GCAGTAGCCTTTACTTCCTCCCAG were designed to amplify and sequence the target region of the genome, and exon-intron boundaries were inferred by comparing genomic and cDNA sequences. We recommend designing target sites at exons because they are more conserved than introns and therefore provide more predictably consistent matches between sgRNAs and genomic targets. We design sgRNAs by scanning for $GGN_{18}NGG$ or $N_{20}NGG$ pattern on the sense or antisense strand of the DNA. Target sequences GGAGTACCGTTACCTGATGA**AGG** and **CCT**CTCTACTTGAAACACFACCA (PAM sequences underlined) were designed to excise a region of 131 bp spanning the functional domains of the DDC enzyme. sgRNA oligos containing T7 promoter, target sequences, and sgRNA backbone were synthesized by a commercial supplier (Integrated DNA Technologies, Inc.). Of note, the PAM sequence is not included in the CRISPR forward primer.

CRISPR forward oligos:

Ddc sgRNA1: GAAATTAATACGACTCACTATA**GGGATCAGCTTTCGTCT GCC**GTTTTAGAGCTAGAAATAGC
Ddc sgRNA2: GAAATTAATACGACTCACTATA**GGAGTACCGTTACCTGA TGA**GTTTTA GAGCTAGAAATAGC
CRISPR universal oligo: AAAAGCACCGACTCGGTGCCACTTTTTCAAGTTG ATAACGGACTAGCCTTATTTTAACTTGCTATTTCTAGCTCTAAAAC

sgRNA Production

sgRNA Template Generation

- With the oligos generated in the preceding step, use High-Fidelity DNA Polymerase PCR Mix (NEB, Cat No. M0530) to generate the template for each sgRNA with CRISPR forward and reverse oligos. We recommend using DEPC-free nuclease-free water (Ambion, Cat No. AM9938).

PCR Reaction		PCR program
High-Fidelity DNA Polymerases PCR	50 µl	
Mix	5 µl	98 °C for 30 s
CRISPR forward oligo (10 µM)	5 µl	35 cycles (98 °C for 10 s; 60 °C for 30 s; 72 °C for 15 s)
CRISPR universal oligo (10 µM)	40 µl	72 °C for 10 min
Nuclease-free water		4 °C hold

- Purify the PCR reaction with MinElute PCR Purification Kit (Qiagen, Cat No. 28004) following the kit instructions and eluting in 15 µl nuclease-free water.
- Dilute 1 µl of this reaction with 9 µl nuclease-free water, and then run on a gel and a fluorometer (e.g., Qubit) to confirm purity, integrity, fragment length, and yield. It is also possible to use gel extraction at this stage if nonspecific products are present.

- The expected size should be around 100 bp, and the expected yield should be around 200 ng/ul.

In Vitro Transcription (IVT)

- Generate sgRNAs by in vitro transcription of the sgRNA PCR template using the T7 MEGAscript Kit (Ambion, Cat. No. AM1334). When producing and handling RNA, it is important to wear gloves and clean equipment and benches with detergent prior to use to avoid RNAse contamination. Pipette tips with filters can also be beneficial to prevent contamination from pipettes.

IVT reaction mix		Incubation and purification
ATP	2 µl	
CTP	2 µl	37 °C overnight incubation
GTP	2 µl	Add 1 µl Turbo DNAse and incubate for 15 min at 37 °C
UTP	2 µl	Add 115 µl ddH_2O and 15 µl ammonium acetate stop solution
10 × reaction buffer	2 µl	
Template	2 µl	
T7 Enzyme Mix	2 µl	
Nuclease-free water	Up to 20 µl	

- Extract sgRNA by adding 150 µl phenol:chloroform:isoamyl alcohol (25:24:1) at pH 6.7 (Sigma-Aldrich, Cat No. P2069), and vortex thoroughly for 30 s.
- Separate phases by centrifugation at 10,000 × g for 3 min at room temperature, and remove the upper phase to a fresh tube.
- Precipitate the RNA by addition of an equal volume (150 µl) of cold isopropanol (Sigma-Aldrich, Cat No. I9516).
- Mix thoroughly, and incubate at −20 °C for greater than 2 h (can be left overnight).
- Collect RNA by centrifugation at 17,000 × g for 30 min at 4 °C.
- Wash pellet twice in 0.5 ml room temperature fresh made 70% ethanol, centrifuging at 17,000 × g for 3 min at 4 °C between each wash.
- Remove the remaining liquid and dry RNA pellet for 3 min at room temperature.
- Resuspend in 30 ul nuclease-free water.
- Measure concentration on a Qubit. The expected concentration should be around 2 µg/ul; sgRNAs can be stored at −80 °C.
- MEGAclear™ Transcription Clean-Up Kit (Ambion, Cat No. AM1908) also works very well for sgRNA purification.

Cas9 Production

Cas9 is typically provided by injection of a plasmid, mRNA, or recombinant protein. We have tried both Cas9 mRNA and protein injections, and both yield similarly efficient mutation rates in butterflies. However, we recommend using commercially available Cas9 protein (PNA Bio, Cat No. #CP01) because it is more stable than Cas9 mRNA and is easier and faster to use.

- Cas9 mRNA is generated by in vitro transcription of the linearized MLM3613 (Addgene plasmid 42,251) plasmid template. The mMessage mMachine T7 Kit (Ambion, Cat No. AM1344) is used to perform in vitro transcription with T7 RNA polymerase, followed by in vitro polyadenylation with the PolyA Tailing Kit (Ambion, Cat No. AM1350). An Agilent Bioanalyzer, or similar instrument, should be used to check the size and integrity of Cas9 mRNA. Note that Cas9 mRNA can show some degree of degradation yet still produce fairly efficient results.

Egg Injection and Survivor Ratio Calculation

- Collect eggs for 2–4 h by placing a host plant leaf into the butterfly cage.
- For thick chorion eggs (e.g., *J. coenia*), dip eggs in 5% benzalkonium chloride for 90 s.
- Cut double-sided tape into several thin strips and fix them to a glass slide.
- Use a paintbrush to line the eggs onto the double-sided tape.
- For high-pressure eggs (e.g., *V. cardui* or *J. coenia*), place the slide in a desiccation chamber for 15 min before injection.
- Mix Cas9 and CRISPR sgRNAs prior to microinjection.

Injection mix		Incubation
Cas9 mRNA or protein (1 μg/μl)	1 μl	Incubate on ice for 20 min
CRISPR sgRNA1 (375 ng/μl)	1 μl	
CRISPR sgRNA2 (375 ng/μl)	1 μl	
Nuclease-free water	2 μl	

- Break the closed tip of the needle with an optimum angle about 30°–40°.
- Load the needle with 0.5 μl injection mix by capillary action or by using by Eppendorf™ Femtotips Microloader Tips (Eppendorf, Cat No. E5242956003).
- One by one inject the eggs with the injector.
- Generally, higher amounts of sgRNA and Cas9 protein will increase mutation rate and decrease egg survival (hatch rate).

Genotyping for Modification

- In order to investigate the efficiency of CRISPR-/Cas9-mediated *Ddc* knockout, we randomly surveyed 81 first instar caterpillars. DNA was extracted according to Bassett et al. (2013) to confirm CRISPR/Cas9 lesions. Generally, place one caterpillar in a PCR tube and mash the caterpillar for 30 s with a pipette tip in 50 μl of squishing butter (10 mM Tris-HCl, pH 8.2, 1 mM EDTA, 25 mM NaCl, 200 μg/ml proteinase K). Incubate at 37 °C for 30 min, inactivate the proteinase K by heating to 95 °C for 2 min, and store in −20 °C for PCR genotyping. Genotyping can also be done with adult butterfly leg DNA by using proteinase K in digestion buffer. We typically use QIAamp DNA Mini Kit (Qiagen, Cat No. 51304) for DNA extraction when genotyping from muscle tissue.
- Design genotyping primers outside of the target region. For *Ddc*, genotyping forward (GCTGGATCAGCTATCGTCT) and reverse primers (GCAGTAGCCTTTACTTCCTCCCAG) were designed to produce a 584 bp PCR fragment in wild-type individuals.
- Mix PCR reagents. PCR fragments containing two sgRNA target sites are expected to produce smaller mutant bands than wild type.

PCR reaction for genotyping		PCR program
Taq DNA Polymerases PCR Mix (NEB)	12.5 μl	98 °C for 1 min
Genotyping F primer (10 μM)	1 μl	35 cycles (98 °C for 10 s; 55 °C for 30 s; 72 °C for 40 s)
Genotyping R primer (10 μM)	1 μl	72 °C for 10 min
DNA template	1 μl	4 °C hold
Nuclease-free water	9.5 μl	

- Recover mutant bands by gel extraction using MinElute Gel Extraction Kit (Qiagen, Cat No. 28604).
- Ligate recovered DNA fragment to T4 vector for TA cloning using a TA cloning kit (Invitrogen, Cat No. K202020).
- Extract plasmid with mutant DNA fragment using QIAprep Miniprep Kit (Qiagen, Cat No. 27104).
- Sequence plasmids and align mutant sequences to wild-type sequences to confirm deletions (Fig. 8.1a in Zhang and Reed, 2016).

References

Auer TO, Duroure K, De Cian A, Concordet J-P, Del Bene F (2014) Highly efficient CRISPR/Cas9-mediated knock-in in zebrafish by homology-independent DNA repair. Genome Res 24:142–153

Bassett AR, Tibbit C, Ponting CP, Liu J-L (2013) Highly efficient targeted mutagenesis of *Drosophila* with the CRISPR/Cas9 system. Cell Rep 4:220–228. doi:10.1016/j.celrep.2013.06.020

Beldade P, Peralta CM (2017) Developmental and evolutionary mechanisms shaping butterfly eyespots. Curr Opin Insect Sci 19:22–29

Böttcher R et al (2014) Efficient chromosomal gene modification with CRISPR/cas9 and PCR-based homologous recombination donors in cultured *Drosophila* cells. Nucleic Acids Res 42:e89–e89

He X et al (2016) Knock-in of large reporter genes in human cells via CRISPR/Cas9-induced homology-dependent and independent DNA repair. Nucleic Acids Res gkw064

Khan SA, Reichelt M, Heckel DG (2017) Functional analysis of the ABCs of eye color in *Helicoverpa armigera* with CRISPR/Cas9-induced mutations. Sci Rep 7

Kolliopoulou A, Swevers L (2014) Recent progress in RNAi research in Lepidoptera: intracellular machinery, antiviral immune response and prospects for insect pest control. Curr Opin Insect Sci 6:28–34

Lewis JJ, van der Burg KR, Mazo-Vargas A, Reed RD (2016) ChIP-Seq-annotated *Heliconius erato* genome highlights patterns of cis-regulatory evolution in Lepidoptera. Cell Rep 16:2855–2863

Li X et al (2015a) Outbred genome sequencing and CRISPR/Cas9 gene editing in butterflies. Nat Commun 6

Li Z et al (2015b) Ectopic expression of ecdysone oxidase impairs tissue degeneration in Bombyx mori. Proc R Soc B 282:20150513

Ling L et al (2015) MiR-2 family targets awd and fng to regulate wing morphogenesis in *Bombyx mori*. RNA Biol 12:742–748

Liu Y et al (2014) Highly efficient multiplex targeted mutagenesis and genomic structure variation in *Bombyx mori* cells using CRISPR/Cas9. Insect Biochem Mol Biol 49:35–42

Ma S et al (2014) CRISPR/Cas9 mediated multiplex genome editing and heritable mutagenesis of *BmKu70* in *Bombyx mori*. Sci Rep 4:4489

Markert MJ et al (2016) Genomic access to monarch migration using TALEN and CRISPR/Cas9-mediated targeted mutagenesis. G3: Genes| Genomes| Genetics 6:905–915

Merlin C, Beaver LE, Taylor OR, Wolfe SA, Reppert SM (2013) Efficient targeted mutagenesis in the monarch butterfly using zinc-finger nucleases. Genome Res 23:159–168

Perry M et al (2016) Molecular logic behind the three-way stochastic choices that expand butterfly colour vision. Nature 535:280–284. doi:10.1038/nature18616

Takasu Y et al (2010) Targeted mutagenesis in the silkworm *Bombyx mori* using zinc finger nuclease mRNA injection. Insect Biochem Mol Biol 40:759–765

Takasu Y et al (2013) Efficient TALEN construction for *Bombyx mori* gene targeting. PLoS One 8: e73458

Tamura T et al (2000) Germline transformation of the silkworm *Bombyx mori* L. using a piggyBac transposon-derived vector. Nat Biotechnol 18:81–84

Terenius O et al (2011) RNA interference in Lepidoptera: an overview of successful and unsuccessful studies and implications for experimental design. J Insect Physiol 57:231–245

Wang Y et al (2013) The CRISPR/Cas system mediates efficient genome engineering in *Bombyx mori*. Cell Res 23:1414–1416

Wei W et al (2014) Heritable genome editing with CRISPR/Cas9 in the silkworm, *Bombyx mori*. PLoS One 9:e101210

Xin Hh et al (2015) Transcription factor Bmsage plays a crucial role in silk gland generation in silkworm, Bombyx Mori. Arch Insect Biochem Physiol 90:59–69

Zeng B et al (2016) Expansion of CRISPR targeting sites in *Bombyx mori*. Insect Biochem Mol Biol 72:31–40

Zhang L, Reed RD (2016) Genome editing in butterflies reveals that *spalt* promotes and *Distal-less* represses eyespot colour patterns. Nat Commun 7. doi:10.1038/ncomms11769

Zhang Z et al (2015) Functional analysis of Bombyx Wnt1 during embryogenesis using the CRISPR/Cas9 system. J Insect Physiol 79:73–79

Zhang L et al (2017) Genetic basis of melanin pigmentation in butterfly wings. Genetics 102632. doi:10.1534/genetics.116.196451

Zhu L, Mon H, Xu J, Lee JM, Kusakabe T (2015) CRISPR/Cas9-mediated knockout of factors in non-homologous end joining pathway enhances gene targeting in silkworm cells. Sci Rep 5:18103

9

Genetic Basis for Bright Color Patterns in Heliconius Butterflies

Chris D. Jiggins

Abstract *Heliconius* wing patterns are an adaptive trait under strong selection in the wild. They are also amenable to genetic studies and have been the focus of evolutionary genetic analysis for many years. Early genetic studies characterised a large number of Mendelian loci with large effects on wing pattern elements in crossing experiments. The recent application of molecular genetic markers has consolidated these studies and led to recognition that a huge range of allelic variation at just a few major loci controls patterns across most of the *Heliconius* radiation. Some of these loci consist of tightly linked components that control different aspects of the phenotype and can be separated by occasional recombination. More recent quantitative analyses have also identified minor-effect loci that influence the expression of these major loci.

Studies of a single locus polymorphism in *Heliconius numata* provide an example of a 'supergene', in which a single major locus controls segregation of a variable phenotype. This supports 'Turner's Sieve' hypothesis for the evolution of supergenes, whereby sequential linked mutations arise at the same locus. In addition, inversion polymorphisms are associated with wing pattern variation in wild populations, which reduce recombination across the supergene locus. This provides direct evidence that the architecture and organisation of genomes can be shaped by natural selection. There is also evidence that patterns of dominance of the alleles at this locus have also been shaped by natural selection. Mimicry therefore provides a case study of how natural selection shapes the genetic control of adaptive variation.

Keywords Mimicry • Heliconius • Convergent evolution • Input-output gene • Developmental pathway • Adaptive radiation

A major research effort in evolutionary biology is devoted to determining the molecular changes in DNA sequences that control adaptive phenotypic changes. By identifying the number and identity of genes controlling traits, and the relative

C.D. Jiggins (✉)
Department of Zoology, University of Cambridge, Downing Street, Cambridge CB2 3EJ, UK
e-mail: c.jiggins@zoo.cam.ac.uk

contribution of individual mutations to changes in the appearance of an organism, we can address a wealth of questions in evolutionary biology including some that were debated by early geneticists, such as the importance of large versus small mutations in evolution. Mimicry patterns in *Heliconius* butterflies have contributed significantly to our understanding of the genetic basis for adaptation over the past 40 years. Here I review what is known of the genetic basis for these bright colour patterns and some of the implications for our understanding of evolution.

9.1 Phenotypic Effects of Major Loci: The Red Locus Optix

The most striking aspect of *Heliconius* wing pattern genetics is that a few major loci control large phenotypic changes (Fig. 9.1). This major locus control of adaptive traits is an emerging pattern in other organisms, but studies of butterflies provided some of the first clear examples (Nadeau and Jiggins 2010) and were already evident in early work (Sheppard et al. 1985). The locus that is best understood at a molecular level and has perhaps the largest phenotypic effect controls red patterns (Table 9.1). Alternate alleles represent regulatory switches controlling expression of the transcription factor *optix*. The most studied red patterns controlled by this locus can be divided into three main elements: the red forewing band, the red ray pattern on the hindwing and the basal patch on the forewing. The latter is known as the 'Dennis' patch, after an individual butterfly that William Beebe named '*Dennis the Menace'*. Once linked genetic markers were identified, it became clear that there is a remarkable degree of homology between species in the control of these elements (Baxter et al. 2008).

This shows that convergent patterns in mimetic species are controlled by the same genetic mechanism. But what about other types of patterns? It turns out that a huge diversity of patterns are controlled by the same genetic loci. For example, this locus also controls orange patches in silvaniform butterflies, *H. hecale* and *H. ismenius* (Huber et al. 2015), and the brown forceps-shaped pattern on the ventral hindwing of *H. cydno* (Naisbit et al. 2003; Chamberlain et al. 2011). In fact, in every species so far investigated genetically, this locus has major phenotypic effects on red and orange pattern elements.

The *optix* locus actually consists of distinct, tightly linked elements. Direct estimation of recombination rates between these has proven difficult, but there are rare natural recombinants. For example in *H. erato*, a single individual with *ray* but not *dennis* was collected in a Peruvian hybrid zone, and similar individuals are known in *H. melpomene* (Mallet 1989). There are also established races that have recombinant genotypes, such as *H. e. amalfreda* and *H. m. meriana* that have *dennis* but not *ray*, while *H. timareta timareta f. contigua* is a form with *ray* but not *dennis*.

Fig. 9.1 Phenotypes from a hybrid zone in Eastern Ecuador
There are three parental races that contribute variation to the hybrid zone, pictured here along the top row *H. m. plesseni*, *H. m. malleti* and *H. m. ecuadorensis*. Three major loci control the wing patterns, *D* controls red/orange pattern elements, *Ac* controls the shape of the forewing band (two spots or one) and *Yb* produces the yellow forewing band. These butterfly hybrids are all from the Neukirchen collection. Scale bar is 1 cm

Table 9.1 Summary of published wing patterning loci

Species	Locus	Phenotypic effect	Reference (corresponding to the caption)
D – Optix – LG18			
H. melpomene	D	Dennis patch	1
	B	Red FW band	1
	R	HW rays	1
	M	Yellow FW band	2
H. erato	Y	Yellow/red FW band	1
	D	Dennis patch	1
	R	HW rays	1
	Wh	White in FW	1
H. cydno	Br	Brown cydno 'C'	3
H. pachinus/heurippa	G	Red HW spots	3, 4
H. hecale	HhBr	HW orange/black	6
H. ismenius	HiBr	HW orange/black	6
Yb – cortex – LG15			
H. melpomene/cydno	Yb	Yellow HW bar	1,3
	N	Yellow FW band	1,3
	Sb	HW white margin	3,5
	Vf	Pale ventral FW band	3
H. erato	Cr	Cream rectangles	1
H. hecale	HhN	FW submarginal spots	6
H. ismenius	HiN	FW submarginal spots	6
H. ismenius	FSpot	FW subapical spots	6
H. ismenius	HSpot	HW marginal spots	6
H. numata	P	All pattern variants	7
Ac – WntA – LG10			
H. melpomene/cydno	Ac	FW band shape	1,3
	C	Broken FW band	1
	S	Shortens FW band	1,8
H. erato	Sd	FW band shape	1,9
	Sd	HW bar	1,9,10
	St	Split FW band	1,9
	Ly	Broken FW band	1,9
	Yl	Yellow FW line	1,11
H. hecale	HhAc	Yellow FW band	6
H. ismenius	HiAc	Yellow FW band	6

(continued)

Table 9.1 (continued)

Species	Locus	Phenotypic effect	Reference (corresponding to the caption)
LG1			
H. melpomene/cydno	K	FW band colour (yellow/white)	3,12
	Khw	HW margin colour (yellow/white)	13
LG13			
H. melpomene	Unnamed	FW band shape	14
H. erato	Ro	Rounded FW band	15
Unknown			
H. melpomene	Or	Orange/red switch	1
H. cydno	L/Wo	Forewing white spots	16
H. cydno/pachinus	Ps	Pachinus 'shutter'	17
H. cydno	Fs	Forewing 'shutter'	17
H. cydno	Cs	Cydno 'shutter'	17

A summary of previously described wing patterning loci and their homology to major effect genes. HW and FW refer to hindwing and forewing respectively. Notes: [1] Sheppard et al. (1985). [2] The *M* locus interacts with *N* to influence the forewing yellow band in *H. melpomene* (Mallet 1989). Unpublished work (Baxter and Mallet pers. Comm.) indicates that *M* is an effect of the *optix* locus. [3] Naisbit et al. (2003). [4] Mavarez et al. (2006). [5] Linares (1996). [6] Huber et al. (2015). [7] The *P* supergene locus in *H. numata* controls all aspects of phenotype. The locus is homologous to *Yb* although it seems likely that the supergene includes several functional loci (Joron et al. 2006).[8] Nijhout (1990). [9] Papa et al. (2013). [10] Mallet (1989). [11] Sheppard et al. (1985) infer that *Yl* and *Sd* are linked, but that *Yl* and *Ly* segregate independently. *Sd* and *Ly* are now known to be the same locus, so it is unclear whether *Yl* is unlinked. Further crosses of Brazilian forms would be needed to test this. [12] Kronforst et al. (2006). [13] Joron et al. (2006). [14] Baxter et al. (2009). [15] The *Ro* locus was mapped to linkage group 13 by means of a hybrid zone association study (Nadeau et al. 2014). [16] L and Wo are linked loci that control forewing white elements in *H. cydno* and may be homologous to *Ac* (Linares 1996). [17] *Ps*, *Fs* and *Cs* from Nijhout (1990) are included for completeness but patterns of segregation and linkage are not known. These may be effects of the *WntA* locus

Recent molecular analysis has confirmed that these phenotypes are indeed recombinants between tightly linked elements located in non-coding DNA near to *optix* (Wallbank et al. 2016). Thus, there are at least three very tightly linked elements that independently control different patches of red on the wing.

9.2 Phenotypic Effects of Major Loci: The Yellow Locus Cortex

This second major locus is similar in many ways to the red locus – it consists of tightly linked elements that similarly control different patches of yellow and white pattern. The *cortex* locus represents a cluster of tightly linked loci located on linkage group 15. These include effects known as *Yb*, *Sb* and *N* in *H. melpomene* and *Cr* in *H. erato* (Sheppard et al. 1985; Mallet 1986). Alleles that produce a yellow band are recessive to the absence of the band, although heterozygotes typically show an alteration in scale morphology in the band region that can be seen in altered reflectance in the otherwise black hindwing. Another allele at the same locus produces a band only on the underside of the hindwing and is present in the west Colombian race *H. m. venustus*. The same genomic region also controls a white hindwing margin found in the west Ecuador races *H. e. cyrbia* and *H. m. cythera* (Jiggins and McMillan 1997; Ferguson et al. 2010).

Many of the coloured patches on *Heliconius* wings are controlled in this very simple one-allele makes one-phenotype manner. However, there are also more complex interaction effects between loci. For example, in East Andean populations of *H. erato*, the yellow hindwing bar results from the joint effects of two loci, *cortex* and *WntA*. Thus, in Peruvian *H. e. favorinus*, recessive alleles at both loci are required for full expression of the hindwing bar (Mallet 1989) (although in Central American *H. erato*, a very similar bar results from a recessive allele at one locus). There is also evidence for rare recombination events between tightly linked loci at this locus. Thus, for example, *Yb* and *Sb* were mapped to within ~1 cM of one another, with two recombinant phenotypes identified in 175 individuals (Ferguson et al. 2010). Similar results are seen in crosses between *H. melpomene rosina* and *H. c. chioneus* (Naisbit et al. 2003).

In summary, these two loci both consist of a set of tightly linked genetic elements that control major phenotypic changes. Each locus controls pattern elements with broadly similar phenotypic effects: yellow and white patches in the case of *cortex* and red and orange patches in the case of *optix*. Patterns of dominance are also predictable, with alleles for red elements dominant, and those for yellow or white elements recessive, giving a dominance series of red > black > white > yellow. In both cases, loci most likely represent tightly linked *cis*-regulatory elements of the same protein-coding gene, with linkage a result of genetic architecture rather than being favoured by selection.

9.3 Phenotypic Effects of Major Loci: The Shape Locus *WntA*

The third major locus is located on linkage group 10 and primarily controls the shape of the forewing elements. For example, in crosses between *H. melpomene rosina* and *H. cydno chioneus,* a recessive allele *ac* places a triangle that forms a white hourglass shape in the main forewing cell of *H. cydno* (Naisbit et al. 2003). In the Ecuadorean *H. m. plesseni*, this locus produces the 'split' forewing band – the largely recessive *H. m. plesseni* allele expresses the more proximal of the two white patches of this form and also influences the shape of the more distal patch (Salazar 2012). This locus likely results from variation in expression of the gene *WntA* (Martin et al. 2012).

A wide variety of loci have previously been described (*St*, *Sd* and *Ly*) which all map to the same genomic location (Papa et al. 2013), corresponding to *WntA*. These loci influence the shape of forewing band elements. In some cases the phenotypic effects of this locus are extremely similar to those seen in *H. melpomene*; thus, for example, in *H. e. notabilis*, which is mimetic with *H. m. plesseni, Sd* also acts to generate the split forewing band phenotype (Salazar 2012). In Amazonian forms, the allele at this locus also generates the broken yellow forewing band (Sheppard et al. 1985; Papa et al. 2013).

9.4 Phenotypic Effects of Other Loci

A further locus, termed *K*, controls the colour change between yellow and white pigments in *H. melpomene*, *H. cydno* and *H. pachinus*. Most strikingly, this locus controls a polymorphism of yellow and white forms in *H. cydno alithea* in western Ecuador. The *K* locus is located on linkage group 1 and is linked with the gene *wingless* (Kronforst et al. 2006). This differs from other loci in that it influences solely colour, with no effect on pattern. There are also a number of minor-effect loci described in the older literature, but in most cases, these have been found to represent allelic effects of the major loci described above. Nonetheless, some of these loci are likely to be distinct. For example, a locus named *Or* described in both *H. melpomene* and *H. erato* controls the switch between red and orange colours (Sheppard et al. 1985). 'Postman' races typically have a bright red forewing band, while Amazonian forms have orange *dennis* and *ray* patterns. Another locus that has been better characterised is *Ro*, which generates a rounded forewing band phenotype such as that seen in *H. e. notabilis* (Salazar 2012; Papa et al. 2013; Nadeau et al. 2014). Some of the most beautiful but poorly characterised are the iridescent blue and green colours that result from structural variation in the wing scales. These traits vary continuously and are difficult to quantify (Jiggins and McMillan 1997). However, while most analysis of *Heliconius* genetics has relied on the scoring of presence/absence of major pattern elements, a better characterisation of these minor-effect loci is gained by a quantitative analysis of pattern segregation.

9.5 Quantitative Analysis

A comprehensive QTL analysis was carried out by Papa et al. using crosses between *H. e. notabilis* and *H. himera* (Papa et al. 2013). This confirmed the subjective finding from generations of earlier researchers that major loci control the segregation of most of the wing variation in crosses. For example, an additive model showed that the *optix* locus controlled 87% of variance in the amount of white versus yellow in the forewing, while the amount of red was best described by an epistatic model in which *optix* explained ~56% of the variation. The sizes of the two forewing spots showed a less skewed distribution of effect sizes and were controlled by several QTL of moderate effect (>5%), some as large in effect as the major locus *WntA*. For example, four QTL together explained 63% of the variance in the 'big spot', one of which was the *WntA* locus. This spot shape analysis therefore suggests a less skewed, more quantitative genetic architecture. Nonetheless, the overall variance explained across the complete set of *H. erato* crosses described in this paper is strongly dominated by large-effect loci.

These QTL analyses of specific wing pattern traits still fail to capture and quantify both segregation of the presence and absence of major pattern elements in the same analysis as quantitative variation in the expression of those traits. More recently, analytical methods have been developed that capture all of the variation in colour and pattern into a single PCA analysis (Huber et al. 2015; Le Poul et al. 2014), which was used to analyse broods of *H. hecale* and *H. ismenius*. All of the significant QTL identified corresponded to the existing major wing patterning loci. More minor QTL did not pass the significance threshold, although some of these additional loci would likely become significant with larger sample sizes. These quantitative analyses therefore support the conclusion that most variations are controlled by a handful of major-effect loci, although their expression is modified by minor-effect loci. In the future, there is a clear need for studies that combine large mapping families with objective methods for pattern analysis to better characterise the distribution of wing patterning variants.

9.6 Non-genetic Effects and Plasticity

There has been considerable interest recently in the role of phenotypic plasticity in evolution, and it has been proposed that plasticity can promote evolutionary novelty, for example, by allowing populations to explore new phenotypes without genetic change (Pfennig et al. 2010; Moczek et al. 2011). However, there is little evidence for phenotypic plasticity in the expression of *Heliconius* wing patterns.

First, most of the variation in wing pattern among hybrid butterflies can be explained by genetic variation at just a handful of major loci. Second, in the wild there is very little phenotypic variation in wing pattern among individuals occurring across a wide range of altitudes and habitats – apart from genetically divergent wing pattern races. Some pigment colours do fade with age, or in stressed individuals, but this is not adaptive plasticity. In summary, while plasticity may play a role in many aspects of *Heliconius* biology, such as learning of behaviour, there is no evidence that it plays a role in wing pattern evolution.

9.7 A Distribution of Effect Sizes?

Early workers used major genes in butterfly mimicry as an argument for major mutations driving evolution, but Fisher countered that mutations with a large effect on the organism will virtually always be deleterious (Fisher 1930). More recently Orr has shown that during an adaptive walk, we expect an exponential distribution of mutational effect sizes (Orr 1998, 2005). Early in the process, there is a high likelihood of mutations that move the population a large distance relative to the optimum. Later on, smaller effect mutations are more probable, that act to 'fine-tune' the adaptation. To some extent this modern view therefore reconciles the two camps.

The theory developed by Orr and others hypothesised a population evolving towards a single adaptive peak. However, the frequency-dependent nature of mimicry and warning colour means that these traits have a different dynamic. If a population of butterflies has a bright warning colour pattern (hereafter the 'mimic'), predators will learn this pattern, and the population will generally be well protected from predation. There may be other butterfly species locally that are perhaps more abundant or more toxic (the 'model') and therefore have a better-protected wing patterns, so the mimic species would gain in fitness by evolving mimicry of the model pattern. However, an individual 'mimic' that deviates from the rest of the population would be selected against, even if it becomes slightly more similar to the model. The two patterns would have to be very similar for predators to generalise between them, in order for gradual evolution towards the model to be possible (Turner 1981). Most current *Heliconius* patterns in different mimicry rings are sufficiently different from one another that gradual convergence seems unlikely. There is a valley of low fitness between the model and mimic which would seem to prevent gradual evolution of mimicry. This difficulty can be overcome if a single mutation causes a large change, sufficient to induce enough similarity to the model in one step that overall fitness is increased. This initial mutation is unlikely to produce a perfect mimic, so subsequent mutations will then be needed to perfect the phenotype. This argument was first outlined by Nicholson (1927) and termed the 'Nicholson two-step model' by John Turner (1977, 1984, 1987). Mimicry may therefore have a different genetic architecture to traits evolving under a single-peak-climbing model (Baxter et al. 2009).

The major locus control of *Heliconius* patterns seems to fit with the predictions of the 'Nicholson two-step model' (Huber et al. 2015; Papa et al. 2013; Turner 1981; Baxter et al. 2009), with a few major loci and additional modifiers of small effect. However, there are a number of reasons to be sceptical of this simple interpretation. First, many races within both *H. erato* and *H. melpomene* differ at several unlinked major-effect loci. For example, hybrid zones in both Peru and Ecuador between races of both *H. melpomene* and *H. erato* differ in at least two major loci (Mallet 1989; Salazar 2012; Nadeau et al. 2014). It is not clear whether a substitution at just one of these loci would be sufficient to gain enough mimetic similarity to provide protection, while the population 'waited' for a subsequent mutation at the second locus. Turner has acknowledged this difficulty but suggested either multiple rounds of 'two-step' evolution or that changes at just one of the loci would be sufficient to confer a fitness advantage (Turner 1977).

Another mismatch between the theory and empirical data is that the data from crossing experiments refers to the phenotypic effects of genetic loci, not separate mutations (Baxter et al. 2009). As pointed out by Fisher (1930), and more recently in dissection of major effect QTL in other organisms (Stam and Laurie 1996; Linnen et al. 2013), major-effect loci can result from accumulation of many mutations at a single locus. It seems likely that single large-effect genetic loci harbour many mutations corresponding to adaptive steps towards the peak. Testing the 'two-step model' therefore becomes a much more challenging problem of separating the order and effect size of individual mutations at a single locus. Nonetheless, mimicry can arise through hybridisation, in which an already well-adapted large-effect allele is acquired from a related species. This represents a clear case of single-step 'major-effect' evolution, so there certainly are at least *some* cases in which large changes are involved (The Heliconius Genome Consortium, 2012). Overall therefore, the 'rugged' adaptive landscape of mimicry likely favours adaptation via large steps as described under a two-step theory, and this might provide some part of the explanation for the major-effect loci involved in *Heliconius* mimicry.

9.8 Supergenes and Polymorphism

The broad picture of wing pattern genetics outlined above applies to most *Heliconius* that have been studied, but there is one species in the genus that has a very different pattern: *H. numata*. Mimicry patterns in *Heliconius numata* are polymorphic, with different morphs mimetic with different species mostly in the genus *Melinaea*. These dramatic differences are controlled by a single genetic locus, with several alternate alleles. Such loci are known as '*supergenes*', which we have defined as '*A genetic architecture involving multiple linked functional genetic elements that allows switching between discrete, complex phenotypes maintained in a stable local polymorphism*' (Thompson and Jiggins 2014). There are two major characteristics of the *Heliconius numata* supergene that maintain an integrated phenotype. First, a lack of recombination – all aspects of the phenotype are inherited as a single non-recombining locus – and second, dominance: alternate alleles show complete dominance relationships such that heterozygote genotypes develop the wing pattern of one or other parent.

The *P* supergene is genetically homologous to the region of the *cortex* locus in *H. melpomene* (Joron et al. 2006). The genetic architecture of 3–4 major loci is ancestral because it is shared by all other species in the genus that have been studied (Huber et al. 2015), so in *H. numata* this locus has 'taken over' control of all aspects of pattern variation (Jones et al. 2012). There are several hypotheses to explain the gradual evolution of tightly linked elements in a supergene. A long-standing hypothesis is that alleles located in different regions of the genome might be translocated into tight linkage (Turner 1967). However, there is no evidence for long-range movement of genes; the gene content of the region is similar in all *Heliconius*. The *P* locus has therefore evolved control of pattern variation normally influenced by genes on different chromosomes, rather than by moving those genes into linkage. The second hypothesis is that sequential mutations might arise in tight linkage with the polymorphic locus and be favoured by selection (Turner 1977; Charlesworth and Charlesworth 1976). Mutations that improve one mimetic form are likely to make things worse for other forms. However, if a new mutation is tightly linked at the *P* locus, then it will always be inherited with the alleles with which it is coadapted. This process has become known as 'Turner's sieve', because it involves sieving of the genetic variation that arises in order to select only linked variants (Turner 1977; Charlesworth and Charlesworth 1976; Turner 1978). The fact that *P* consists of linked elements suggests that these must have arisen through multiple sequential mutations.

Once linked elements have arisen, theory predicts that selection can act to further reduce recombination between them (Turner 1967; Charlesworth and Charlesworth 1976; Kirkpatrick and Barton 2006). Mathieu Joron and his group have identified large genomic inversions (400 kb) that segregate in polymorphic populations around the *P* locus (Fig. 9.2). Alternate gene arrangements are fully associated with wing pattern phenotypes in natural populations and show strong linkage disequilibrium in natural populations. Effectively, there is a block of about 400 kb of DNA sequence that is inherited in complete association with different wing pattern forms (Joron et al. 2011). Similar inversions have been seen in complex polymorphisms in other species – notably a behavioural and plumage polymorphism in the white-throated sparrow, a social polymorphism in fire ants and a behavioural polymorphism in the ruff, a wading bird (Thompson and Jiggins 2014; Huynh et al. 2011; Wang et al. 2013; Küpper et al. 2015; Lamichhaney et al. 2015). In all cases, inversions lock together inheritance of a large part of one chromosome. Perhaps more similar to the *Heliconius numata* case is *Papilio polytes,* in which a very localised inversion around the *Dsx* gene controls a wing pattern mimicry polymorphism (Kunte et al. 2014; Nishikawa et al. 2015). These examples all suggest that the evolution of inversions to reduce recombination between coadapted alleles may be a common phenomenon.

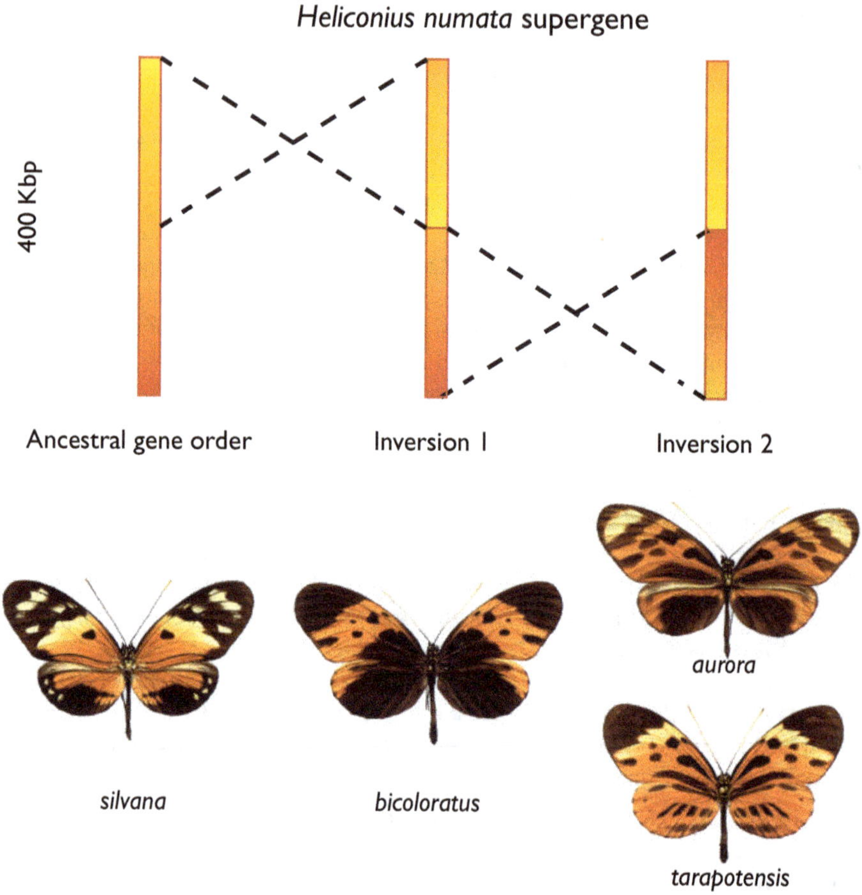

Fig. 9.2 Structural variation associated with the *Heliconius numata* supergene
At least two genetic inversions are associated with the *H. numata* supergene. The ancestral gene order, which matches that in *H. melpomene* and *H. erato* is shown on the left and is associated with ancestral phenotypes such as *H. n. silvana*. Two sequentially derived inversions are associated with dominant alleles and are shown in the middle and right. Redrawn from (Joron et al. 2011)

The second aspect of a supergene that ensures mimicry is a strong pattern of dominance (Llaurens et al. 2015). Alternate alleles show complete dominance, with an allelic series between morphs (Le Poul et al. 2014; Joron et al. 2011; Brown 1976). Remarkably, one heterozygote genotype is distinct but appears to have been stabilised because of its effective mimicry of a different species (Le Poul et al. 2014). In most *Heliconius*, there are predictable rules for dominance. Red/orange pattern elements are generally dominant over black, while yellow/white pattern elements are recessive. The complete dominance of alleles across the entire wing surface in *H. numata* therefore represents a derived state that apparently overturns typical 'rules of inheritance'. Dominance has been optimised by natural selection.

These patterns of dominance could be controlled by mutations within the supergene itself or unlinked loci acting to control dominance at *P*. Although there is evidence for both of these processes, recent analysis provides strong evidence for evolution of dominance at the *P* locus itself. Patterns of dominance *between* derived and ancestral alleles show unusual patterns of dominance in which the typical dominance patterns are overruled. In contrast, *among* derived alleles, patterns of dominance follow the typical colour hierarchy seen in other *Heliconius* species (Le Poul et al. 2014). These patterns suggest that dominance is a property of the alleles themselves, rather than the genetic background. This will be a fascinating system in which to explore mechanisms underlying the evolution of dominance.

9.9 Conclusions

The extraordinary diversity of wing patterns among the *Heliconius* butterflies has provided insights into the diversification of animal form and its genetic control. An important discovery has been the repeated role of just a handful of loci in diversification of not just convergent mimetic patterns but also diverse and novel phenotypes. Nonetheless, there is still a need for better quantitative analysis of patterns that will reveal the distribution of loci controlling adaptation. These patterns parallel discoveries in other systems, for example, sticklebacks, where similarly there are a few loci with major effects on phenotype (Colosimo et al. 2005; Chan et al. 2010), but many traits are also influenced by more polygenic control (Peichel and Marques 2017).

I have reviewed our understanding of wing patterning based on genetic crossing experiments, but have not considered in detail the developmental basis for pattern diversity, which has recently been reviewed elsewhere (Jiggins et al. 2017).

References

Baxter SW, Johnston SE, Jiggins CD (2009) Butterfly speciation and the distribution of gene effect sizes fixed during adaptation. Heredity 102:57–65. doi:10.1038/hdy.2008.109

Baxter SW, Papa R, Chamberlain N, Humphray SJ, Joron M, Morrison C, ffrench-Constant RH, McMillan WO, Jiggins CD (2008) Convergent evolution in the genetic basis of Müllerian mimicry in *Heliconius* butterflies. Genetics 180:1567–1577. doi:10.1534/genetics.107.082982

Brown KS (1976) An illustrated key to the silvaniform Heliconius (Lepidoptera: Nymphalidae) with descriptions of new subspecies. Trans Am Entomol Soc 102:373–484

Chamberlain NL, Hill RI, Baxter SW, Jiggins CD, Kronforst MR (2011) Comparative population genetics of a mimicry locus among hybridizing Heliconius butterfly species. Heredity 107:200–204. doi:10.1038/hdy.2011.3

Chan YF et al (2010) Adaptive evolution of pelvic reduction in sticklebacks by recurrent deletion of a Pitx1 enhancer. Science 327:302–305. doi:10.1126/science.1182213

Charlesworth D, Charlesworth B (1976) Theoretical genetics of batesian mimicry II. Evolution of supergenes. J Theor Biol 55:305–324. doi:10.1016/S0022-5193(75)80082-8

Colosimo PF et al (2005) Widespread parallel evolution in sticklebacks by repeated fixation of ectodysplasin alleles. Science 307:1928–1933. doi:10.1126/science.1107239

Ferguson L et al (2010) Characterization of a hotspot for mimicry: assembly of a butterfly wing transcriptome to genomic sequence at the HmYb/Sb locus. Mol Ecol 19:240–254. doi:10.1111/j.1365-294X.2009.04475.x

Fisher RA (1930) The Genetical theory of natural selection, 1st edn. Oxford University Press, Oxford

Huber B et al (2015) Conservatism and novelty in the genetic architecture of adaptation in Heliconius butterflies. Heredity 114:515–524. doi:10.1038/hdy.2015.22

Huynh LY, Maney DL, Thomas JW (2011) Chromosome-wide linkage disequilibrium caused by an inversion polymorphism in the white-throated sparrow (*Zonotrichia albicollis*). Heredity 106:537–546. doi:10.1038/hdy.2010.85

Jiggins CD, McMillan WO (1997) The genetic basis of an adaptive radiation: warning colour in two Heliconius species. Proc R Soc Biol Sci 264:1167–1175. doi:10.1098/rspb.1997.0161

Jiggins CD, Wallbank RWR, Hanly JJ (2017) Waiting in the wings: what can we learn about gene co-option from the diversification of butterfly wing patterns? Philos Trans R Soc B 372:20150485. doi:10.1098/rstb.2015.0485

Jones RT, Salazar PA, ffrench-Constant RH, Jiggins CD, Joron M (2012) Evolution of a mimicry supergene from a multilocus architecture. Proc R Soc Lond B Biol Sci 279:316–325. doi:10.1098/rspb.2011.0882

Joron M et al (2006) A conserved supergene locus controls colour pattern diversity in Heliconius butterflies. PLoS Biol 4:1831–1840. doi:10.1371/journal.pbio.0040303

Joron M et al (2011) Chromosomal rearrangements maintain a polymorphic supergene controlling butterfly mimicry. Nature 477:203–206. doi:10.1038/nature10341

Kirkpatrick M, Barton N (2006) Chromosome inversions, local adaptation and speciation. Genetics 173:419–434. doi:10.1534/genetics.105.047985

Kronforst MR, Young LG, Kapan DD, McNeely C, O'Neill RJ, Gilbert LE (2006) Linkage of butterfly mate preference and wing color preference cue at the genomic location of wingless. Proc Natl Acad Sci U S A 103:6575–6580. doi:10.1073/Pnas.0509685103

Kunte K, Zhang W, Tenger-Trolander A, Palmer DH, Martin A, Reed RD, Mullen SP, Kronforst MR (2014) doublesex is a mimicry supergene. Nature 507:229–232. doi:10.1038/nature13112

Küpper C et al (2015) A supergene determines highly divergent male reproductive morphs in the ruff. Nat Genet 48:79–83. doi:10.1038/ng.3443

Lamichhaney S et al (2015) Structural genomic changes underlie alternative reproductive strategies in the ruff (*Philomachus pugnax*). Nat Genet 48:84–88. doi:10.1038/ng.3430

Le Poul Y, Whibley A, Chouteau M, Prunier F, Llaurens V, Joron M (2014) Evolution of dominance mechanisms at a butterfly mimicry supergene. Nat Commun 5:5644. doi:10.1038/ncomms6644

Linares M (1996) The genetics of the mimetic coloration in the butterfly *Heliconius cydno weymeri*. J Heredity 87(2):142–149

Linnen CR, Poh Y-P, Peterson BK, Barrett RDH, Larson JG, Jensen JD, Hoekstra HE (2013) Adaptive evolution of multiple traits through multiple mutations at a single gene. Science 339:1312–1316. doi:10.1126/science.1233213

Llaurens V, Joron M, Billiard S (2015) Molecular mechanisms of dominance evolution in Müllerian mimicry. Evolution 69:3097–3108. doi:10.1111/evo.12810

Mallet J (1986) Hybrid zones of Heliconius butterflies in Panama and the stability and movement of warning color clines. Heredity 56:191–202

Mallet J (1989) The genetics of warning color in Peruvian hybrid zones of *Heliconius erato* and Heliconius melpomene. Proc R Soc Lond Ser B-Biol Sci 236:163–185

Martin A et al (2012) Diversification of complex butterfly wing patterns by repeated regulatory evolution of a Wnt ligand. Proc Natl Acad Sci 109:12632–12637. doi:10.1073/pnas.1204800109

Mavarez J, Salazar CA, Bermingham E, Salcedo C, Jiggins CD, Linares M (2006) Speciation by hybridization in Heliconius butterflies. Nature 441(7095):868–871

Moczek AP, Sultan S, Foster S, Ledón-Rettig C, Dworkin I, Nijhout HF, Abouheif E, Pfennig DW (2011) The role of developmental plasticity in evolutionary innovation. Proc R Soc Lond B Biol Sci 278:2705–2713. doi:10.1098/rspb.2011.0971

Nadeau NJ, Jiggins CD (2010) A golden age for evolutionary genetics? Genomic studies of adaptation in natural populations. Trends Genet 26:484–492. doi:10.1016/j.tig.2010.08.004

Nadeau NJ et al (2014) Population genomics of parallel hybrid zones in the mimetic butterflies, H. melpomene and *H. erato*. Genome Res 24:1316–1333. doi:10.1101/gr.169292.113

Naisbit RE, Jiggins CD, Mallet J (2003) Mimicry: developmental genes that contribute to speciation. Evol Dev 5:269–280

Nijhout HF (1990) A comprehensive model for colour pattern formation in butterflies. Proc Royal Soc B Biol Sci 239(1294):81–113

Nicholson AJ (1927) A new theory of mimicry in insects. Aust Zool 5:10–104

Nishikawa H et al (2015) A genetic mechanism for female-limited Batesian mimicry in Papilio butterfly. Nat Genet 47:405–409. doi:10.1038/ng.3241

Orr HA (1998) The population genetics of adaptation: the distribution of factors fixed during adaptive evolution. Evolution 52:935–949. doi:10.2307/2411226

Orr HA (2005) The genetic theory of adaptation: a brief history. Nat Rev Genet 6:119–127. doi:10.1038/nrg1523

Papa R, Kapan DD, Counterman BA, Maldonado K, Lindstrom DP, Reed RD, Nijhout HF, Hrbek T, McMillan WO (2013) Multi-allelic major effect genes interact with minor effect QTLs to control adaptive color pattern variation in *Heliconius erato*. PLoS One 8:e57033. doi:10.1371/journal.pone.0057033

Peichel CL, Marques DA (2017) The genetic and molecular architecture of phenotypic diversity in sticklebacks. Philos Trans R Soc B 372:20150486. doi:10.1098/rstb.2015.0486

Pfennig DW, Wund MA, Snell-Rood EC, Cruickshank T, Schlichting CD, Moczek AP (2010) Phenotypic plasticity's impacts on diversification and speciation. Trends Ecol Evol 25:459–467. doi:10.1016/j.tree.2010.05.006

Salazar PCA (2012) Hybridization and the genetics of wing colour-pattern diversity in Heliconius butterflies. Cambridge University, Cambridge

Sheppard PM, Turner JRG, Brown KS, Benson WW, Singer MC (1985) Genetics and the evolution of Muellerian mimicry in Heliconius butterflies. Philos Trans R Soc Lond Ser B Biol Sci 308:433–610. doi:10.2307/2398716

Stam LF, Laurie CC (1996) Molecular dissection of a major gene effect on a quantitative trait: the level of alcohol dehydrogenase expression in *Drosophila melanogaster*. Genetics 144:1559–1564

The Heliconius Genome Consortium (2012) Butterfly genome reveals promiscuous exchange of mimicry adaptations among species. Nature 487:94–98. doi:10.1038/nature11041

Thompson MJ, Jiggins CD (2014) Supergenes and their role in evolution. Heredity 113:1–8. doi:10.1038/hdy.2014.20

Turner JRG (1967) On supergenes. I. The evolution of supergenes. Am Nat 101:195–221

Turner JRG (1977) Butterfly mimicry – genetical evolution of an adaptation. In: Hecht MK, Steere WC, Wallace B (eds) Evolutionary biology. Plenum Press, New York, pp 163–206

Turner JRG (1978) Why male butterflies are non-mimetic: natural selection, sexual selection, group selection, modification and sieving. Biol J Linn Soc 10:385–432. doi:10.1111/j.1095-8312.1978.tb00023.x

Turner JRG (1981) Adaptation and evolution in Heliconius – a defense of Neodarwinism. Annu Rev Ecol Syst 12:99–121

Turner JRG (1984) Mimicry: the palatability spectrum and its consequences. In: Vane-Wright RI, Ackery PR (eds) The biology of butterflies. Academic Press, London, pp 141–161

Turner JRG (1987) The evolutionary dynamics of batesian and muellerian mimicry: similarities and differences. Ecol Entomol 12:81–95. doi:10.1111/j.1365-2311.1987.tb00987.x

Wallbank RWR et al (2016) Evolutionary novelty in a butterfly wing pattern through enhancer shuffling. PLoS Biol 14:e1002353. doi:10.1371/journal.pbio.1002353

Wang J, Wurm Y, Nipitwattanaphon M, Riba-Grognuz O, Huang Y-C, Shoemaker D, Keller L (2013) A Y-like social chromosome causes alternative colony organization in fire ants. Nature 493:664–668. doi:10.1038/nature11832

10

Batesian Mimicry in Females: Molecular Mechanism and Evolutionary Process

Haruhiko Fujiwara

Abstract Mimicry is an important evolutionary trait involved in prey-predator interactions. In a swallowtail butterfly *Papilio polytes*, only mimetic-form females mimic the unpalatable butterfly, *Pachliopta aristolochiae*, but it remains unclear how this female-limited polymorphic Batesian mimicry is generated and maintained. To explore the molecular mechanisms, we determined two whole genome sequences of *P. polytes* and its related species *P. xuthus* for comparison. The genome projects revealed a single long-autosomal inversion outside *doublesex* (*dsx*) between mimetic (*H*) and non-mimetic (*h*) chromosomes (Chr25) *in P. polytes*. The inversion site was just same as the mimicry locus *H* identified by linkage mapping. The gene synteny around *dsx* among Lepidoptera suggests that *H*-chromosome originates from *h*-chromosome. The 130 kb inverted region includes three genes, *doublesex* (*dsx*), *UXT*, and *U3X*, all of which were expressed from *H*-chromosome, but rarely from *h*-chromosome, indicating that these genes in *H*-chromosome are involved in the mimetic trait as supergene. Amino acid sequences of Dsx were substituted at over 13 sites between *H*- and *h*-chromosomes. To certify the functional difference of Dsx, we performed electroporation-mediated knock-down and found that only female *dsx* from *H*-chromosome (*dsx_H)* induced mimetic patterns but simultaneously repressed non-mimetic patterns on female wings. We propose that *dsx_H* switches the coloration of predetermined patterns in female wings and that female-limited polymorphism is tightly kept by chromosomal inversion. In this chapter, I will introduce the above results and discuss about the molecular mechanism and evolutionary process underlying the female-limited Batesian mimicry in *P. polytes*.

Keywords Batesian mimicry • Female-limited polymorphic mimicry • *Papilio polytes* • Whole genome sequence • *Doublesex* • Supergene • Chromosomal inversion • Electroporation-mediated functional analysis • Wing coloration pattern

---

H. Fujiwara (✉)
Department of Integrated Biosciences, Graduate School of Frontier Sciences, The University of Tokyo, Kashiwa, Chiba 277-8562, Japan
e-mail: haruh@edu.k.u-tokyo.ac.jp

10.1 Research Background

One of the most essential problems in evolutionary biology is to elucidate the molecular basis of various and adaptive morphological phenotypes in living organisms. The morphological diversity plays an important role in adaptation to the surrounding environment in many cases (Darwin 1872). Insects at the bottom of the food chain have been continuously attacked by the predators and thus developed various defense strategies to avoid predation during evolution (Ruxton et al. 2005).

Among the various strategies used by butterflies to avoid predators, some butterflies have become unpalatable and inform predators to their toxicity by exhibiting the conspicuous wing patterns. Some unpalatable butterflies share similar wing patterns to provide mutualistic protection called Mullerian mimicry (Müller 1878). In contrast, some palatable butterflies have evolved Batesian mimicry, in which they resemble unpalatable model to protect them from predators (Bates 1862; Brower 1958; Uesugi 1996). Multiple loci are involved in the expression of Mullerian mimicry phenotypes in *Heliconius* butterflies (Jiggins et al. 2005; Kapan et al. 2006), whereas the phenotypes of Batesian mimicry species reported so far are determined by a single locus (Clarke and Sheppard 1959, 1962, 1972).

It is noteworthy that the Common Mormon butterfly, *Papilio polytes*, shows a female-limited Batesian mimicry (Clarke and Sheppard 1972). The females have two forms: non-mimetic female (also called *cyrus*) which wing patterns are almost identical to monomorphic males and mimetic female (also called *polytes*) which resembles wing patterns of the distasteful butterfly, the Common Rose, *Pachliopta aristolochiae* (Fig. 10.1). This female-limited dimorphism is controlled by a single autosomal locus *H*, and the mimetic phenotype (genotype: *HH* or *Hh*) is dominant (Clarke and Sheppard 1972). However, how the female-limited Batesian mimicry is generated or how the female dimorphism is maintained is largely unknown.

There are two models for the *H* gene: a conceptual "supergene" that comprises a series of the neighboring genes tightly linked to each other (Clarke and Sheppard

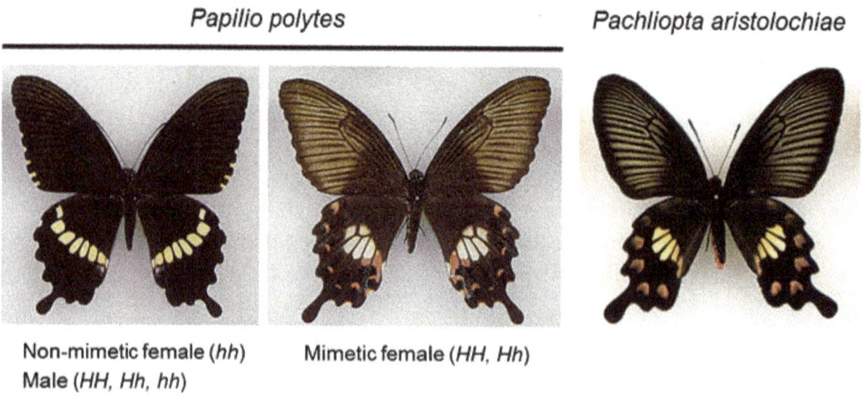

Fig. 10.1 *P. polytes* and model butterfly, *Pachliopta aristolochiae*

1960, 1972) and a single regulatory gene that controls downstream, unlinked genes affecting the color pattern. It is hypothesized that a supergene unit is created by recombination events and fixed by inhibitory effects of a chromosomal inversion on recombination (Nijhout 2003; Joron et al. 2011), although the mechanism underlying this hypothesis has remained obscure.

Recently, we found that drastic changes of gene networks not only in red but also pale-yellow regions can switch wing color patterns between non-mimetic and mimetic female of *P. polytes* (Nishikawa et al. 2013). It is presumed that these pigmentation processes involved in Batesian mimicry of *P. polytes* should be downstream of the *H* gene. To elucidate the evolutionary processes of this mimicry comprehensively, it is important to clarify the *H* locus and its structure and function. More recently, Kunte et al. (2014) and our group have identified the *H* gene locus and revealed its structure, independently (Nishikawa et al. 2015).

10.2 *Papilio* Genome Projects Reveal the *H* Locus and Chromosomal Inversion Near *dsx*

To reveal the *H* locus and its flanking structure, we first determined the whole genome sequences of *P. polytes* and *P. xuthus* for comparison (Nishikawa et al. 2015). We have prepared the *P. polytes* genome DNA (Ishigakijima Island strain in Japan) from one inbred female (genotype, *H*/*h*) after four generations of laboratory inbreeding and the *P. xuthus* genome DNA from a male captured in the field near Tokyo, Japan. We used a whole genome shotgun approach with next-generation sequencing platform. Filtered paired-end reads (135.2 Gb pairs for *P. polytes* and 73.8 Gb pairs for *P. xuthus*) were assembled using Platanus (version 1.2.1) (Kajitani et al. 2014) with some mate-pair libraries sequenced by Illumina Hiseq2000 and Hiseq2500. Consequently, we obtained 3873 and 5572 scaffolds, with an N50 of 3.7 Mb and 6.2 Mb pairs, spanning 227 Mb and 244 Mb pairs of the genome sequences for *P. polytes* and for *P. xuthus*, respectively.

In validating resulting assembled scaffolds, we noticed that there were two independent scaffolds including *dsx* in *P. polytes* while only one scaffold including *dsx* in *P. xuthus*. Because these two scaffolds in *P. polytes* were significantly different in sequences and the genome DNA was prepared from one heterozygous (*Hh*) mimetic female, we assumed that each haplotype (*H* or *h*) was highly diverged around the *dsx* locus in the two independent scaffolds. To survey such heterozygous regions in the whole genome of *P. polytes*, we picked windows in which the coverage depth was $\leq$350, which is approximately half the homozygous peak of 600. After clustering overlapping windows, we found 15 highly diverse (identity of $\leq$90%) and long ($\geq$100 kbp) heterozygous regions; 14 were mapped on heterozygous sex chromosome-1 (ZW) and one on chromosome-25 near *dsx* (Nishikawa et al. 2015). In the heterozygous region near *dsx*, we detected an approximately 130 kbp autosomal inversion (Fig. 10.2b). Strikingly, in the whole genome data of

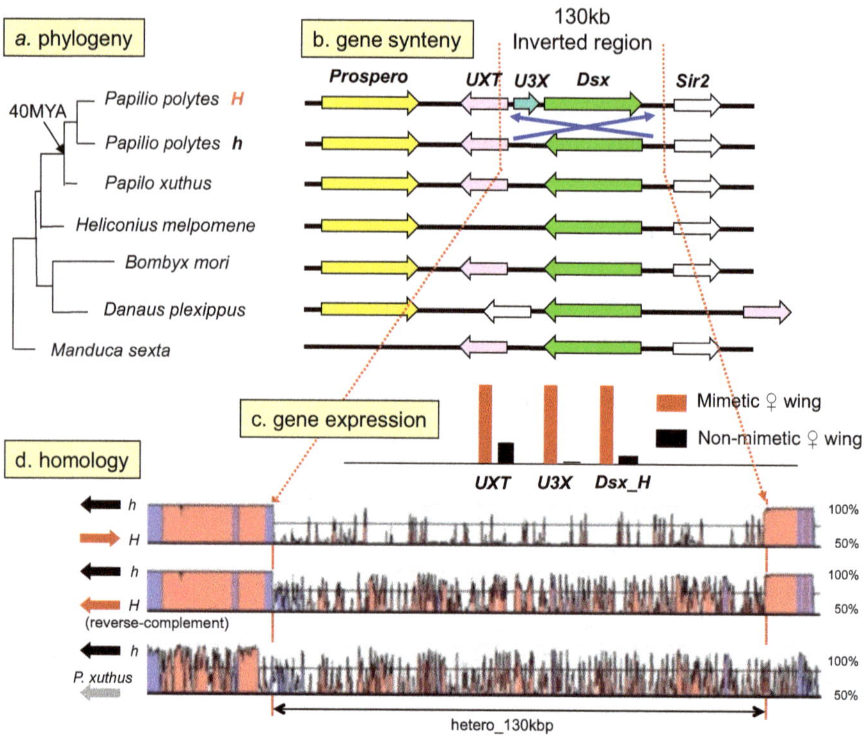

Fig. 10.2 Schematic view of the *H* locus in *P. polytes*. (**a**) Phylogenetic tree of Lepidoptera based on Dsx amino acid sequences. (**b**) Gene structure around *dsx* for the *h* and *H* alleles and gene synteny among Lepidoptera. (**c**) Expression level of genes within inverted region of the *H* locus, at early pupal stage in hind wings of mimetic (*red*) and non-mimetic (*black*) females. (**d**) Sequence homology around the *H* locus between *H* and *h* alleles and between *h* and *P. xuthus*

P. polytes, we could not find a long heterozygous region other than in the sex chromosomes (*Z*/*W*) which include many various repetitive sequences. Thus, the putative *H* locus region located on the chromosome-25 is thought to be only a long and unique heterozygous site among the whole autosomal chromosomes, which structure is maintained by reduced recombination due to the chromosomal inversion.

10.3 Linkage Mapping of the *H* Locus

To identify the mimicry locus *H*, we also performed the linkage mapping using non-mimetic type of *P. polytes* in Minamidaitōjima Island in Japan and mimetic type of *Papilio alphenor* in the Philippines (Nishikawa et al. 2015) (this work was performed mainly by Dr. H. Hori). After analyzing 84 F2 backcrossed females with mimetic phenotype in heterozygote of *H* (*Hh*) and 69 of non-mimetic females (*hh*) using amplified fragment length polymorphism (AFLP) and restriction fragment length polymorphism (RFLP) markers, we mapped the mimicry locus in *P. polytes* within 800 kbp genomic region containing 41 genes between two markers designed in *kinesin* and *intermediate* on chromosome-25. The association between the region and mimicry phenotype in natural populations was further examined using single nucleotide polymorphisms (SNPs) in 54 wild-caught females (Nishikawa et al. 2015). Consequently, eight SNPs in *dsx* showed significantly higher association (chi-squared test of independence, $P < 10^{-10}$) but none outside the gene. This is consistent with the result of the association study by K. Kunte et al. (2014) using laboratory-reared *P. polytes alphenor*. It is noteworthy that the *H* locus revealed by linkage mapping coincides completely with the long heterozygous region revealed by whole genome sequencing. This means that a genetic locus responsible for some polymorphic trait with a long heterozygous region can be identified only by genome sequencing without linkage mapping.

10.4 Detailed Structure of a Long Heterozygous Region Linked to the *H* Locus

Gene prediction and RNA-seq mapping showed that most of the inverted region of the *H* locus was occupied by *dsx* and the intron/exon structure was reversed in the *h*- and *H*-chromosomes, suggesting that a simple inversion occurred near both ends of *dsx* (Kunte et al. 2014; Nishikawa et al. 2015). Sequence comparison of the inverted region between *H* and *h* showed low-level homology not only directly but also in reverse (Fig. 10.2d). However, it is remarkable that some scattered regions including exons (shown by blue) for *dsx* were highly conserved (Fig. 10.2d).

To estimate the evolutionary process of the chromosomal inversion between *H*- and *h*- chromosomes, we further compared the gene synteny around *dsx* of *P. polytes*, with those of other Lepidoptera (Fig. 10.2a, b). We found that all tested genomes (*Papilio* species, *Heliconius*, *Bombyx* and *Manduca*) except *Danaus* have the same oriented synteny as the *h*-chromosome of *P. polytes*. Only in *H*-chromosome of *P. polytes*, *dsx* resides in the reverse orientation. These observations suggest that the *H*-chromosome may have originated from *h*-chromosome by a single inversion. Based on the gene synteny, we speculate that different types of inversion may have occurred near *dsx* in the *Danaus* genome independently. When comparing the inverted region (named hetero_130kbp) with a corresponding region of *P. xuthus*, the homology between *h* and *P. xuthus* was a little bit lower than that between *h* and *H* (Fig. 10.2d). This fact and phylogenetic tree suggest that the chromosomal inversion may have occurred after the branch of *P. polytes* and *P. xuthus* (Fig. 10.2a, d).

To clarify structural features of the inverted region of the *H* locus, we identified the exact place for the chromosomal inversion (Nishikawa et al. 2015). We have detected the sharp decline of the sequence conservation at both ends of inverted region between *H*- and *h*-chromosomes and considered these as putative breakpoints (Fig. 10.2d). The left breakpoint which closes on *Prospero* resides on about 700 bp downstream of the sixth exon of *dsx* in *h*-chromosome but on about 14.6 kbp upstream of the first exon of *dsx* in *H*-chromosome. The right breakpoint which closes on *Sir2* resides on about 8.9 kb upstream of the first exon of *dsx* in *h*-chromosome but on about 1.1 kb downstream of *dsx* in *H*-chromosome. Compared with *dsx* in *h*-chromosome (*dsx_h*), *dsx* in *H*-chromosome (*dsx_H*) was longer in the second, fourth, fifth, and sixth introns and sixth exon. Just outside of both breakpoints, in contrast, more than 99% homology was carried on between the *h*- and *H*-sequences (Fig. 10.2d). These structures implied that many mutations and several insertion and deletion events may have accumulated in the inverted region for *H* after the inversion and were maintained by suppression of recombination between two chromosomes.

10.5 Dimorphic Dsx Structure Associated with the *H* and *h* Alleles

The fact that a complete *dsx* was encoded inside of the inversion region indicates a possible involvement of the gene on the mimetic phenotype. RNA-seq assembly from mimetic (*HH*) and non-mimetic female (*hh*) revealed three types of female isoforms of *dsx* (F1, F2, and F3) in wings (Fig. 10.3). Dsx isoforms are generated by alternative splicing between the third and the forth exon both on the *h* and *H* alleles. Translational stop codon appeared in the fourth exon for F1 and for F3 and in the third exon for F2. Amino acid differences among isoforms were restricted merely in the C-terminal region (4–23 amino acids); three isoforms shared the first 244 amino acids including *dsx* DNA-binding motif and oligomerization domain (Fig. 10.3). Although there were 13–15 amino acid changes in three *dsx* isoforms between *H* and *h* alleles (Fig. 10.3), most substitutions occurred around the DNA-binding motif and dimerization domain (An et al. 1996). The comparison of *dsx* sequences among Lepidoptera showed that only five amino acids were specifically changed in *dsx_H* of *P. polytes* (Fig. 10.3, *). Recently, we have revealed dimorphic structure of Dsx sequences in another polymorphic, female-limited Batesian mimic species *P. memnon*, which shows different sites of amino acid changes between mimetic and non-mimetic alleles, in comparison with *P. polytes* (Komata et al. 2016). This finding suggests parallel evolution of the mimicry locus in two *Papilio* species, and further researches are necessary to clarify the structural features of Dsx involved in the mimicry traits.

Respective differences of amino acid sequence for F1, F2, and F3 between *dsx_H* and *dsx_h* are 2, 0, and 1, respectively (Fig. 10.3). This indicates that at least the C-terminal region of F2 may not be involved in the specific function of *dsx_H* on the mimetic phenotype. Furthermore, sequence comparison of these

```
                                              *
MVSVGAWRRRSPDECDDRNEPGASSSGVPRAPPNCARCRNHRLKVELKGHKRYCKYRYCTCE
                           T         A                  (DNA binding domain)
                                                              *        *
KCRLTADRQRVMAMQTALRRAQAQDEARARAAEH GHQPPGIELERGEPPIVKAPRSPVVPAPLPP
                                        QP                  M    L       L   PA
            *
RSLGSSSCDSVPGSPGVSPFAPPPPSVPPPPIMPPLLPPQQP(intron1)AVSLETLVENCHRLLEKF
    A     E                                   P               (oligmerization domain)

HYSWEMMPLVLVIFNYAGSDLDEASRKIDE (intron2)GKLIVNEYARKHNLNIFDGLELRNSTR
               L
                         *
F1 (intron3) HDRTKVAKFEI
                     E    K
F2  QYGL

F3 (intron3)QKMLSEINNISGVVSSSLKLFCE
                               M
```

Fig. 10.3 Amino acid sequences of Dsx for the *H* and *h* alleles. The N-terminal 244 amino acid sequence of Dsx which is common to female-specific isoforms (F1, F2, and F3) is shown on *top*. Each sequence of three isoforms for C-terminal region is shown at the *bottom*. The *h* allele sequence (Dsx_h) is shown. The amino acids substituted in Dsx_H are shown in *blue* below the sequence. * indicates an amino acid residue which is substituted only in Dsx_H among Lepidoptera

isoforms with those in other Lepidoptera revealed highly conserved structure except the C-terminal amino acid in F1 isoform. These observations suggest that no special isoform of *dsx_H* seems to be involved in the mimetic wing coloration, although it needs further evidence to show this possibility.

In males which show merely non-mimetic phenotype, we found only one isoform of *dsx* which skips exons 3 and 4 included in all female isoforms, implying the importance of exons 3 and 4 for the mimicry. In these regions of three female isoforms, however, there was only one amino acid (the C-terminal end of F1) changed specifically in *dsx_H*, as described above. The male-specific isoform of *dsx_H* was scarcely expressed in prepupal to pupal wings, suggesting that male *dsx_H* is not involved in the mimetic phenotype (Fig. 10.4).

10.6 Expression Profiles of Genes Around the Inverted Region of *H* Locus

To clarify the transcribed regions around the inverted regions, we mapped reads of RNA-seq to *h* and *H* alleles and found that three independent transcripts near left breakpoints, *ubiquitously expressed transcript* (*UXT*, transcriptional regulator) (Schroer et al. 1999), *unknown-3-exons (U3X*, long noncoding RNA emerged in *H*), and unknown transcript downstream of *Prospero*. These genes were highly expressed in wings of mimetic females (*HH* or *Hh*) compared with that in wings in

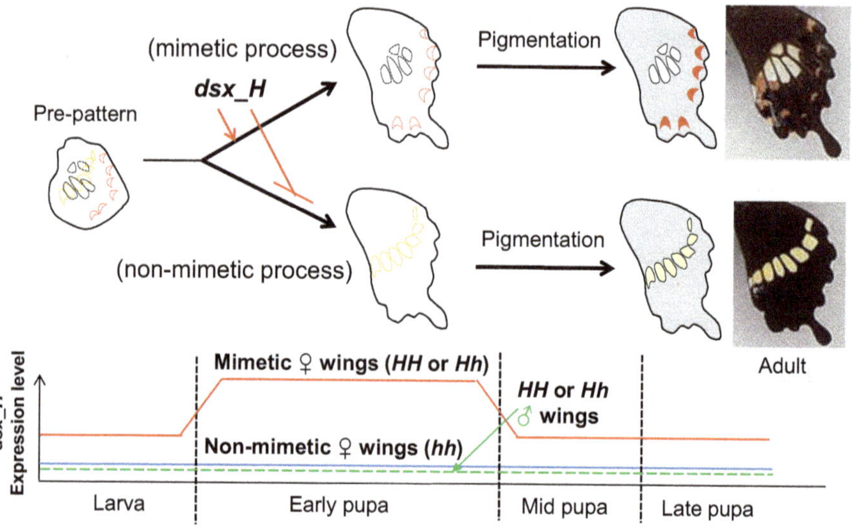

Fig. 10.4 Hypothetical model of Dsx_H function on mimetic and non-mimetic wing coloration patterns. The expression patterns of *dsx_H* in wings of mimetic (*red*) and non-mimetic (*blue*) females and males (*dotted green*) during larval to pupal stages are shown below

non-mimetic females (*hh*) (Nishikawa et al. 2015) (Fig. 10.2c), suggesting a possible involvement of these genes in the mimicry. The 5′ untranslated region (UTR) structure and transcriptional start site for *UXT* were altered by the inversion event between *H* and *h*, while the open reading frame was the same. A newly emerged gene *U3X* was found merely in the heterozygous region of the *H*-chromosome in the whole genome of *P. polytes*. The downstream sequence of *Prospero* was differently expressed between *h*- and *H*-chromosomes while located outside of inverted regions. These facts demonstrate that the chromosomal inversion affects not only the genome structure but also the expression of neighboring genes drastically probably through changes of gene regulatory elements.

We found that there seemed no significant differences in expression level of each isoform (F1, F2, and F3) of *dsx* between mimetic and non-mimetic wings in P1–2, P4–5, and P10.5 stages (Nishikawa et al. 2015). However, the expression level of *dsx_H* in mimetic female wings was quite higher than in non-mimetic wings in early pupal stages, but *dsx_h* did not show such expression profiles. RNA-seq analyses support the results that *dsx_H* was dominantly expressed in *Hh* female wings (Nishikawa et al. 2015). Differential expression level between *dsx_h* and *dsx_H* becomes significant on female wings at P2 stage when the patterning of wing pigmentation may be determined (Nishikawa et al. 2013) (Fig. 10.4). In contrast to *Hh* male, *dsx_H* was scarcely expressed, while *dsx_h* was dominantly expressed both in wandering and early pupal stages. In the report of Kunte et al. (2014), however, the expression pattern of *dsx_H* was upregulated at late pupal stage, which was different with our result.

The comparison of promoter regions of *dsx_H* and *dsx_h* showed highly nucleotide conservation near the transcriptional start site, but the conservation gradually reduced in more upstream regions (Nishikawa et al. 2015). Some of the nucleotide differences in the regulatory regions or intron regions between *dsx_H* and *dsx_h* may be responsible for the specific regulation of *dsx_H* in the female wings. The above results suggest that not only amino acid substitution but also regulatory changes for female *dsx_H* are possibly involved in the mimetic phenotype.

10.7 Functional Analysis of *dsx*

To verify the *dsx* function on the mimetic wing pattern formation, we performed the functional analysis with electroporation-mediated siRNA incorporation optimized for pupal wings, which enables mosaic analysis by knocking down target genes (Ando and Fujiwara 2013; Yamaguchi et al. 2011; Fujiwara and Nishikawa 2016). First, to confirm the validity of this newly established method, we knocked down *tyrosine hydroxylase* (*TH*) that is involved in melanin synthesis and found that the black pigmentation in adult wings was clearly repressed in the siRNA incorporated region. When injecting Universal Negative Control siRNA which is used generally as a negative control, no phenotypic change was observed.

Using this method, we injected siRNA designed to knock down *dsx_H* but not *dsx_h* into the whole hind wings of mimetic female and applied electroporation. This treatment caused the mimetic wing pattern switching to non-mimetic wing patterns. Furthermore, electroporation of *dsx_H* siRNA into part of early pupal hind wings of mimetic females also resulted in severe repression of mimetic wing patterns in the siRNA incorporated region; the peripheral red spots became the small pale orange ones; the central white marking mostly disappeared in the mosaic area. By this treatment, the ectopic white patterns for non-mimetic females emerged in the predicted position (Nishikawa et al. 2015). These results indicated that *dsx_H* not only induces the mimetic wing patterns but also simultaneously represses the emergence of the non-mimetic wing patterns. In contrast, *dsx_h* siRNA in mimetic females did not influence wing phenotype. When knocking down both *dsx_H* and *dsx_h* expression by siRNA which was designed in the common region between *H* and *h* (*dsx_H/h*), the same phenotype was observed as *dsx_H* siRNA alone (Nishikawa et al. 2015). These results implied that *dsx_h* is not involved both in mimetic and non-mimetic wing pattern formation. This strongly suggests that only *dsx_H* is involved in the mimicry phenotypes. It is noteworthy that *H/H* homozygous individuals are viable. This means that *dsx_H* should have basic functions for sexual differentiation in addition to the wing coloration.

It is unexpected that *dsx_H* not only induces the formation of mimetic color patterns but also represses non-mimetic patterns (Fig. 10.4). We previously showed that white (pale-yellow in mimetic) regions on mimetic and non-mimetic female wings of *P. polytes* are composed of different pigments (Nishikawa et al. 2013). Additionally, kynurenine/N-beta-alanyldopamine (NBAD) synthesis and *Toll* signaling genes were upregulated in the red spots of mimetic wings of *P. polytes* (Nishikawa et al. 2013; Rembold and Umebachi 1984). In addition to these observations, in the mimetic female wings, the positions of white bands in non-mimetic females are altered to the central area, which pattern resembles that of the model species *Pachliopta aristolochiae*. Therefore, both region-specific patterns and the synthesis of pale-yellow and red pigments in the mimetic female wings should be switched by *dsx_H*. We observed that chemical properties of pale-yellow pigments in mimetic females are similar to the model butterfly. Thus, various changes controlled by *dsx_H* lead to the successful mimicry of mimetic females of *P. polytes*.

More noteworthy, the appearance of non-mimetic-type white bands in mimetic pupal wings by knockdown *dsx_H* suggests that the pigmentation pattern is preset at least in early pupal stages. We propose a possible model for the functional role of *dsx_H* on the color pattern formation (Fig. 10.4). We assume that *dsx_H* acts as the pigmentation selector for the mimetic pattern. In late larval to early pupal stages, both mimetic and non-mimetic color patterns are predetermined by pattern formation genes other than *dsx*. Moreover, *dsx_H* merely selects the pigmentation processes for the mimetic pattern and represses the non-mimetic pattern in the fate-determined wings. In heliconid butterflies, the black region at the center of the forewings is determined by *WntA* (Martin et al. 2012), and the forewing band pattern is determined by *optix* (Reed et al. 2011). In early pupal wings of *Bicyclus anynana*, the *Distal-less* (*Dll*) homeobox gene is involved in positive regulation of focal differentiation and eyespot signaling (Monteiro et al. 2006; Brakefield et al. 1996). In addition, the red spots in the margin of hind wings are often observed among many butterflies in *Papilionidae*, irrelevant to males or females. From these facts, we speculated that the wing patterns of *P. polytes* may be determined initially by other genes than *dsx*.

10.8 Evolution of Female-Limited Batesian Mimicry

We revealed fine structure of the chromosomal inversion linked to female-limited Batesian mimicry of *Papilio polytes*, which has been studied mainly by classical genetics or from ecological points of views to date. Female-limited Batesian mimicry is widely observed among *Papilio* species (Kunte 2009) and may be controlled by a similar *dsx*-mediated system used in *P. polytes*. The characteristic of female-limited mimicry of *P. polytes* is the intraspecies polymorphisms and mimetic and non-mimetic phenotypes, which molecular background is explained well by the existence of two differentiated chromosomes, *h*- and *H*-chromosomes. The structure and function of *dsx_H* (and maybe neighboring genes) specialized for mimetic phenotype may be tightly kept through reduced recombination due to chromosomal inversion just outside of *dsx* for a long time in natural populations.

A question to be answered is when the chromosomal inversion or the differentiation of *dsx_H* and *dsx_h* occurred. The comparison of gene synteny among Lepidoptera indicates that *H*-chromosome of *P. polytes* may be originated from the *h*-type chromosome which structure is common in many species. Therefore, the chromosomal inversion might have occurred after the branch of *P. polytes* and *P. xuthus* 40 million years ago (Zakharov et al. 2004) (Figs. 10.1 and 10.2a), and thereafter the sequence difference between *dsx_H* and *dsx_h* had been fixed and accumulated. Indeed, the rate of single nucleotide variations (SNVs) and phylogenetic analysis of *dsx* showed considerable rate of nucleotide substitutions in the inverted region, suggesting that *dsx_H* has a high evolutionary rate and may evolve a new function under positive selection pressure. On the other hand, we observed lower homology in the whole hetero_130 kbp region between the *H*- and *h*-chromosomes of *P. polytes*, which level seemed similar to that between the *P. polytes* and *h*-chromosomes (Fig 10.2d). This indicates the chromosomal inversion might have occurred in very ancient age, and *dsx_H* function was refined by repeated and accumulated mutations during evolution. This evolutionary scenario needs to be certified by further detailed analysis of genome structure in other closely related species.

References

An W, Cho S, Ishii H, Wensink PC (1996) Sex-specific and non-sex-specific oligomerization domains in both of the doublesex transcription factors from *Drosophila melanogaster*. Mol Cell Biol 16:3106–3111

Ando T, Fujiwara H (2013) Electroporation-mediated somatic transgenesis for rapid functional analysis in insects. Development 140:454–458. doi:10.1242/dev.085241

Bates HW (1862) Contributions to an insect fauna of the Amazon valley. Lepidoptera: Heliconidae Transactions of the Linnean Society of London 23:495–566

Brakefield PM, Gates J, Keys D, Kesbeke F, Wijngaarden PJ, Monteiro A, French V, Carroll SB (1996) Development, plasticity and evolution of butterfly eyespot patterns. Nature 384:236–242

Brower JVZ (1958) Experimental studies of mimicry in some North American butterflies. Part II. *Battus philenor* and *Papilio troilus*, *P. polyxenes* and *P. glaucus*. Evolution 12:123–136

Clarke CA, Sheppard PM (1959) The genetics of Papilio Dardanus, Brown I. Race cenea from South Africa. Genetics 44:1347–1358

Clarke CA, Sheppard PM (1960) Super-genes and mimicry. Heredity 14:175–185

Clarke CA, Sheppard PM (1962) The genetics of the mimetic butterfly *Papilio glaucus*. Ecology 43:159–161

Clarke CA, Sheppard PM (1972) The genetic of the mimetic butterfly *Papilio polytes*. Philos Trans R Soc Lond Ser B Biol Sci 263:431–458

Darwin CR (1872) The origin of species by means of natural selection, or the preservation of favoured races in the struggle for life, 6th edn. John Murray, London

Fujiwara H, Nishikawa H (2016) Functional analysis of genes involved in color pattern formation in Lepidoptera. Curr Opin Insect Sci 17:16–23. doi:10.1016/j.cois.2016.05.015

Jiggins CD, Mavarez J, Beltrán M (2005) A genetic linkage map of the mimetic butterfly Heliconius melpomene. Genetics 171:557–570

Joron M, Frezal L, Jones RT, Chamberlain NL, Lee SF, Haag CR, Whibley A, Becuwe M, Baxter SW, Ferguson L, Wilkinson PA, Salazar C, Davidson C, Clark R, Quail MA, Beasley H, Glithero R, Lloyd C, Sims S, Jones MC, Rogers J, Jiggins CD, French-Constant RH (2011) Chromosomal rearrangements maintain a polymorphic supergene controlling butterfly mimicry. Nature 477:203–206. doi:10.1038/nature10341

Kajitani R, Toshimoto K, Noguchi H, Toyoda A, Ogura Y, Okuno M, Yabana M, Harada M, Nagayasu E, Maruyama H, Kohara Y, Fujiyama A, Hayashi T, Itoh T (2014) Efficient de novo assembly of highly heterozygous genomes from whole-genome shotgun short reads. Genome Res 24:1384–1395. doi:10.1101/gr.170720.113

Kapan DD, Flanagan NS, Tobler A, Papa R, Reed RD, Gonzalez JA, Restrepo MR, Martinez L, Maldonado K, Ritschoff C, Heckel DG, McMillan WO (2006) Localization of Mullerian mimicry genes on a dense linkage map of *Heliconius erato*. Genetics 173:735–757

Komata S, Lin CP, Iijima T, Fujiwara H, Sota T (2016) Identification of doublesex alleles associated with the female-limited Batesian mimicry polymorphism in Papilio memnon. Sci Report 6:34782. doi:10.1038/srep34782

Kunte K (2009) The diversity and evolution of batesian mimicry in Papilio swallowtail butterflies. Evolution 63:2707–2716. doi:10.1111/j.1558-5646.2009.00752.x

Kunte K, Zhang W, Tenger-Trolander A, Palmer DH, Martin A, Reed RD, Mullen SP, Kronforst MR (2014) Doublesex is a mimicry supergene. Nature 507:229–232. doi:10.1038/nature13112

Martin A, Papa R, Nadeau NJ, Hill RI, Counterman BA, Halder G, Jiggins CD, Kronforst MR, Long AD, McMillan WO, Reed RD (2012) Diversification of complex butterfly wing patterns by repeated regulatory evolution of a Wnt ligand. Proc Natl Acad Sci U S A 109:12632–12637. doi:10.1073/pnas.1204800109

Monteiro A, Glaser G, Stockslager S, Glansdorp N, Ramos D (2006) Comparative insights into questions of lepidopteran wing pattern homology. BMC Dev Biol 6:52

Müller F (1878) Über die Vortheile der Mimicry bei Schmetterlingen. Zool Anz 1:54–55

Nijhout HF (2003) Polymorphic mimicry in Papilio Dardanus: mosaic dominance, big effects, and origins. Evol Dev 5:579–592

Nishikawa H, Iga M, Yamaguchi J, Saito K, Kataoka H, Suzuki Y, Sugano S, Fujiwara H (2013) Molecular basis of wing coloration in a Batesian mimic butterfly, Papilio polytes. Sci Report 3:3184. doi:10.1038/srep03184

Nishikawa H, Iijima T, Kajitani R, Yamaguchi J, Ando T, Suzuki Y, Sugano S, Fujiyama A, Kosugi S, Hirakawa H, Tabata S, Ozaki K, Morimoto H, Ihara K, Obara M, Hori H, Itoh T, Fujiwara H (2015) A genetic mechanism for female-limited Batesian mimicry in Papilio butterfly. Nat Genet 47:405–409. doi:10.1038/ng.3241

Reed RD, Papa R, Martin A, Hines HM, Counterman BA, Pardo-Diaz C, Jiggins CD, Chamberlain NL, Kronforst MR, Chen R, Halder G, Nijhout HF, McMillan WO (2011) Optix drives the repeated convergent evolution of butterfly wing pattern mimicry. Science 333:1137–1141. doi:10.1126/science.1208227

Rembold H, Umebachi Y (1984) The structure of papiliochrome II, the yellow wing pigment of the Papilionid butterflies. In: Schlossberger HG, Kochen W, Linzen B, Steinhart H (eds) Progress in Tryptophan and Serotonin Research. Walter de Gruyter, Berlin, pp 743–746

Ruxton GD, Sherratt TN, Speed M (2005) Avoiding attack: the evolutionary ecology of crypsis, warning signals and mimicry. Oxford Univ. Press, Oxford

Schroer A, Schneider S, Ropers H, Nothwang H (1999) Cloning and characterization of UXT, a novel gene in human Xp11, which is widely and abundantly expressed in tumor tissue. Genomics 56:340–343

Uesugi K (1996) The adaptive significance of Batesian mimicry in the swallowtail butterfly, Papilio polytes (Insecta, Papilionidae): associative learning in a predator. Ethology 102:762–775

Yamaguchi J, Mizoguchi T, Fujiwara H (2011) siRNAs induce efficient RNAi response in *Bombyx mori* embryos. PLoS One 6:e25469. doi:10.1371/journal.pone.0025469

Zakharov EV, Caterino MS, Sperling FA (2004) Molecular phylogeny, historical biogeography, and divergence time estimates for swallowtail butterflies of the genus Papilio (Lepidoptera: Papilionidae). Syst Biol 53:193–215

Part IV
Adaptation and Ecological Perspectives

11

Unpalatable Chemical Elements in Butterflies and their Chemical Ecology

Ritsuo Nishida

Abstract A number of butterfly species are toxic or unpalatable against predators by developing mechanisms either to biosynthesize such noxious elements de novo or to acquire directly from the poisonous host plants for their own defense. Most of these "poisonous butterflies" exhibit aposematically colored wing patterns that are often associated either with Batesian or Müllerian mimicry species to form a "mimicry ring." This review focuses on unpalatable chemical elements potentially operating in three typical mimicry rings: (1) the tiger *Danaus* mimicry ring, (2) the *Idea* mimicry ring, and (3) the red-bodied swallowtail mimicry ring, in association with their mimetic wing patterns. Female-limited polymorphisms are a common feature of the Batesian mimicry but not in Müllerian mimicry, because such diversification is unfavorable for the models. I present here some unique cases of sexual dimorphisms within the putative Müllerian mimicry complexes. A *Danaus chrysippus*-mimicking nymphalid, *Argyreus hyperbius*, is a typical example of the female-limited dimorphic mimics. However, *A. hyperbius* were found to be poisonous with toxic cyanogenic glycosides (linamarin and lotaustralin). Likewise, a pipevine swallowtail, *Atrophaneura alcinous*, which sequesters toxic aristolochic acids, exhibits sexually dimorphic color patterns (male, black; female, smoky brown). A sympatric diurnal zygaenid moth, *Histia flabellicornis*, is mimetic to *A. alcinous* males rather than the females and stores cyanogenic glycosides. The moth is regarded as a Müllerian ally that may have stabilized the wing coloration mutually with those of *A. alcinous* males. On the contrary, a diurnal "swallowtail moth," *Epicopeia hainesii*, mimics the brighter wing color of *A. alcinous* females. Possible adaptation mechanisms on these paradoxical mimicry patterns are discussed.

Keywords Batesian mimicry • Müllerian mimicry • Aposematism • Unpalatability • Sequestration • Defense substance • Pharmacophagy • Sexual dimorphism • Sexual selection • Mimicry ring

---

R. Nishida (✉)
Discipline of Chemical Ecology, Kyoto University, Sakyo-ku, Kyoto 606-8502, Japan
e-mail: ritz@kais.kyoto-u.ac.jp

11.1 Introduction

The monarch butterfly, *Danaus plexippus* (Danainae: Nymphalidae), is an exemplar of the model/mimic relationships in butterflies. The larvae feed on milkweed plants (Asclepiadoideae of the family Apocynaceae) and selectively accumulate toxic cardenolides (cardiac glycosides, CGs) from the host plants and pass over to the adults (Reichstein et al. 1968). CGs are powerful heart poison and induce emesis to predatory birds (Brower 1984). If a "hungry blue jay" ingested a monarch butterfly, the sequestered CGs in the butterfly body tissues strongly induced emesis (Brower 1969). After such a noxious experience, the bird would never eat the butterflies with the same wing pattern again. Predatory birds avoid the butterflies primarily because they experience an obnoxious "bitter" taste and/or emesis after ingestion. This results in a visually conditioned aversion toward prey with similar appearance, allowing for the evolution of Batesian mimicry (Brower 1969). Therefore, the monarch is considered to be a typical poisonous model butterfly. The viceroy, *Limenitis archippus* (Nymphalinae, Nymphalidae), was initially thought to be a typical example of Batesian mimicry. However, the viceroy was shown to be fairly unpalatable to some predators (Ritland and Brower 1991; Prudic et al. 2007). Moreover, monarch can sometimes be palatable (or nontoxic) particularly if the larvae grew up on the nontoxic or low-CG milkweed plants (Brower 1969). Thus, the situation between the models and mimics may be interchangeable as in this case, and if both are unpalatable mimicking to each other, it forms an association of the Müllerian mimicry type, rather than Batesian (Rothschild 1979; Huheey 1984).

Unpalatability or distastefulness is strongly linked to the toxicity of the sequestered defense substances, as in human taste (Brower 1984), and thus, the predator effectively avoids toxic prey before ingestion. Visually oriented, avian predators are the most effective selective pressure to contribute to the formation of mimicry butterflies in conjunction with the distastefulness of the model species. The evolution of mimicry is highly dependent on both the visual and chemosensory (gustatory/olfactory) physiology of the predatory animals in addition to the intrinsic toxicity of the model butterfly (Brower 1984; Nishida 2002). However, knowledge on the chemistry of defensive agents stored in each species within the mimicry rings is often scarce (Trigo 2000; Nishida 2002). I highlight chemical backgrounds of acquired defensive elements among some aposematic Asian butterfly species in the following three typical/putative mimicry rings:

(1) The tiger *Danaus* mimicry ring
(2) The *Idea* mimicry ring
(3) The red-bodied swallowtail mimicry ring

The female-limited mimetic dimorphisms are a common feature of the Batesian mimics as in the *Papilio polytes* of the red-bodied swallowtail mimicry ring (Turner 1978). By contrast, Müllerian mimics lack strong sexual dimorphism, presumably due to density-dependent selection, where the divergence of a morph is disadvantageous in the toxic model (Mallet and Joron 1999). However, some exceptional cases of sexual dimorphisms can be seen within the putative Müllerian mimicry complexes in the above mimicry rings (1) and (3). My particular attention is focused on those sexually dimorphic species where both sexes were confirmed to harbor noxious chemicals – to evaluate their adaptive characteristics within the mimicry rings.

11.1.1 Tiger Danaus Mimicry Ring

Although the monarch butterfly, *Danaus plexippus*, and other milkweed butterflies sequester CGs from the *Asclepias* hosts in the body tissues and become unpalatable to bird predators, some of the butterfly populations often lack CGs as mentioned above. Similar to the American *Danaus* spp., the Old World milkweed butterfly, *Danaus chrysippus* (so-called plain tiger, widely distributed from Africa to tropical Asia) is assumed to be "poisonous" with its conspicuous appearance with a black apex and white subapical spots on the forewing in blight tawny-orange background coloration, likely as a typical model for various mimicry species (Smith 1973) (Figs. 11.1 and 11.2). *D. chrysippus* feeds exclusively on the Asclepiadoideae and presumed to sequester toxic cardenolides from its host milkweed plants during the

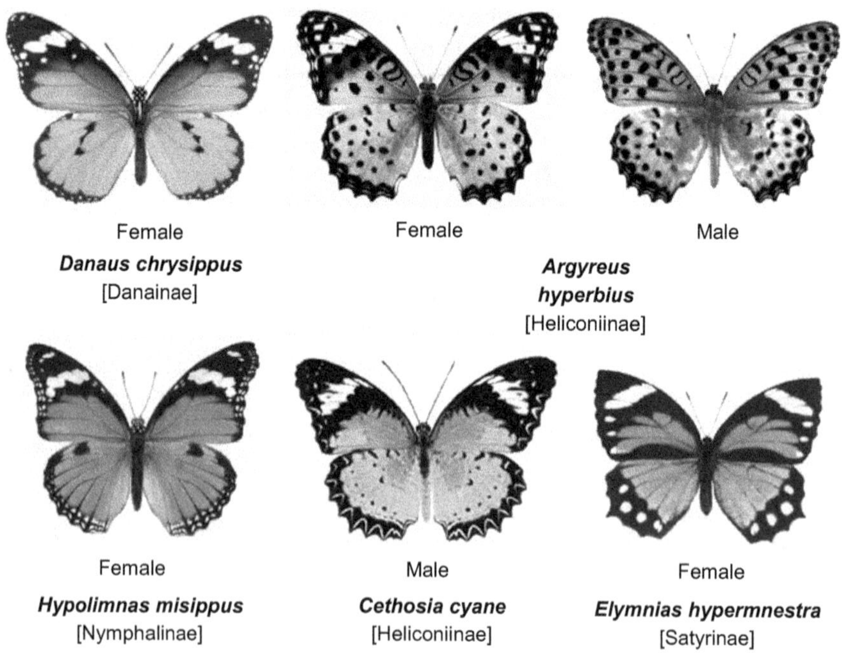

Fig. 11.1 The tiger Danaus mimicry ring. Note that some species in this figure may not occur in the same habitat

Fig. 11.2 (**a**) *Danaus chrysippus* female, (**b**) *Argyreus hyperbius* female, (**c**) *A. hyperbius* male

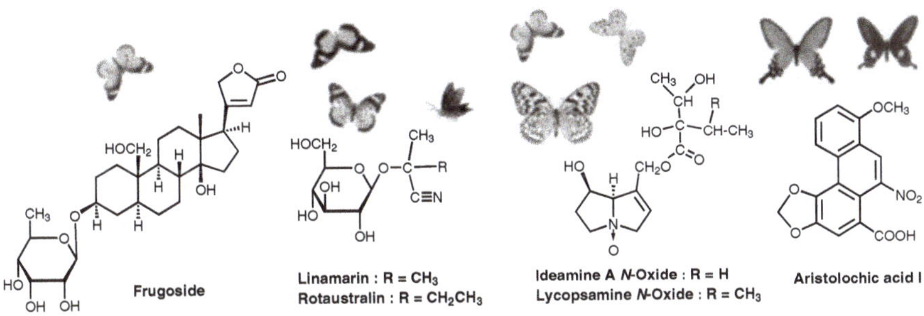

Fig. 11.3 Typical defensive substances in aposematic butterflies

larval stage. However, they are known to be poor and inconsistent sequesterers of CGs (Schneider et al. 1975; Rothschild et al. 1975; Mebs et al. 2005).

In order to confirm the toxicity of a local population of *D. chrysippus* in Okinawa, we examined CG contents in their body tissue. A relatively polar CG, frugoside (Fig. 11.3), was characterized as the major sequestrate in *D. chrysippus* adults raised on *Asclepias curassavica* (Wada et al. unpublished). While calotropin and calactin are CGs known as typical sequestrates in *D. plexippus* raised on *A. curassavica* (A), these less polar CGs were not found as prominent components in *D. chrysippus*, suggesting a selective or differential accumulation of CG species as in the case between *D. plexippus* and *D. gilippus* (Cohen 1985; Mebs et al. 2012). In addition, males of *D. chrysippus* frequently visit plants containing poisonous pyrrolizidine alkaloids (PAs) and pharmacophagously sequester the alkaloids in the body tissues both as a defense substance and pheromone precursors (Edgar et al., 1979; Boppré, 1986) (see also Sect. 11.2). Since *D. chrysippus* males in Okinawa frequently visit PA-containing plants such as *Eupatorium* (Asteraceae) and *Heliotropium* (Boraginaceae), they are assumed to be protected dually both by CGs and PAs. Nevertheless, a more detailed spatiotemporal dimensional survey of sequestered defensive elements is needed further to clarify the "palatability spectrum," because the distribution of *D. chrysippus* is widespread throughout the African-Asian tropical and subtropical regions up to Okinawa in association with their local mimics as discussed below.

There are a variety of butterfly species that closely resemble wing color patterns of *D. chrysippus*, particularly within Nymphalidae to form a "tiger Danaus mimicry ring" in Asian regions (Fig. 11.1):

Danaus genutia (striped tiger) (Danainae: Nymphalidae): This butterfly resembles *D. chrysippus* but with conspicuous black stripes on both fore- and hindwings. These two coinhabiting *Danaus* species, widely distributed throughout India, South Asia, and Japan, are regarded as Müllerian mimics, as both species feed on the milkweed subfamily Asclepiadoideae, although the ability of sequestration of CGs from their host *Cynanchum liukiuense* in Okinawa is unknown. Males of these two species show strong affinity to PA-containing plants and are considered to obtain alkaloids both as defensive measure as well as sex pheromone precursors.

Hypolimnas misippus (danaid eggfly) (Nymphalinae: Nymphalidae): Females are known polymorphic and the plain tiger mimic occurs sympatrically with *D. chrysippus* in Asia (Gordon and Smith, 1998). Since the larvae feed on the "presumably innocuous plant" *Portulaca oleracea* (Portulacaceae), this butterfly is thought to be a palatable mimic.

Elymnias hypermnestra (common palmfly) (Satyrinae: Nymphalidae): Some populations are sexually dimorphic. While the males mimic *Euploea* species, the females mimic *Danaus* spp. (Yata and Morishita 1985). The larvae feed on the palm family Arecaceae (Ackery and Vane-Wright 1984).

Cethosia cyane (leopard lacewing) (Heliconiinae: Nymphalidae): There are several lacewing species mimicking the wing patterns of *D. chrysippus*. Their larvae feed on Passifloraceae, the subfamily of which is known to biosynthesize toxic cyanogenic glycosides (CNs) (Nahrstedt and Davis 1985).

Argyreus hyperbius (Indian fritillary) (Heliconiinae: Nymphalidae): This species is sexually dimorphic and the females have similar patterns to *D. chrysippus* but with additional black spots scattered on the dorsal wings (Fig. 11.2). The female has been considered to be a Batesian mimic (Su et al. 2015). The larvae feed on Violaceae plants similar to many other related species in the tribe Argynnini in Japan. Since the toxicity of this butterfly is unknown, we examined the possible defensive substance in the body as described below.

Among the species listed above, only females are mimetic to *D. chrysippus* in *H. misippus*, *A. hyperbius*, and *E. hypermnestra*, whereas males are considered to be pre-existing morphs at least in the case of the former two, as typical examples of sexual dimorphism. Female-limited wing pattern dimorphism in the butterfly mimicry rings seems restricted to Batesian mimicry, whereas Müllerian mimics lack strong sexual dimorphism (Mallet and Joron 1999). An increase of palatable mimetic pattern relative to a model would weaken the mimetic advantage in Batesian mimicry by negative frequency-dependent selection, whereas an increase of unpalatable mimetic pattern becomes more favorable in Müllerian mimicry by positive density-dependent selection (Turner 1978; Mallet and Joron 1999). If this theory is applicable, both *H. misippus* and *A. hyperbius* would be palatable Batesian mimics of the model *Danaus*. This would also be supported by chemical constituents in their host plants; both *Portulaca* and *Viola* are presumably "nontoxic" (the former often listed as a local edible wild vegetable). However, the chemical analysis of *A. hyperbius* butterfly bodies revealed the presence of highly toxic cyanogenic glycosides (CNs), linamarin and lotaustralin, both in males and females (raised on wild *Viola yedoensis* as well as cultivated varieties of pansy, *V. tricolor* during the larval stage) in substantial quantities (contents of total CNs: females, 300–400 μg; males 100–250 μg/body) (Nakade et al. unpublished) (Fig. 11.3). Since these CNs were not detected in the host *Viola* plants, *A. hyperbius* seems to biosynthesize these toxic compounds from amino acids, as in *Heliconius* butterflies (Nahrstedt and Davis 1985). It is also known that some of monomorphic *Cethosia* spices store linamarin and lotaustralin (Nahrstedt and Davis 1985).

Here *A. hyperbius* butterfly is determined to be poisonous despite its sexual dimorphism in wing pattern, to form the *Danaus* (male + female) – *Cethosia* (male + female) – *Argyreus* (female) Müllerian mimicry ring. It is noted that a number of *Cethosia* lacewings exhibit a black-spotted wing pattern similar to that of *A. hyperbius* rather than that of *D. chrysippus* which entirely lacks such "leopard"--spotted patterns, suggesting a closer reciprocal interaction between these two taxa within the Heliconiinae in the South Asian regions. We still do not know whether the other sexually dimorphic nymphalids, *H. misippus* and *E. hypermnestra*, store any unpalatable allelochemicals or not.

11.1.2 Idea *Butterfly Mimicry Ring*

A giant butterfly, *Idea leuconoe* (mangrove tree nymph) (Danainae: Nymphalidae), is a primitive danaine species, having a wingspan of 120–140 mm with black markings and veins on the white wing background (Fig. 11.4). Although the butterfly is not aposematic in coloration, compared to other toxic butterflies, it is highly conspicuous in the subtropical forests flying slowly and gracefully under the high predatory avian pressure such as the blue rock thrush, *Monticola solitarius*. The butterfly (originated in Okinawa) accumulates large quantities of PAs (ideamines A, B, and C, lycopsamine, parsonsine) as *N*-oxides from the host *Parsonsia laevigata* (Apocynaceae) during the larval stage (Nishida et al. 1991; Kim et al. 1994) (Fig. 11.3). The total alkaloids often exceed 3 mg/insect, suggesting a high unpalatability to potential predators.

Although males of most danaine butterflies show strong affinity to PAs during adult stage and accumulate PAs by foraging pharmacophagously from various plant sources to obtain defensive agents, as well as pheromone precursors, by visiting PA-containing plants (Ackery and Vane-Wright 1984), *I. leuconoe* and some other related *Idea* species in Southeast Asia likely obtain PAs not pharmacophagously but directly from the apocynad hosts during the larval stage. *I. leuconoe* is highly adapted to PAs not only by sequestering PAs for defense (allomone) but also for specific oviposition cues (kairomone) in females (Honda et al. 1997) and precursors

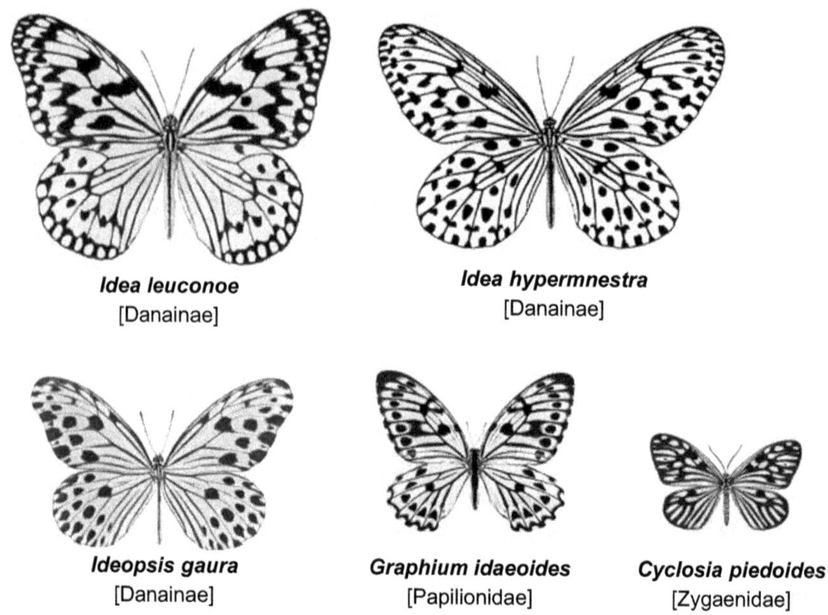

Fig. 11.4 The *Idea* mimicry ring. Note that some species in this figure may not occur in the same habitat

of volatile sex pheromone in males (Nishida et al. 1996). Such a tight linkage with PAs seems to represent an ancestral feature of the Danainae, from which the secondary colonization of the Danainae on PA-free plants (such as Asclepiadaceae) might have taken place with retention of the PA-mediated allomonal and pheromonal systems via pharmacophagous acquisition of the precursors from nonhost PA plants (Edgar 1984; Nishida 2002). *I. hypermnestra* (Fig. 11.4) and *I. lynceus*, both sympatric in Peninsular Malaysia, sequester a series of PA *N*-oxides, tentatively assigned from the hemolymph of the wild adults in the spectrometric analysis (Nishida unpublished). These two species are regarded as the Müllerian mimics in the rainforest habitat.

Beside these two giant *Idea* species, a smaller danaine species, *Ideopsis gaura*, often share the same habitat in Southeast Asia (Fig. 11.4). Its larvae feed on the Apocynaceae, but sequestration of toxic elements from the host is unknown. *I. gaura* males frequently visit PA-containing *Eupatorium* flowers (Asteraceae). A PA *N*-oxide of lycopsamine (or its stereoisomers) was detected by mass spectrometry from the body extracts of *I. gaura* males (captured in Penang, Malaysia) (Nishida unpublished). Thus, these three coinhabiting danaines seem to form a Müllerian mimicry association within the subfamily Danainae.

A Southeast Asian satyrine species, *Elymnias kuenstleri* (Nymphalidae), is also known to mimic *Idea* spp., which is analogous to the relationship between *Elymnias hypermnestra* and *D. chrysippus* in the tiger Danaus mimicry ring (Fig. 11.1). There are several swallowtail butterfly species mimetic to *Idea* spp. or *Ideopsis gaura*, such as *Graphium idaeoides* (Fig. 11.4), *G. delessertii*, and a whitish morph of

Papilio memnon (Papilionidae), which are assumed to be Batesian mimics of *Idea* spp.

The wing pattern of a day-flying *Cyclosia pieriodes* (false *Idea* moth, Zygaenidae) (Fig. 11.4) resembles that of *Idea* spp. or *Ideopsis gaura*. Although the defense substance has not been examined, it is likely that the moth stores some CNs, as many other unpalatable zygaenids are known to biosynthesize/store linamarin and lotaustralin (Holzkamp and Nahrstedt 1994; Nishida 1994). The wingspans between the putative model giant *Idea* spp. (or medium-size *I. gaura*) and minute *C. pieriodes* may be beyond comparison for the insectivorous birds. However, the exquisite wing color pattern may be a more significant factor to develop mimicry beyond their size in the avian vision, probably involving a psychological effect (Wickler 1968; Halpin et al. 2013).

11.1.3 Red-Bodied Swallowtail Mimicry Ring

As quoted by Wallace, the common rose, *Pachliopta aristolochiae*, and many other "red-bodied swallowtails" in the tribes Troidini (Papilionidae) are known as unpalatable models for various Batesian mimicry complexes particularly in the genus *Papilio* (Uésugi 1996) (Fig. 11.5). These troidines feed selectively on the pipevine family, Aristolochiaceae, and sequester toxic aristolochic acids (AAs) (Fig. 11.3) in the body tissues (Euw et al. 1968; Nishida and Fukami 1989; Wu et al. 2000). One of the pipevine swallowtails, *Atrophaneura alcinous*, from mainland Japan exhibits sexual dimorphism to some degree, in that the dorsal wings of males are typically jet black, while the females' are gray or smoky brown with some variations in the degree of darkness; the underside hindwings have long tails and a row of pink or orange spots at the edge in both sexes; the lateral sides of the abdomen are red with a black spot on each segment in both sexes (Fig. 11.5b, e, f).

A day-flying swallowtail moth, *Epicopeia hainesii* (Epicopeiidae) (Fig. 11.5c, d), has a red abdomen and a similar color pattern on the wings with long swallowtail projections as those in *A. alcinous*, although the body size is much smaller than that of *A. alcinous* (Fig. 11.5d–f and 11.6). The brighter gray or smoky brown-color tone and the black veins on the dorsal wings of *E. hainesii* are strikingly similar to those of *A. alcinous* females but not much of males. It suggests a unique case of mimicry in that the moth adopts one of the sexes of the toxic model. Both *A. alcinous* and *E. hainesii* are considered to be more or less sympatric in forests and grasslands in the middle to southern part of Japan. However, *E. hainesii* is distributed further to northern Japan, including Hokkaido, where its putative model *A. alcinous* is absent (Inoue, 1978). In the absence of a model species, there may be no protection from predation (Prudic and Oliver, 2008), unless other defensive measures are present. Since *E. hainesii* is warningly colored with red body even without the presence of the model swallowtail, it strongly implies that this moth is also unpalatable,

Fig. 11.5 The red-bodied butterflies and moths. (**a**) *Pachliopta aristolochiae* female, (**b**) *Atrophaneura alcinous* female, (**c**) *Epicopeia hainesii* female, (**d**) *Histia flabellicornis* female secreting defensive foams containing linamarin (*orange arrow*). (**e**) *A. alcinous* female, (**f**) *A. alcinous* male, (**g**) *E. hainesii* male, (**h**) *H. flabellicornis* male

although their palatability is unknown. The larval host, *Cornus controversa* (Cornaceae), is not known to contain typical noxious elements such as CGs, PAs, and CNs. The moth does not seem to secrete glandular defense substance. Preliminary chemical analysis of the whole-body extracts of *E. hainesii* revealed some unidentified polar compounds on the thin-layer chromatography, which did not match with CNs such as linamarin and lotaustralin (Nishida unpublished). Further studies are needed to substantiate possible unpalatability of *E. hainesii*. There is a

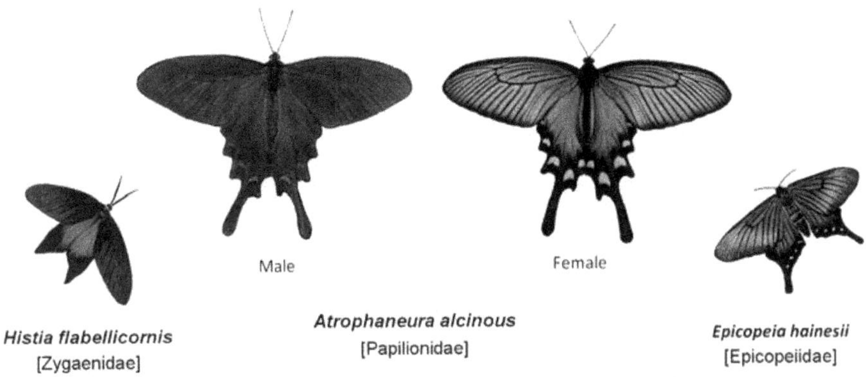

Fig. 11.6 The red-bodied swallowtail mimicry ring. Note that some species in this figure may not occur in the same habitat

spectacular array of red-bodied *Atrophaneura/Pachliopta–Epicopeia* mimicry rings distributed in East Asia further to examine in view of Batesian/Müllerian mimicry associations.

Histia flabellicornis moth (Zygaenidae) is a red-bodied diurnal moth resembling *A. alcinous* with black forewing and metallic blue hindwing coloration (larval host: *Bischofia javanica*, Euphorbiaceae) (Fig. 11.5d, h). *H. flabellicornis* is subdivided into many subspecies, and their distribution from Southeast Asia to Taiwan and Okinawa (but not inhabited in Japan's mainland) overlaps that of the red-bodied swallowtails (*Atrophaneura* and *Pachliopta*). The adults secrete toxic linamarin and hydrogen cyanide (HCN) as a foam when disturbed, suggesting the moth to be a Müllerian mimicry ally with *A. alcinous* in Okinawa (Nishida 1994). Although the degree of mimesis is imperfect in size, shape, and coloration, the wing color of *H. flabellicornis* relatively resembles males of *A. alcinous* but less resembles the females. This situation is contrary to the *A. alcinous–E. hainesii* mimicry association pattern, where the mimesis is biased to the females. The metallic blue burnish characteristic to *H. flabellicornis* hindwings may appear less similar to the *A. alcinous* male wing coloration. However, under the strong subtropical sunlight, dorsal wings of *A. alcinous* males (Fig. 11.5f) in Iriomote Island (Okinawa) shine in a metallic blue as in *H. flabellicornis* wings (Fig. 11.5h). *A. alcinous* males may have developed an implementation of structural coloration interacting with *H. flabellicornis* through visual selection by avian predators. Thus, even though the Müllerian association between these two species may be loose, the mimicking pattern may represent a rare case of sexual dimorphism in the toxic model, which might have stabilized the wing morph only in male. As illustrated in Fig. 11.6, the above two cases exemplify a symmetrical relationship of sexual dimorphism in *A. alcinous*, interacting adaptively with either one of the (putative) Müllerian mimic moths (*H. flabellicornis* and *E. hainesii*).

11.2 Discussion

As presented in those three mimicry rings, the chemistry of the unpalatable elements stored in the butterflies is the critical factor to understand the nature of aposematic wing coloration and mimicry – otherwise a conclusion of adaptive coloration must be tentative (Nishida 2002). Besides, "unpalatability" itself is hard to determine (Mallet and Joron 1999) due to the variable ecological and physiological milieus of both the preys and predators. The quality and quantities of defensive substances in presumptive model butterflies often depend on allelochemicals from the host plants, which may result in a variation in palatability, individually or among local communities, which complicates the mimicry association (Nishida 2002). Thus, Batesian and Müllerian mimicries may be interchangeable depending on the "palatability spectrum" of the prey butterflies as well as the "predatory spectrum" of the potential selective pressure (Huheey 1984; Turner 1984). The predatory spectra may involve variations in chemosensory and toxicological sensitivity/susceptibility of the predators. The sensory sensitivity in insectivorous animals may be different greatly among the species, and such variation in response to chemicals or preferences may often be seasonal particularly during highly predacious period and abundance of preys such as breeding seasons (Fink and Brower 1981; Glendinning and Brower 1990). We certainly need to know both prey/predatory spectra in time and space to understand the overall scheme of the existing mimicry rings (Joron and Mallet 1998).

Two instances of sex-limited dimorphisms were presented here in the Müllerian mimicry complexes – the tiger Danaine mimicry ring (*Argyreus hyperbius*) and the red-bodied swallowtail mimicry ring (*Atrophaneura alcinous*). In a toxic model, the diversification of its wing morphology would decrease their fitness due to a "number-dependent selection" as originally stated by Müller (Turner 1978; Mallet and Joron 1999). This widely accepted rule is not supported by the cases noted above. Such paradoxical examples are also known in *D. chrysippus* in Africa (Owen et al. 1994) and *Heliconius numata* in South America (Brown and Benson 1974).

In the *A. alcinous* (male)–*H. flabellicornis* (both sexes) co-mimicking association of the red-bodied mimicry ring, the defensive allelochemicals (AAs and CNs) fulfill the requirement of potential unpalatability that would reciprocally enhance their fitness through Müllerian mimicry. In the case of the *A. alcinous* (female)–*E. hainesii* (both sexes) co-mimicking association, the chemistry of the latter is still unclear, although the moth is suspected to be a Müllerian mimic as mentioned above. The root cause triggering female-limited wing pattern dimorphism in *A. alcinous* may be attributed to an escape from an overload of potential Batesian mimics, similar to the case of polymorphism in *D. chrysippus* in Africa under the high population density of palatable *Hypolimnas* (Smith et al. 1993). Several Rutaceae-feeding black swallowtails such as *Papilio macilentus* and *P. protenor* (females of the former are highly mimetic to *A. alcinous* males) often share the same microhabitat spatiotemporally in the middle part of Japan. Potential higher predation on females due to behavioral vulnerability may underlie behind this mechanism (Ohsaki 1995 2005). Thus, in addition to the chemistry, further studies are warranted especially on the geographical variations of wing color in both the butterflies and moths but also on the potential interspecific impacts from co-occurring Batesian mimics.

In the *A. hyperbius* (female)–*D. chrysippus* (both sexes) co-mimicking association, both are protected by unpalatable elements (CNs, CGs, and PAs). The black-spotted wing pattern in *A. hyperbius* males resembles that of the co-occurring woodland fritillaries such as *Argynnis paphia* (silver-washed fritillary) and *Argyronome ruslana*. We have recently shown that adults (both sexes) of these fritillary species sequester CNs in their body in substantial quantities (Nakade, Naka et al. unpublished), likely to form a "leopard fritillary mimicry ring," apparently excluding *A. hyperbius* females from the ring. These chemical evidences support that sexual dimorphism is feasible not only in Batesian mimicry but also in Müllerian mimicry, if the other sexual morph is consolidated by a member of other mimetic toxic species, probably to receive better protection than staying in the same color patterns. This would invoke arguments on the sexual selection vs. natural selection. Certainly, additional work is needed to support this hypothesis.

Another type of sexually restrictive feature in the mimicry syndromes can be seen in the specific exocrine organs developed on the wings and/or body, which disseminate odoriferous substances, often accompanied by morphological differences between the sexes. Troidine males such as *A. alcinous* emit characteristic odors from androconial scales on the inner margins of the hindwings (Honda 1980), which represents a structural–chemical dimorphism. Although sex-pheromonal roles of these male-specific scents are mostly unknown, the conspicuous smell of *A. alcinous* males may be attributed to an "odor aposematism," alerting the presence of systemic toxins (AAs) to enhance its visual aposematism (Nishida and Fukami 1989).

In many *Danaus* butterfly species, males pharmacophagously accumulate specific alkaloids from PA-containing plants during adult stage and use these chemicals for their own defense as well as precursors of sex pheromone disseminated from male-specific hairpencil organs (Boppré 1984). In this case, only males become a model with defensive PAs, possibly dually protected also by CGs originating from the larval host plant, representing a sexual "chemical dimorphism" by the sequestered defense substances. In this process, females may also be indirectly protected by visual automimicry, even if they do not have any defensive chemicals (Brower 1969). In the queen butterfly, *Danaus gilippus*, a portion of male's PAs may be transferred to females during copulation as a nuptial gift potentially to protect the female and eventually to eggs and thus to ensure protection of the male's progeny (Eisner and Meinwald 1995).

In contrast, a primitive danaine, *Idea leuconoe*, acquires PAs from the apocynad host during the larval stage (both sexes), and males convert a portion of sequestered PAs to volatile pheromone components to entice females (Nishida et al. 1996). In this hairpencil pheromone system, females assess the quality of protectiveness of a male during the precopulatory behavior (Nishida 2002). PA-derived pheromone

Fig. 11.7 *Idea leuconoe* female in a typical acceptance posture, visually and chemically stimulated by a paper model of male wings with a paper disk treated with the hairpencil extract (male pheromone). The female is arrested, curling her abdomen downward

system thus plays a decisive role in sexual selection of danaine butterflies (Eisner and Meinwald 1987; Nishida 2002). In indoor behavioral tests of *I. leuconoe*, both virgin males and females visually responded to a "fluttering" paper model of *I. leuconoe* (wing image printed on the copy paper) as shown in Fig. 11.7 (Nishida et al. 1996). When the model was scented with crude extracts of hair pencils from mature males (0.1 male equivalent/paper disk), receptive females usually landed on the nearby herbage with abdomen-curling posture (Fig. 11.7). Interestingly, if a fluttering "solid white model + pheromone" was presented instead of the "patterned model + pheromone," many virgin females approached and touched the model, some occasionally exhibiting the typical mate-acceptance posture, whereas none even approached a fluttering "solid black model + pheromone" (Nishida, unpublished). This clearly indicates that the females can select males by both visual and chemical cues and in lesser degrees by vision. If this is the case in nature, non-mimetic males would also be selected by a female by chance, in that the androconial chemical stimulus is sufficient to induce her acceptance, although *Idea* butterflies are basically monomorphic. This suggests an importance of chemical factors together with visual components, which then might facilitate the evolution of a new wing pattern through sexual selection (cf. Turner 1984; Krebs and West 1988).

In addition, males of *Idea* spp. also manipulate the conspicuous white/yellow eversible tufts of hair pencils on the dorsum of the abdomen for defense, by extruding them reflexively to emit a strong odor of phenol and/or *p*-cresol together with other volatiles whenever frightened by predators (Schulz and Nishida 1996). This exemplifies a male-specific aposematic scent operating concurrently with aposematic coloration of defensive organs (Nishida et al. 1996). Interestingly, one of the Idea-mimicking swallowtails, Graphium delessertii, exhibits a conspicuous yellow mark on the inner edge of the hindwings, as though an Idea male in captivity were displaying hair pencils. This could possibly evoke an additional psychological impression upon avian predators. In this case, only males become a template for the mimic.

Batesian or Müllerian mimicry – whichever the case – the evolutionary convergence of morphological characters between the models and mimics seems to have been greatly assisted by chemical elements including wing pigments and subject, concomitantly, to both natural and sexual selections.

Acknowledgments Special thanks are due to the late Miriam Rothschild for the discussion on this subject. I thank Hiroki Takamatsu, Aya Nakade, Atsushi Wada, and Hajime Ono of Kyoto University and Hideshi Naka of Tottori University for the project on *Argyreus hyperbius*. I am grateful to H. Frederik Nijhout and Atsushi Honma for reviewing the manuscript.

References

Ackery PR, Vane-Wright RI (1984) Milkweed butterflies. British Museum (Natural History), London

Boppré M (1984) Chemically mediated interactions between butterflies. In: Vane-Wright RI, Ackery PR (eds) The biology of butterflies. Academic, London, pp 259–275

Boppré M (1986) Insects pharmacophagously utilizing defensive plant chemicals (pyrrolizidine alkaloids). Naturwissenschaften 73:17–26

Brower LP (1969) Ecological chemistry. Sci Am 220:22–29

Brower LP (1984) Chemical defence in butterflies. In: Vane-Wright RI, Ackery PR (eds) The biology of butterflies. Academic, London, pp 109–134

Brown KS, Benson WW (1974) Adaptive polymorphism associated with multiple Müllerian mimicry in *Heliconius numata* (Lepid.: Nymph.) Biotropica 6:205–228

Cohen JA (1985) Differences and similarities in cardenolide contents of queen and monarch butterflies in Florida and their ecological and evolutionary implications. J Chem Ecol 11:85–103

Edgar JA (1984) Parsonsieae: ancestral larval foodplants of the Danainae and Ithomiinae. In: Vane-Wright RI, Ackery PR (eds) The biology of butterflies. Academic, London, pp 91–93

Edgar JA, Boppré M, Schneider D (1979) Pyrrolizidine alkaloid storage in African and Australian danaid butterflies. Experientia 35:1447–1448

Eisner T, Meinwald J (1987) Alkaloid-derived pheromones and sexual selection in Lepidoptera. In: Prestwich GD, Blomquist GJ (eds) Pheromone biochemistry. Academic, Orlando, pp 251–269

Eisner T, Meinwald J (1995) The chemistry of sexual selection. Proc Natl Acad Sci U S A 92:50–55

Euw JV, Reichstein T, Rothschild M (1968) Aristolochic acid-I in the swallowtail butterfly *Pachliopta aristolochiae* (Fabr.) (Papilionidae). Isr J Chem 6:659–607

Fink LS, Brower LP (1981) Birds can overcome the cardenolide defense of monarch butterflies in Mexico. Nature 291:67–70

Glendinning JI, Brower LP (1990) Feeding and breeding responses of five mice species to overwintering aggregations of the monarch butterfly. J Anim Ecol 59:1091–1112

Gordon IJ, Smith DAS (1998) Genetics of the mimetic African butterfly *Hypolimnas misippus*: hindwing polymorphism. Heredity 80:62–69

Halpin C, Skelhorn J, Rowe C (2013) Predators' decisions to eat defended prey depend on the size of undefended prey. Anim Behav 85:1315–1321

Holzkamp G, Nahrstedt A (1994) Biosynthesis of cyanogenic glucosides in the Lepidoptera: Incorporation of [U-^{14}C]-2-methylpropanealdoxime, 2S-[U-^{14}C]-methylbutanealdoxime and D,L-[U-^{14}C]-*N*-hydroxyisoleucine into linamarin and lotaustralin by the larvae of *Zygaena trifolii*. Insect Biochem Molec Biol 24:161–165

Honda K (1980) Odor of a papilionid butterfly: odoriferous substances emitted by *Atrophaneura alcinous alcinous* (Lepidoptera: Papilionidae). J Chem Ecol 6:867–873

Honda K, Hayashi N, Abe F et al (1997) Pyrrolizidine alkaloids mediate host-plant recognition by ovipositing females of an Old World danaid butterfly, *Idea leuconoe*. J Chem Ecol 23:1703–1713

Huheey JE (1984) Warning coloration and mimicry. In: Bell WJ, Cardé RT (eds) Chemical ecology of insects. Chapman and Hall, London, pp 257–297

Inoue H (1978) Genus *Epicopeia* Westwood from Japan, Korea and Taiwan, Lepidoptera: Epicopeiidae. Tyo to Ga (Trans Le Soc Jpn) 29:69–75

Joron M, Mallet J (1998) Diversity in mimicry: paradox or paradigm. Trends Ecol Evol 13:461–466

Kim CS, Nishida R, Fukami H, Abe F, Yamauchi T (1994) 14-Deoxyparsonsianidine *N*-oxide: A pyrrolizidine alkaloid sequestered by the giant danaine butterfly, *Idea leuconoe*. Biosci Biotech Biochem 58:980–981

Krebs RA, West DA (1988) Female mate preference and the evolution of female-limited Batesian mimicry. Evolution 42:1101–1104

Mallet J, Joron M (1999) Evolution of diversity in warning color and mimicry: polymorphisms, shifting balance, and speciation. Annu Rev Ecol Syst 30:201–233

Mebs D, Reuss E, Schneider M (2005) Studies on the cardenolide sequestration in African milkweed butterflies (Danaidae). Toxicon 45:581–584

Mebs D, Wagner MG, Toennes SW et al (2012) Selective sequestration of cardenolide isomers by two species of *Danaus* butterflies (Lepidoptera: Nymphalidae: Danainae). Chemoecology 22:269–272

Nahrstedt A, Davis RH (1985) Biosynthesis and quantitative relationships of the cyanogenic glucosides, linamarin and lotaustralin, in genera of the Heliconiini (Insecta: Lepidoptera). Comp Biochem Physiol 82B:745–749

Nishida R (1994/1995) Sequestration of plant secondary compounds by butterflies and moths. Chemoecology 5(6):127–138

Nishida R (2002) Sequestration of defensive substances from plants by Lepidoptera. Annu Rev Entomol 47:57–92

Nishida R, Fukami H (1989) Ecological adaptation of an Aristolochiaceae-feeding swallowtail butterfly, *Atrophaneura alcinous*, to aristolochic acids. J Chem Ecol 15:2549–2563

Nishida R, Kim CS, Fukami H, Irie R (1991) Ideamine *N*-oxides: Pyrrolizidine alkaloids sequestered by a danaine butterfly, *Idea leuconoe*. Agric Biol Chem 55:1787–1797

Nishida R, Schulz S, Kim CS et al (1996) Male sex pheromone of the giant danaine butterfly, *Idea leuconoe*. J Chem Ecol 22:949–972

Ohsaki N (1995) Preferential predation of female butterflies and the evolution of Batesian mimicry. Nature 378:173–175

Ohsaki N (2005) A common mechanism explaining the evolution of female-limited and both-sex Batesian mimicry in butterflies. J Anim Ecol 74:728–734

Owen DF, Smith DAS et al (1994) Polymorphic Müllerian mimicry in a group of African butterflies: a reassessment of the relationship between *Danaus chrysippus, Acraea encedon* and *Acraea encedana* (Lepidoptera: Nymphalidae). J Zool 232:93–108

Prudic KL, Oliver JC (2008) Once a Batesian mimic, not always a Batesian mimic: mimic reverts back to ancestral phenotype when the model is absent. Proc R Soc B 275:1125–1132

Prudic KL, Khera S et al (2007) Isolation, identification, and quantification of potential defensive compounds in the viceroy butterfly and its larval host-plant, Carolina willow. J Chem Ecol 33:1149–1159

Reichstein T, Jv E, Parsons JA, Rothschild M (1968) Heart poison in the monarch butterfly. Science 161:861–866

Ritland DB, Brower LP (1991) The viceroy is not a Batesian mimic. Science 350:497–498

Rothschild M (1979) Mimicry, butterflies and plants. Symb Bot Upsal 22:82–99

Rothschild M, Jv E, Reichstein T et al (1975) Cardenolide storage in *Danaus chrysippus* (L.) with additional notes of *D. plexippus* (L.) Proc Roy Soc London (B) 190:1–31

Schneider D, Boppré M, Schneider H et al (1975) A pheromone precursor and its uptake in male *Danaus* butterflies. J Comp Physiol 97:245–256

Schulz S, Nishida R (1996) The pheromone system of the male danaine butterfly, *Idea leuconoe*. Bioorg Med Chem 4:341–349

Smith DAS (1973) Batesian mimicry between *Danaus chrysippus* and *Hypolimnas misippus* (Lepidoptera) in Tanzania. Nature 242:129–131

Smith DAS, Owen DF, Gordon IJ et al (1993) Polymorphism and evolution in the butterfly *Danaus chrysippus* L. (Lepidoptera: Danainae). Heredity 71:242–251

Su S, Lim M, Kunte L (2015) Prey from the eyes of predators: Color discriminability of aposematic and mimetic butterflies from an avian visual perspective. Evolution 69:2985–2994

Trigo JR (2000) The chemistry of antipredator defense by secondary compounds in neotropical Lepidoptera: facts, perspectives and caveats. J Braz Chem Soc 11:551–561

Turner JRG (1978) Why male butterflies are non-mimetic: natural selection, sexual selection, group selection, modification and sieving. Biol J Linn Soc 10:385–432

Turner JRG (1984) Mimicry: the palatability spectrum and its consequences. In: Vane-Wright RI, Ackery PR (eds) The biology of butterflies. Academic, London, pp 141–161

Uésugi K (1996) The adaptive significance of Batesian mimicry in the swallowtail butterfly, *Papilio polytes* (Insecta, Papilionidae): Associative learning in a predator. Ethology 102:762–775

Wickler W (1968) Mimicry in plants and animals. Wiedenfeld and Nicholson, London

Wu TS, Leu YL, Chan YY (2000) Aristolochic acids as a defensive substance for the aristolochiaceous plant-feeding swallowtail butterfly, *Pachliopta aristolochiae interpositus*. J Chin Chem Soc 47:221–226

Yata O, Morishita K (1985) Butterflies of the South East Asian Islands, vol II. Pieridae and Danaidae. Plapac Co Ltd, Tokyo

12

Mimetic Swallowtail Butterfly and Population Dynamics Model

Toshio Sekimura, Noriyuki Suzuki and Yasuhiro Takeuchi

Abstract Based on recent progresses of both experiment and mathematical analysis, we present an extension of the model for population dynamics of the mimetic swallowtail butterfly *Papilio polytes* in Sakishima Islands, Japan (Sekimura et al. J Theor Biol 361:133–140, 2014). The model includes four major variables, that is, population densities of three kinds of butterflies (two female forms f. *cyrus* and f. *polytes* and the unpalatable butterfly *Pachliopta aristolochiae*) and their predator. In this extension, we introduce difference in the predation rate between two forms f. *cyrus* and f. *polytes*. We still assume that both the benefit of mimicry for the mimic f. *polytes* and the cost for the model are dependent on their relative frequencies, i.e., the mortality of the mimic by predation decreases with increase in frequency of the model, while the mortality of the model increases as the frequency of the mimic increases. Taking the density-dependent effect by carrying capacity into account, we set up an extended model system consisting of three ordinary differential equations (ODEs), analyze it mathematically, and provide computer simulations that confirm the analytical results.

Keywords *Papilio polytes* • Batesian mimicry • Population dynamics • Mathematical model • Computer simulations • Relative abundance of the mimic • Sakishima Islands

---

T. Sekimura (✉)
Department of Biological Chemistry, Graduate School of Bioscience and Biotechnology, Chubu University, Kasugai, Aichi 487-8501, Japan
e-mail: sekimura@isc.chubu.ac.jp

N. Suzuki • Y. Takeuchi
Department of Physics and Mathematics, College of Science and Engineering, Aoyama Gakuin University, Sagamihara, Kanagawa 252-5258, Japan

12.1 Introduction

P. polytes is a mimetic swallowtail butterfly species widely distributed across India and Southeast Asia, including Southeast of China, the Philippines, Taiwan, and the Ryukyu Islands of Japan (Clarke and Sheppard 1972). *P. polytes* exhibits the female limited polymorphism, that is, the female is polymorphic, whereas the male is monomorphic and exhibits a white bar on the black hindwing. In the Ryukyu Islands located in the southwest of Japan, the female of *P. polytes* has two different forms, the mimetic form f. *polytes* and the nonmimetic form f. *cyrus* resembling the monomorphic male in appearance. The form f. *polytes* mimics the unpalatable butterfly *Pachliopta aristolochiae* as a mimetic model, which has a large white area in the center and a row of submarginal red spots on the black hindwing (Fig. 12.1). Mimicry in the female of *P. polytes* is known to be Batesian mimicry.

Yamauchi (1994) built a population dynamic model of Batesian mimicry, in which two populations of both model and mimic species were considered. The dynamic model has two components, growth at intrinsic growth rate and carrying capacity and reduction by predation. The probability of a predator catching prey on an encounter was assumed to depend on the frequency of the mimic. He applied the dynamic model to field records of butterflies in Ryukyu Islands, and his model has successfully explained some features, e.g., multiple dynamic equilibria between the model and the mimic in the field. However, his model did not account for realistic

Fig. 12.1 The female limited polymorphism of the mimetic butterfly *Papilio polytes* and the model butterfly *Pachliopta aristolochiae*. *Top left*: *P. polytes* f. *cyrus* (nonmimetic form). *Top right*: *P. polytes* f. *polytes* (mimetic form). *Bottom left*: *P. polytes* male. *Bottom right*: the model *Pachliopta aristolochiae*

model-mimic systems such as polymorphism in *P. polytes* and intraspecific competition between the mimic f. *polytes* and the non-mimic f. *cyrus*.

Sekimura et al. (2014) presented a mathematical model for population dynamics of *P. polytes* observed in the Sakishima Islands (i.e., the Miyako Islands and the Yaeyama Islands), which are the southernmost island group of the Ryukyu Islands, Japan (Fig. 12.2). The model system consists of three ordinary differential equations (ODEs), in which variables are population densities of three butterflies, the unpalatable butterfly *P. aristolochiae* and two female forms of *P. polytes*. The model was constructed on the basis of field data in the islands and also experimental data in the laboratory. Using mathematical analysis and computer simulations of the system equations, they clarified the logical relationship hidden behind field data on the population dynamics of the mimetic butterfly *P. polytes* in Sakishima Islands. In particular, they discussed both temporal change in the relative abundance (RA) since 1975 in Miyako-jima Island and variation in the RA in Sakishima Islands.

Before going into details of our extended model and results, we summarize field data briefly in Sect. 12.2 and describe main features of mimicry of *P. polytes* in Sect. 12.3.

12.2 Field Records of *Papilio polytes* Observed in Sakishima Islands

12.2.1 Observation of Temporal Change in the Population of the Mimetic Female of P. polytes *in Miyako-jima Island*

Uesugi (1992) observed temporal change of the relative abundance of the mimic f. *polytes* in all the females of *P. polytes* for 14 years from 1975 to 1989 after the establishment of the model *P. aristolochiae* in the land. Here, the relative abundance (RA) of the mimic denotes the population ratio of f. *polytes* to all the two forms (f. *cyrus* and f. *polytes*) and is defined as follows:

$$\mathrm{RA} = \frac{\text{population of the mimic f. } \textit{polytes}}{\text{population of the non-mimic f. } \textit{cyrus} + \text{population of the mimic f. } \textit{polytes}} \times 100\ (\%)$$

The observational result revealed that the RA increases with the date like the sigmoidal curve, and in 1985 about 10 years after starting observation, the RA reached at a saturated value (or the value of equilibrium) of about 50% (Fig. 12.5a).

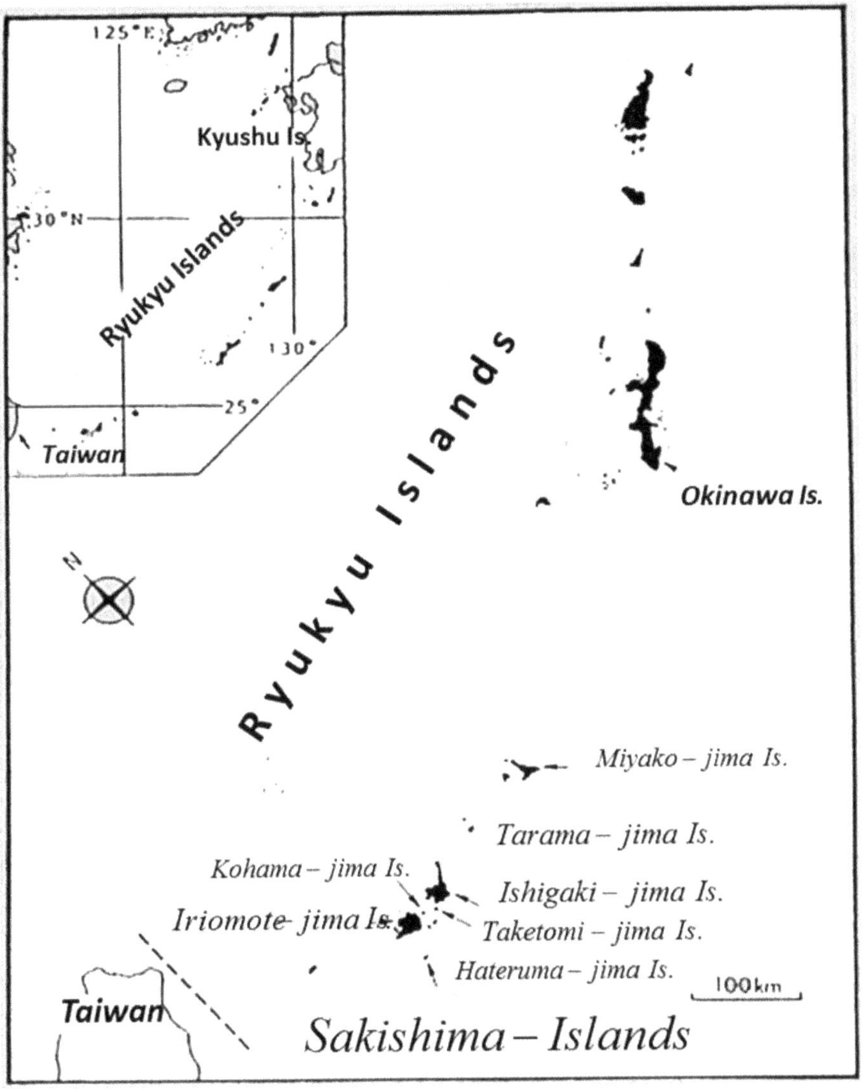

Fig. 12.2 Map of the Sakishima Islands, Japan. The Sakishima Islands are located at the southeast end of Japan and part of the Ryukyu Islands, which include both the Miyako Islands and the Yaeyama Islands. The Miyako Islands include Miyako-jima Is., Tarama-jima Is., etc., and the Yaeyama Islands include Ishigaki-jima Is., Hateruma-jima Is., Iriomote-jima Is., Taketomi-jima Is., Kohama-jima Is., etc.

12.2.2 Variation in the Relative Abundance (RA) in Sakishima Islands

Uesugi (1992) also observed variation in the RA in Sakishima Islands. In order to investigate the relationship between two populations of the model *P. aristolochiae* and the mimic f. *polytes*, he recorded three populations of the model and two female forms of *P. polytes* in seven islands from island to island in the Sakishima Islands in 1982, 14 years after the establishment of the model in the Yaeyama Islands. The horizontal axis in Fig. 12.5b shows the advantage index (AI) of Batesian mimicry, which is defined to be the population ratio of the model to all the related butterflies as follows:

$$\mathrm{AI} = \frac{\text{population of the model}}{\text{population of the model} + \text{populations of f. } \textit{cyrus} \text{ and f. } \textit{polytes}} \times 100\ (\%)$$

The vertical axis in Figs. 12.5a and 12.5b is the relative abundance (RA), which denotes the population rate of the form f. *polytes* to all two forms (f. *cyrus* and f. *polytes*).

Field data, which is shown by solid circles in Fig. 12.5b, clearly shows the positive correlation between the AI and the RA, which means that the higher the population ratio of the model butterfly in an island, the higher the ratio of the mimetic female to all females in the island.

12.3 Extended Mathematical Model for Population Dynamics of *P. polytes*

We first summarize fundamental facts on the mimicry, which will allow us to design the content of the mathematical model of three variables, that is, three kinds of populations of the model *P. aristolochiae*, the mimic f. *polytes*, and the non-mimic f. *cyrus*.

12.3.1 Fundamental Facts on the Mimicry of P. polytes

12.3.1.1 Difference in Predation Risk Between Two Forms f. *polytes* and f. *cyrus*

The butterfly *P. polytes* has two female forms: one is the non-mimic f. *cyrus* that resembles the monomorphic male with a white bar on the black hindwing, while the other is the mimic f. *polytes* that resembles the unpalatable butterfly species *P. aristolochiae* (Fig.12.1). The mimetic form f. *polytes* is considered to have an advantage over f. *cyrus* with respect to protection from predation, when it lives in sympatry with the model *P. aristolochiae*. In reality, Uesugi (1996) examined the idea positively by learning experiments. Unexperienced birds, brown-eared bulbuls, *Hypsipetes amaurotis oryeri* as predators, were first trained to take food from two feeders in a cavity and then offered *P. aristolochiae* in one of the feeders. After experiencing an uncomfortable encounter with this butterfly, the birds reduced the frequency of taking regular food from the feeder where the butterfly had been placed. On the other hand, Ohsaki (1995) paid attention to the rates of beak marks by predators on wings of both palatable and unpalatable butterflies. By analyzing the number of beak marks on butterfly wings caught in Borneo, he found that by comparing the species *P. polytes* and the model species *P. aristolochiae*, nonmimetic females were selectively attacked, while males, mimetic females, and models were attacked less. Thus, the mimic f. *polytes* is considered to gain benefit of reduced predation risk by living in sympatry with the model *P. aristolochiae*.

However, we should note an experimental result on survival rates of the mimetic form f. *polytes*, the nonmimetic form f. *cyrus*, and male in Taketomi-jima Is., where the model *P. aristolochiae* and the mimetic female f. *polytes* are both absent (Uesugi 1997). Uesugi (1997) used a releasing and re-catching method, that is, he released at first 300 butterflies into the field and then caught again released butterflies day after day for a week in the field. The result was remarkable, that is, the survival rate of the mimetic f. *polytes* was statistically significantly lower than those of the nonmimetic f. *cyrus* and male. This means that the mimetic f. *polytes* has received an apparent disadvantage for survival in Taketomi-jima Is. in comparison with the nonmimetic f. *cyrus* and male.

12.3.1.2 Males Prefer the Non-mimic f. *cyrus* to the Mimic f. *polytes*?

The mimetic female f. *polytes* might be less distinguishable than the nonmimetic female f. *cyrus* by the male, because the form f. *polytes* resembles a different species *P. aristolochiae*, while the form f. *cyrus* resembles the male of the same species. This means that f. *cyrus* could get more chances to mate with the male than f. *polytes*. In reality, Uesugi (1997) counted mating times of young butterflies of both f. *cyrus* and f. *polytes* just after emergence. The result was striking to show that f. *polytes* had no approach from the male and no count, while f. *cyrus* could mate with the male. From the viewpoint of making offspring, the non-mimic f. *cyrus* has an advantage over the mimic f. *polytes*. Thus, viewing from the preference of the male to females in the species, the mimicry is not always beneficial.

12.3.1.3 Physiological Life Span of Two Forms f. *cyrus* and f. *polytes*

Ohsaki (2005) measured physiological life spans of flying three types (the male, the non-mimic f. *cyrus*, and the mimic f. *polytes*) in Itami City Museum of Insects, Japan, and found that the order of the life span length is as follows: f. *cyrus* > the male > f. *polytes*. The problem is why the life span of the mimic f. *polytes* is the shortest. To explain the reason, Ohsaki proposed a hypothesis that the mimic f. *polytes* pays an additional cost for producing red colored pigments (carotenoid) by activating biochemical reaction networks in cells. The hypothesis coincides with the result that the bigger the red colored area on the wing that the butterfly has, the shorter the physiological life span that the butterfly has.

On the other hand, by using *P. polytes* caught in Okinawa Island, Kinjyo (2000) showed that the physiological life span of f. *cyrus* and f. *polytes* is dependent on the feeding condition. For example, there is no statistically significant difference in the life span when both f. *cyrus* and f. *polytes* are fed in a good condition, under which honey is given every day (i.e., the mean life span, 23.5 days for f. *cyrus* and 22 days for f. *polytes*, respectively). When butterflies are fed in other conditions, under which plain water or no honey is given, mean life spans of the non-mimic f. *cyrus* become 8.8 days (with plain water) and 5.3 days (with no food), and those of the mimic f. *polytes* become 7.8 days (with plain water) and 4.6 days (with no food), respectively. This result shows that in nature, when environmental condition or food conditions (e.g., long rainy weather, dry weather) become worse, the life span of both f. *cyrus* and f. *polytes* becomes shorter rapidly. In any case, the result shows that the life span of f. *cyrus* is somewhat longer than that of f. *polytes*.

For the last several years, Sekimura measured the life span of both f. *cyrus* and f. *polytes* in the laboratory and found that in a regular condition at 25 °C room temperature, there was no statistically significant difference in the life span of both f. *cyrus* and f. *polytes* under different feeding conditions (Fig. 12.3).

12.3.2 Mathematical Model of Three ODEs for Population Dynamics of P. polytes *with Intraspecific Competition*

In order to analyze field data summarized in Sect. 12.2, we present here a mathematical model for population dynamics of *P. polytes*. The model system consists of three ordinary differential equations (ODEs), in which variables are population densities of three butterflies, the unpalatable butterfly *P. aristolochiae* and two female forms of *P. polytes*. The male of *P. polytes* is not included directly in the system but included indirectly through parameter values such as intraspecific competition coefficients.

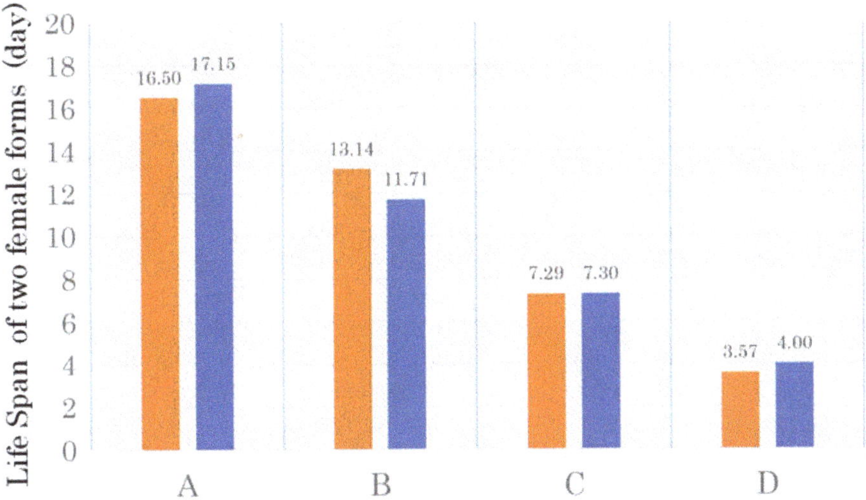

Fig. 12.3 Phylogenetic life span of two female forms f. *cyrus* and f. *polytes* under the feeding condition. The horizontal axis indicates the feeding condition. "A" is the condition under which Calpis (or Calpico) water (i.e., Japanese milk-based soft drink) is given twice a day; "B" is the condition under which the Calpis water with doubling dilution is given twice a day; "C" is the condition under which only water is given twice a day; "D" is the condition under which nothing is given after hatching. The left side (*orange-colored*) column of each condition corresponds to the average life span of f. *cyrus*, and the right side (*blue colored column*) is that of f. *polytes*. The number upon each column indicates the average life span. For example, the number 16.50 upon the left column of the condition "A" is the average life span (days) of 10 (f. *cyrus*) individuals. Numbers of individuals used in the experiment are as follows: numbers of (f. *cyrus*) individuals, 10 for the feeding condition "A," 7 for "B," 14 for "C," and 14 for "D," respectively, and numbers of (f. *polytes*) individuals: 13 for the feeding condition "A," 7 for "B," 10 for "C," and 10 for "D," respectively. According to our statistical analysis of the data, there is no statistically significant difference in the physiological life span between two forms f. *cyrus* and f. *polytes* for all feeding conditions A–D. In contrast, the data analysis has made it clear that there exists statistically significant effect of the feeding condition (*A–D*) on difference in the life span

We denote four population densities of the model *P. aristolochiae*, the mimic f. *polytes*, the non-mimic f. *cyrus*, and predator as n_1, n_2, n_3, and p, respectively. Changes of the densities in time become the following three ODEs:

$$\frac{\mathrm{d}n_1}{\mathrm{d}t} = n_1\left\{r_1\left(1 - \frac{n_1}{K_1}\right) - \alpha\left(\frac{n_2}{n_1 + n_2}\right)p\right\} \tag{12.1}$$

$$\frac{\mathrm{d}n_2}{\mathrm{d}t} = n_2\left\{r_2\left(1 - \frac{n_2 + a_{23}n_3}{K_2}\right) - \beta_2\left(\frac{n_2}{n_1 + n_2}\right)p\right\} \tag{12.2}$$

$$\frac{\mathrm{d}n_3}{\mathrm{d}t} = n_3\left\{r_3\left(1 - \frac{n_3 + a_{32}n_2}{K_3}\right) - \beta_3 p\right\} \tag{12.3}$$

where the population density of predator p is given as a fixed parameter value. In the system Eqs. (12.1), (12.2), and (12.3), we evaluated the density effect or saturation effect by using carrying capacities K_1, K_2, K_3 $(=K_2)$ for *P. aristolochiae*, f. *polytes*, and f. *cyrus,* respectively. We assume that K_2, K_3 have the same value since both f. *polytes* and f. *cyrus* are females in the same species. Growth rates of three butterflies are denoted by r_1, r_2, r_3, respectively. In Eqs. (12.2) and (12.3), we introduced an intraspecific competition between f. *polytes* and f. *cyrus* through competition coefficients (a_{23}, a_{32}). The intraspecific competition effect includes competitions for resources such as nectar and indirectly the male as noted in Sect. 12.3.1.2. The term $(n_2/(n_1+n_2))$ multiplied by p in Eqs. (12.1) and (12.2) represents the effect of mimicry, that is, the negative density effect implying that the increase in the density of f. *polytes* n_2 causes the decrease in both densities of *P. aristolochiae* and f. *polytes*, n_1 and n_2. Parameters α, β_2, β_3 represent difference in the predation rate among *P. aristolochiae*, *P. polytes* f. *polytes*, and f. *cyrus,* respectively, and it would be reasonable to assume the inequality $\alpha < \beta_2, \beta_3$, because *P. aristolochiae* is an unpalatable butterfly species. In Sect. 12.3.1.1, we noted the result on beak marks by predators showing that nonmimetic females were selectively attacked, while males, mimetic females, and model butterflies were attacked less. This fact is evaluated mathematically in the second term of Eqs. (12.2) and (12.3) by multiplying p by $(n_2/(n_1+n_2))$ (<1) for the mimic f. *polytes*, while by 1 for the non-mimic f. *cyrus*.

12.4 Mathematical Analysis of the System Equations and Computer Simulations

12.4.1 Mathematical Analysis

Based on discussions in Sect. 12.3.1, we consider and analyze mathematically the following three cases of the system Eqs. (12.1), (12.2), and (12.3) classified by growth rate and predation rate of two forms f. *polytes* and f. *cyrus*: (a) case 1, $r_2 < r_3$ and $\beta_2 = \beta_3 (=\beta)$; (b) case 2, $r_2 = r_3$ and $\beta_2 > \beta_3$; and (c) case 3, $r_2 = r_3$ and $\beta_2 = \beta_3 (=\beta)$.

12.4.1.1 Case 1: $r_2 < r_3$ and $\beta_2 = \beta_3 (=\beta)$

This is just the case that was analyzed in our previous paper (Sekimura et al. 2014), where following inequalities were assumed, $r_2 < r_3$, i.e., (growth rate of f. *polytes*) < (growth rate of f. *cyrus*), and $\alpha < \beta$, i.e., (predation rate of *P. aristolochinae*) < (predation rate of f. *polytes*).

We put one more assumption on the survival of f. *cyrus*, $r_3 > \beta p$, i.e., the growth rate of f. *cyrus* is larger than the predation rate. Note that the population n_3 tends to zero when $r_3 \leq \beta p$, since $dn_3/dt \leq 0$.

We summarize here main analytical results and computer simulations by solving the system Eqs. (12.1), (12.2), and (12.3).

[Result C1-1] The change of relative abundance (RA) of the mimic in the female of *P. polytes* with respect to the carrying capacity K_1 of the model is positive, that is,

$$\frac{d}{dK_1}\left(\frac{n_2}{n_2+n_3}\right) = \frac{d(RA)}{dK_1} > 0 \tag{12.4}$$

[Result C1-2] The change of the advantage index (AI) with respect to the carrying capacity K_1 of the model is positive:

$$\frac{d}{dK_1}\left(\frac{n_1}{n_1+n_2+n_3}\right) = \frac{d(AI)}{dK_1} > 0 \tag{12.5}$$

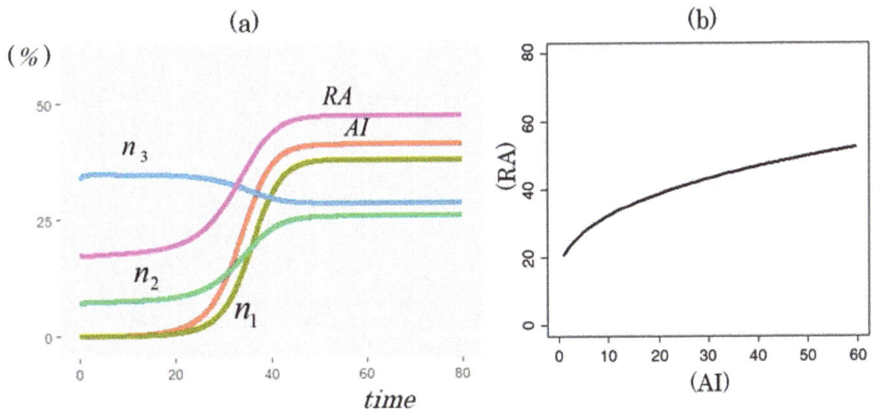

Fig. 12.4 Simulation results on temporal changes of all related quantities
(**a**) The *blue line* denotes the population density n_3 of the non-mimic f. cyrus, the *green line* is the density n_2 of the mimic f. *polytes*, the *purple line* is the RA, the *orange line* is the AI, and the *ocher line* denotes the density n_1 of the model *P. aristolochiae*, respectively. Parameter values used in numerical simulations are as follows, and the stability condition for the positive equilibrium is satisfied with these parameter values:
$r_1 = 0.5, r_2 = 1.0, r_3 = 2.0; K_1 = 50, K_2 = 50; a_{23} = {}^{23}0.5, a_{32} = 0.35; p = 0.5; \alpha = 0.6, \beta_2 = \beta_3 = 1.0; n_{10} = 0.01, n_{20} = 7.2, n_{30} = 34$ (initial values)
(**b**) Numerical simulation results of the positive dependence of the RA on the AI. Parameter values used in the numerical simulation are all the same as in Fig. 12.4a

From inequalities (12.4) and (12.5), we see that the ratio of change of the relative abundance (RA) to change of the advantage index (AI) with respect to the carrying capacity K_1 of the model is also positive:

$$\frac{d(RA)}{dK_1} / \frac{d(AI)}{dK_1} > 0 \tag{12.6}$$

The inequality (12.6) provides the analytical evidence for the field record on the positive dependence of the RA on the AI in Sakishima Islands in Fig.12.5b (Uesugi 1992) noted in Sect. 12.2.2.

Parameter values used in numerical simulations of Figs. 12.4, 12.5a and 12.5b are chosen as follows, so as to satisfy the existence and stability condition for the equilibrium of all the three population densities $E_{123} = (n_1, n_2, n_3)$, $r_1 = 0.5$, $r_2 = 1.0$, $r_3 = 2.0$, $K_1 = 50$, $K_2 = 50$, $a_{23} = 0.5$, $a_{32} = 0.35$, $p = 0.5$, $\alpha = 0.6$, $\beta = 1.0$, $n_{10} = 00.1$, $n_{20} = 7.2$, and $n_{30} = 34$, where n_{i0} denotes the initial value of n_i.

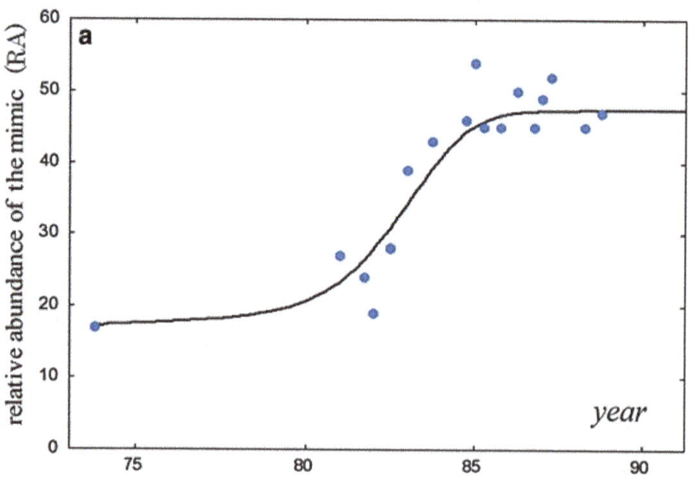

Fig. 12.5a Temporal change in the population of the mimetic female of *P. polytes* on Miyako-jima Island and its numerical simulation results
Solid circles (●) show temporal change in the relative abundance (RA) of the mimetic form f. *polytes* to all the females of *P. polytes* for 14 years after the establishment of the model *P. aristolochiae* on Miyako-jima Is. in 1975. The *solid line* shows numerical simulation results of temporal change in the RA by use of the Eqs. (12.1), (12.2), and (12.3). Parameter values used in numerical simulations are all the same as in Fig. 12.4a, and the stability condition for the positive equilibrium is satisfied with these parameter values:
$r_1 = 0.5, r_2 = 1.0, r_3 = 2.0; K_1 = 50, K_2 = 50; a_{23} = {}^{23}0.5, a_{32} = 0.35; p = 0.5; \alpha = 0.6, \beta_2 = \beta_3 = 1.0; n_{10} = 0.01, n_{20} = 7.2, n_{30} = 34$ (initial values)

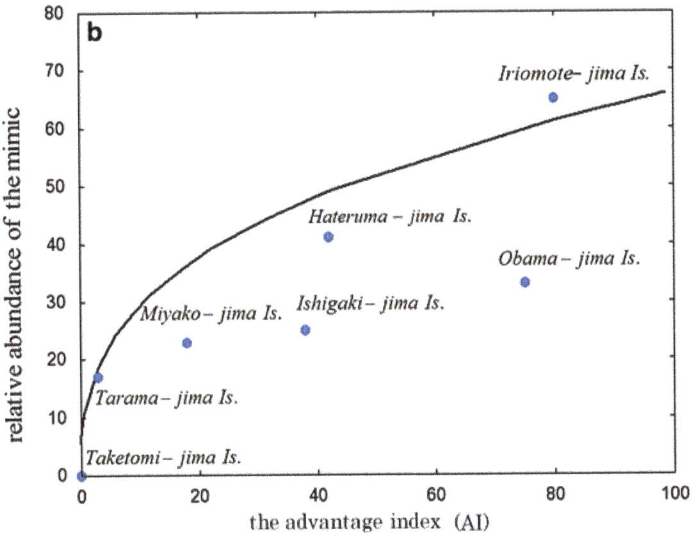

Fig. 12.5b Variation in the relative abundance (RA) in Sakishima Islands and its numerical simulation results
The horizontal axis shows the advantage index (AI), which is defined as the number ratio of the model to all the related butterflies (the model and two female forms of *P. polytes*). The vertical axis is the relative abundance (RA). *Solid circle* (●) represents the RA corresponding to the AI on each Island of 7 Islands (i.e., Taketomi-jima Is., Tarama-jima Is., Miyako-jima Is., Ishigaki-jima Is., Hateruma-jima Is., Obama-jima Is., and Iriomote-jima Is.) in Sakishima Islands (From Uesugi 1992). The *solid line* (i.e., Fig. 12.4b) represents numerical simulation results by use of the Eqs. (12.1), (12.2), and (12.3), which clearly show the positive dependence of the RA on the AI with a convexity. The inequality (6), which is an analytical result of the system equations, provides the theoretical basis on the positive dependence. Parameter values used in the numerical simulation are all the same as in Fig. 12.5a.

12.4.1.2 Case 2: $r_2 = r_3$ and $\beta_2 > \beta_3$

This is the case corresponding to experimental results noted in the last paragraphs of Sects. 12.3.1.3 and 12.3.1.1, that is, (a) there is no statistically significant difference in the life span of both f. *cyrus* and f. *polytes* (i.e., $r_2 = r_3$), and (b) the survival rate of mimetic f. *polytes* is statistically significantly lower than that of nonmimetic f. *cyrus* (i.e., $\beta_2 > \beta_3$) (this result was obtained by an experiment in Taketomi-jima Is. (Uesugi 1997), where the model *P. aristolochiae* and the mimetic female f. *polytes* are both absent) (Uesugi 1991, 1992).

[Result C2-1] First consider the system Eqs. (12.1), (12.2), and (12.3) under the condition $n_{10} = 0, n_{20} > 0$ and $n_{30} > 0$. The setting of initial parameter values corresponds to the situation of Taketomi-jima Is. The uniqueness of the solution of the system implies that $n_1(t) = 0, n_2(t) > 0$ and $n_3(t) > 0$ for any $t > 0$. Then the solution $n_2(t)$ and $n_3(t)$ satisfies

$$\frac{dn_2}{dt} = n_2\left\{r_2 - \beta_2 p - r_2\left(\frac{n_2 + a_{23}n_3}{K_2}\right)\right\} \tag{12.7}$$

$$\frac{dn_3}{dt} = n_3\left\{r_3 - \beta_3 p - r_3\left(\frac{n_3 + a_{32}n_2}{K_3}\right)\right\}, \tag{12.8}$$

which is a traditional Lotka-Volterra competition model. Suppose that $r_i > \beta_i p$ for $i = 2, 3$ (otherwise species i always tends to zero). It is easy to show that:

1. $n_2(t) \to 0$ for any $n_{20} > 0$ and $n_{30} > 0$, when

$$a_{23} > \frac{1 - \beta_2 p/r_2}{1 - \beta_3 p/r_3} \quad \text{and} \quad a_{32} < \frac{1 - \beta_3 p/r_3}{1 - \beta_2 p/r_2} \tag{12.9}$$

2. $n_2(t) \to 0$ for some $n_{20} > 0$ and $n_{30} > 0$, when

$$a_{23} > \frac{1 - \beta_2 p/r_2}{1 - \beta_3 p/r_3} \quad \text{and} \quad a_{32} > \frac{1 - \beta_3 p/r_3}{1 - \beta_2 p/r_2}. \tag{12.10}$$

Mathematically we can prove that the equilibrium point $E_3 = \left(0, \frac{K_3}{r_3}(r_3 - \beta_3 p)\right)$ is globally stable for system (12.7) and (12.8) under the condition (12.9). Also E_3 and $E_2 = \left(\frac{K_2}{r_2}(r_2 - \beta_3 p), 0\right)$ are locally stable under (12.10). Note that under the parameter values for case 1,

$$a_{23} < \frac{1 - \beta_2 p/r_2}{1 - \beta_3 p/r_3} \quad \text{and} \quad a_{32} > \frac{1 - \beta_3 p/r_3}{1 - \beta_2 p/r_2} \tag{12.11}$$

are satisfied, and f. *polytes* and f. *cyrus* can coexist at the positive equilibrium point when *P. aristolochiae* is absent.

[Result C2-2] Now let us choose the parameter values for case 2 as $r_1 = 0.5$, $r_2 = 1.0, r_3 = 1.0, K_1 = 50, K_2 = 50, a_{23} = 1, a_{32} = 0.3$, $p = 0.5, \alpha = 0.6, \beta_2 = 0.8, \beta_3 = 0.7, n_{10} = 0.01, n_{20} = 1, and\ n_{30} = 34$. This case corresponds to an expectation of the population dynamics of butterflies after introducing the model *P. aristolochiae* into Taketomi-jima Is. The different choices between this and case 1 are $r_2 = r_3 = 1.0$, $\beta_2 > \beta_3$, and the smaller initial value for f. *polytes*, since the above parameters satisfy condition (12.9) and only f. *cyrus* remains when *P. aristolochiae* is absent. Figure 12.6a shows that the introduction of the model yields the stable coexistence of *P. aristolochiae*, f. *polytes*, and f. *cyrus*. Further Fig. 12.6b shows the positive dependence of the RA and the AI.

[Result C2-3] The above results imply that the qualitative properties obtained for case 1 still hold true for case 2.

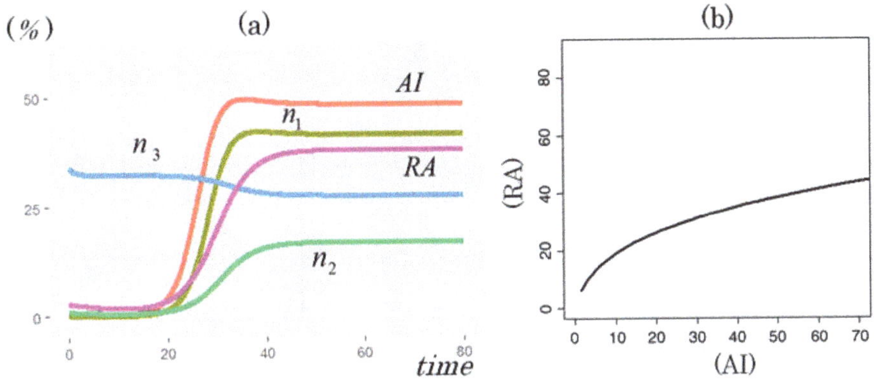

Fig. 12.6 Simulation results on temporal changes of all related quantities
(**a**) The *blue line* denotes the population density n_3 of the non-mimic f. *cyrus*, the *green line* is the density n_2 of the mimic f. *polytes*, the *purple line* is the RA, the *orange line* is the AI, and the *ocher line* denotes the density n_1 of the model *P. aristolochiae*, respectively. Parameter values used in numerical simulations are as follows, and the stability condition for the positive equilibrium is satisfied with these parameter values:
$r_1 = 0.5, r_2 = 1.0, r_3 = 1.0; K_1 = 50, K_2 = 50; a_{23} = {}^{23}1, a_{32} = 0.3; p = 0.5; \alpha = 0.6, \beta_2 = 0.8,$ $\beta_3 = 0.7; n_{10} = 0.01, n_{20} = 1, n_{30} = 34$ (initial values)
(**b**) Numerical simulation results of the positive dependence of the RA on the AI. Parameter values used in the numerical simulation are all the same as in Fig. 12.6a

12.4.1.3 Case 3: $r_2 = r_3$ and $\beta_2 = \beta_3$ $(=\beta)$

Now consider case 3. We adopt the parameter values $a_{23} = 1.1$, $a_{32} = 0.1$, $\beta = 1$ and the same values for the remaining as case 2. Similar case 2, the parameters satisfy condition (12.9). Figure 12.7 shows the similar property as Fig. 12.6.

Finally, we note mathematical results in a compact way in Sect. 12.4 as follows, Let us define the right-hand sides of the inequality in (12.9), (12.10), and (12.11) as

$$a_{23}^c = \frac{1 - \beta_2 p/r_2}{1 - \beta_3 p/r_3} \quad \text{and} \quad a_{32}^c = \frac{1 - \beta_3 p/r_3}{1 - \beta_2 p/r_2}. \tag{12.12}$$

It is easy to check that $a_{23}^c = a_{32}^c = 1$ for case 3 and $a_{23}^c < 1, a_{32}^c > 1$ for case 1 and 2.Since the condition $a_{23} > a_{23}^c$ implies that f. *polytes* goes extinct under the competition with f. *cyrus* $(n_2(t) \to 0)$ when *P. aristolochiae* is absent, both cases 1 and 2 enlarge the possibility for the extinction of f. *polytes* when *P. aristolochiae* is absent. The results obtained in Sect. 12.4 show that f. *polytes* can coexist with f. *cyrus* under the invasion of *P. aristolochiae* even f. *polytes* has disadvantageous property as case 1 (relatively small intrinsic growth rate) or case 2 (relatively small survival rate under that predation). Note that the same result can be obtained the case where $\frac{\beta_2}{r_2} > \frac{\beta_3}{r_3}$.

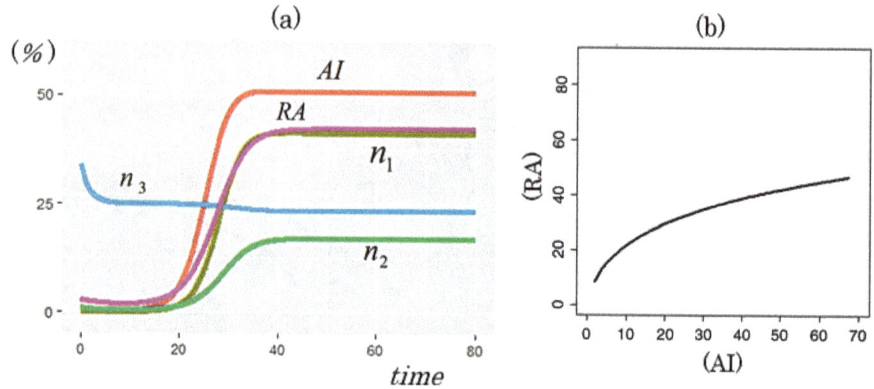

Fig. 12.7 Simulation results on temporal changes of all related quantities
(**a**) The *blue line* denotes the population density n_3 of the non-mimic f. *cyrus*, the *green line* is the density n_2 of the mimic f. *polytes*, the *purple line* is the RA, the *orange line* is the AI, and the *ocher line* denotes the density n_1 of the model *P. aristolochiae*, respectively. Parameter values used in numerical simulations are as follows, and the stability condition for the positive equilibrium is satisfied with these parameter values:

$r_1 = 0.5, r_2 = 1.0, r_3 = 1.0; K_1 = 50, K_2 = 50; a_{23} = {}^{23}1.1, a_{32} = 0.1; p = 0.5; \alpha = 0.6, \beta_2 = \beta_3 = 1.0; n_{10} = 0.01, n_{20} = 1, n_{30} = 34$ (initial values)
(**b**) Numerical simulation results of the positive dependence of the RA on the AI. Parameter values used in the numerical simulation are all the same as in Fig. 12.7a

12.5 Summary and Discussions

Based on new experimental results, we presented an extension of the model for population dynamics of the mimetic butterfly *P. polytes* in Sakishima Islands, Japan (Sekimura et al. 2014). We introduced here difference in the predation rate between two female forms f. *polytes* and f. *cyrus* by parameters β_2, β_3, respectively. The new model system (12.1), (12.2), and (12.3) still includes three major effects: (a) self-density effect by carrying capacity; (b) mimetic effect, that is, the probability that a predator attacks prey on an encounter is proportional to the relative frequency of the mimic among the model and the mimic; and (c) intraspecific competition between two forms f. *polytes* and f. *cyrus* for resources such as nectar and, indirectly, the male. As to growth rate and predation rate, we took account two cases ($r_2 < r_3$) and ($r_2 = r_3$) into consideration and assumed the inequality $\alpha < \beta_2, \beta_3$ among predation rates of the model *P. aristolochiae*, f. *polytes*, and f. *cyrus*, respectively.

Using mathematical analysis and computer simulations of the system equations, we extended the possibility of the logical relationship on the population dynamics of the mimetic butterfly *P. polytes* in Sakishima Islands. In particular, we discussed both temporal change in the relative abundance (RA) and variation in the RA in Sakishima Islands. We estimated conditions for existence of equilibrium solutions of butterfly populations by making a comparison between experimental results and mathematical analyses of the model equations. Our results show that one of key factors to understand field data is the carrying capacity K_1 of the model in each island. The positive dependence of the RA on the AI originates from the result that changes of both the relative abundance (RA) and the advantage index (AI) with respect to the carrying capacity K_1 are positive.

The results in Sect. 12.4 have shown that both cases with respect to production rate: ($r_2 < r_3$) and ($r_2 = r_3$) could be possible to reproduce experimental data on the population dynamics of the mimetic butterfly *P. polytes* in Sakishima Islands. The first case: ($r_2 < r_3$) means that the mimicry of *P. polytes* requires a kind of genetic change in production rate of the mimetic form f. *polytes* to reproduce the data. On the other hand, the second case ($r_2 = r_3$) means that in order to reproduce the data, changes in ecological factors such as intraspecific competition coefficients (a_{23}, a_{32}) and predation rate β_2, β_3 are required for the mimicry of *P. polytes* without any genetic change. We think that it is not enough to determine at the moment which case is the real case that occurred in Sakishima Islands, because experiments on butterflies in both the field and the laboratory are somewhat subtle. We hope that much reliable experiments will be done to understand the reality in the future.

Finally, we hope that the mathematical analysis and computer simulations in the paper provide the theoretical basis on the female limited polymorphism of *P. polytes* in Sakishima Islands.

References

Clarke CA, Sheppard PM (1972) The genetics of the mimetic butterfly *Papilio Polytes* L. Proc R Soc Lond B 263:431–458

Kinjyou A (2000) Mimetic relationship in swallowtail butterflies in the Ryukyus. Nat Insects 36 (12):24–27 (In Japanese)

Ohsaki N (1995) Preferential predation of female butterflies and the evolution of batesian mimicry. Nature 378:173–175

Ohsaki N (2005) A common mechanism explaining the evolution of female-limited and both-sex Batesian mimicry in butterflies. J Anim Ecol 74:728–734

Sekimura T, Fujihashi Y, Takeuchi Y (2014) A model for population dynamics of the mimetic butterfly *Papilio polytes* in the Sakishima Islands, Japan. J Theor Biol 361:133–140

Uesugi K (1991) Temporal changes in records of the mimetic butterfly *Papilio polytes* with establishment of its model *Pachiliopta aristolochiae* in the Ryukyu Islands. Jpn J Ent 59 (1):183–198

Uesugi K (1992) Polymorphism of the mimetic butterfly *Papilio Polytes*, L. Insectalium 22:4–10. (In Japanese)

Uesugi K (1996) The adaptive significance of Batesian mimicry in the butterfly, *Papilio Polytes* (Insecta, Papilionidae): Associative learning in a predator. Ethology 102:762–775

Uesugi K (1997) Iden 51(2):68–71 (In Japanese)

Yamauchi A (1994) A population dynamic model of the Batesian mimicry. Res Popul Ecol 53:295–315

13

Tribe Junoniini and Seasonal Polyphenism

Jameson W. Clarke

Abstract Seasonal polyphenism in insects is the phenomenon whereby multiple phenotypes can arise from a single genotype depending on environmental conditions during development. Many butterflies have multiple generations per year, and environmentally induced variation in wing color pattern phenotype allows them to develop adaptations to the specific season in which the adults live. Elements of butterfly color patterns are developmentally semiautonomous allowing for detailed developmental and evolutionary changes in the overall color pattern. This developmental flexibility of the color pattern can result in extremely diverse seasonal phenotypes in a single species. In this study, we asked the following questions: (a) How do wing phenotype elements such as shape and pattern vary between seasonal forms? (b) Can this variation be explained phylogenetically? (c) If so, what are the various pattern development strategies used to achieve crypsis in the dry season form? To answer these questions, we used high-resolution images to analyze pattern element variation of 34 seasonally polyphenic butterfly species belonging to the tribe Junoniini (Lepidoptera: Nymphalidae). We show that forewing shape and eyespot size both vary seasonally and that the methods by which phenotype elements change in the dry season forms are different in different clades and may therefore have independent and diverse evolutionary origins.

Keywords Polyphenism • Seasonal polyphenism • Shape polyphenism • Color pattern • Pattern evolution • Pattern element • Junoniini • *Junonia* • *Precis*

13.1 Introduction

Seasonal polyphenism in insects is the phenomenon whereby multiple phenotypes can arise from a single genotype depending on environmental conditions during development. Brakefield and Shapiro more formally define seasonal polyphenism

---

J.W. Clarke (✉)
Department of Biology, Duke University, Durham, NC 27705, USA
e-mail: jameson.clarke@duke.edu

as the expression of a repeating pattern of changing phenotypes under the control of some environmental factor (Brakefield et al. 1996). Many butterflies have multiple generations per year, and environmentally induced variation in wing color pattern allows them to develop adaptations to the specific season in which the adults live. For example, when predators and prey are both plentiful during the spring or wet season, large striking eyespots may serve to deter or deflect predators (Prudic et al. 2015). In contrast, during the autumn or dry season when prey are scarce, having large striking eyespots might increase the chances of being detected by a predator, and benefit may be obtained by a more cryptic coloration (Brakefield and Larsen 1984).

Because it is potentially adaptive to have specialized forms for predictable environmental heterogeneity, seasonal polyphenism is not uncommon and is most often seen in families Hesperiidae, Lycaenidae, Pieridae, and of course Nymphalidae (Brakefield and Larsen 1984). The elements of butterfly color patterns are developmentally semiautonomous allowing for detailed developmental and evolutionary changes in the overall color pattern (Nijhout 1991). This developmental flexibility of the color pattern can result in extremely diverse seasonal phenotypes within and among species. Seasonal forms of some species, such as *Precis octavia*, can be so different that they were thought to be a distinct species prior to laboratory experiments that demonstrated that alternative color patterns could be induced by rearing the larvae under varying conditions of temperature and photoperiod (McLeod 1968).

Although the genetic, developmental, and hormonal control of seasonal polyphenism are becoming increasingly understood, there are relatively few studies that examine the evolution of the pattern elements of seasonal forms (Rountree and Nijhout 1995; Monteiro et al. 2015; Oostra et al. 2011). Therefore we asked the following questions: (a) How do wing phenotype elements such as shape and pattern differ between seasonal forms? (b) Can this variation be explained phylogenetically? (c) If so, what are the various pattern strategies used to achieve crypsis in the dry season form? To answer these questions, we analyzed pattern element variation of 34 seasonally polyphenic butterfly species belonging to tribe Junoniini.

The tribe Junoniini (Lepidoptera: Nymphalidae: Nymphalinae) is one of six major tribes in the subfamily Nymphalinae and is comprised of 85 species in 6 genera: *Hypolimnas* (26 spp.), *Precis* (17 spp.), *Salamis* (3 spp.), *Yoma* (2 spp.), *Protogoniomorpha* (2 spp.), and *Junonia* (35 spp.) (Kodandaramaiah 2009). The tribe is estimated to have evolved approximately 30–40 million years ago originating in Africa and spreading throughout Asia and Oceania primarily (Wahlberg et al. 2005; Kodandaramaiah and Wahlberg 2007). Most of the genera diverged approximately 25 million years ago with the exception of *Yoma* and *Protogoniomorpha* which split approximately 5 million years later (2006). Species belonging to this tribe are noted for being swift fliers, having medium to large body sizes, and exhibiting striking polyphenic forms such as the model organism *Junonia coenia* making the tribe an ideal target taxon for this study (Win et al. 2016).

We show that forewing shape and eyespot size both vary seasonally in the Junoniini and that the methods by which phenotype elements change in the dry season forms are different in different clades and may therefore have independent and diverse evolutionary origins.

13.2 Methods

Phenotypic variation in the seasonal forms of the Junoniini was measured and assessed using high-resolution digital images and image processing software. Specimens were acquired from the Museum of Natural History in London and the Smithsonian National Museum of Natural History. Images received from the London Museum were on 35 mm photographic film and were digitized using an EPSON Perfection V600 flatbed digital scanner. Images of the Smithsonian specimens were captured at a fixed distance from the specimen using a Nikon CoolPix P600 16.1 megapixel digital camera. All images were taken with a millimeter ruler for scale.

Specimens were selected for analysis to maximize phylogenetic coverage of tribe Junoniini by including representatives from each of the major genera (*Hypolimnas*, *Junonia*, *Precis*, *Protogoniomorpha*, *Salamis*, and *Yoma*) as well as two out-groups (*Anartia* and *Kallimoides*). Selected Junoniini specimens were labeled and organized into the following functional groups corresponding to major clades within the tribe's phylogeny: upper *Junonia*, Asian *Junonia*, lower *Junonia*, *Yoma*, *Precis*, and *Hypolimnas* (Fig. 13.1a). The lower *Junonia* was treated as a clade for comparative purposes despite its status as an unresolved polytomy. Species were selected that had at least four replicates of comparable quality. To eliminate confounding effects of regional variation and sexual dimorphism, specimens were sexed and selected from the same geographic region. If the species had distinct seasonal phenotypes, these were confirmed by the date of collection and comparison to previous descriptions in the literature. The resulting sample was comprised of 34 species from Africa, Asia, Oceania, and North America.

Size and position of pattern elements of the ventral hind wing were measured by generating landmark points in the image processing software ImageJ v1.51g (Schneider et al. 2012). The ventral surface was chosen because butterflies most often have their wings folded showing the ventral pattern when not in flight. The hind wing was chosen to maximize the amount of visible wing surface since it overlaps in front of the forewing when viewed from the ventral side of the animal. For each image, landmark distances were converted from pixels to millimeters using a 20 mm scale. To normalize the size and relative distances between pattern elements according to the size of the wing on which they appear, a proxy for wing size was generated using the perimeter of a triangle formed using the distances between the root node of the venation system, the Cu2 vein terminus, and the Rs vein terminus, abbreviated RCR triangle (Fig. 13.1b – e).

For the Sc+R1, Rs, M1, M2, M3, and Cu1 cells of the ventral hind wing, the following measurements were taken:

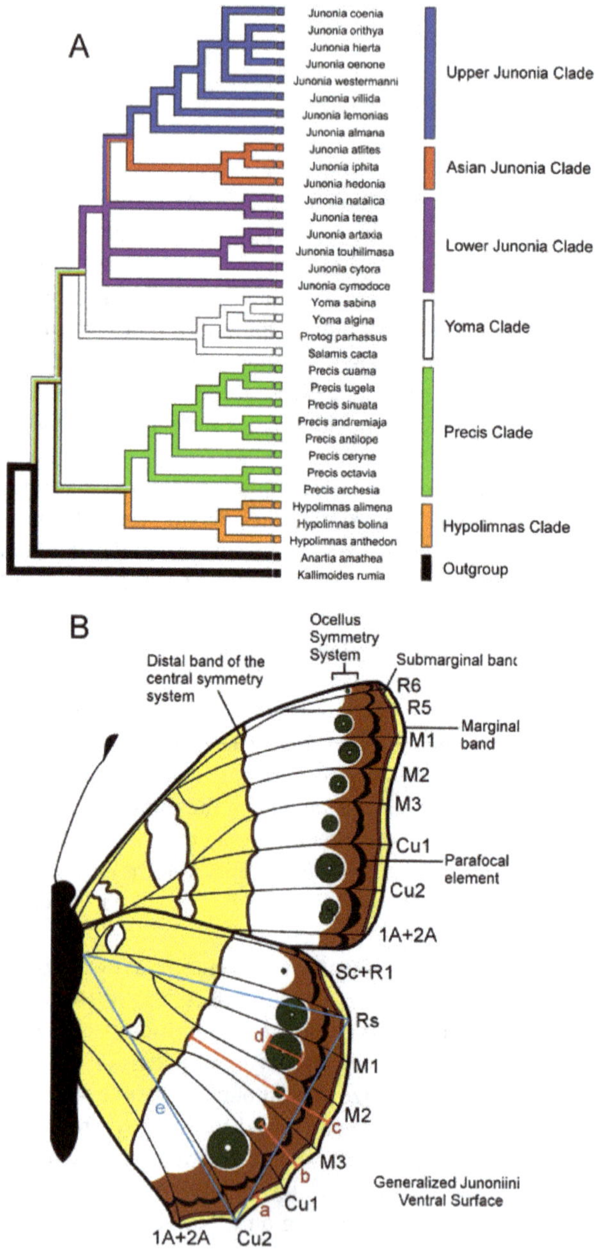

Fig. 13.1 (**a**) Phylogeny of the Junoniini (Lepidoptera: Nymphalidae: Nymphalinae) species used in this study grouped by reference clade. Tree topology based on Kodandaramaiah and Wahlberg (2007). (**b**) Diagram of measurements on a generalized Junoniini ventral wing surface including (*a*) submarginal band proximity [NSP], (*b*) eyespot proximity [NEP], (*c*) central symmetry system

Fig. 13.2 Qualitative categorization of four forewing apex shape classifications observed in the dry season form. Note the increasing acuteness of the angle formed by the margins immediately flanking the M1 terminus as the apex increases in angularity from low to high. Note also that the M1 terminus angle is not present in the leaflike morphotype

(a) [NSP] normalized submarginal band proximity – the distance between the wing margin and the submarginal band divided by the RCR triangle (Fig. 13.1b – a)
(b) [NEP] normalized eyespot proximity – the distance between the wing margin and the focus of the eyespot divided by the RCR triangle (Fig. 13.1b – b)
(c) [NCP] normalized central symmetry system proximity – the distance between the wing margin and the distal band of the central symmetry system divided by the RCR triangle (Fig. 13.1b – c)
(d) [NED] normalized eyespot diameter – the longest distance that intersects with the focus of an eyespot between the distal and the proximal borders of the eyespot divided by the RCR triangle (Fig. 13.1b – d)

The morphology of the forewing apex was categorized into four classifications: leaflike, low angularity, medium angularity, and high angularity (Fig. 13.2). To

Fig. 13.1 (continued) proximity [NCP], and (*d*) eyespot diameter [NED]. Measurements were size-normalized using (*e*) the perimeter of a triangle connecting the Rs and Cu2 termini to the root of the venation system [RCR triangle]. (Tree topology from Kodandaramaiah and Wahlberg 2007)

ensure the consistency of the shape classifications, specimens were independently classified by an outside researcher and by the authors. Each classification converged on the same subdivision of morphotypes.

All measurements, including absolute differences in measurements for each seasonal form, were entered into a character matrix using the software suite Mesquite version 3.04 (Maddison and Maddison 2015). To contrast pattern element data between seasonal forms, parsimony character state reconstructions were mapped to an existing tree topology based on a molecular phylogeny for tribe Junoniini (Kodandaramaiah and Wahlberg 2007) and mirrored to draw comparisons between seasonal forms.

13.3 Results

13.3.1 Variation by Pattern Element

Variation in pattern elements was dependent on the type of pattern element being measured. The normalized submarginal band proximity (NSP) to the margin of the wing did not vary significantly for any wing cell between seasonal forms or across clades. The normalized central symmetry system proximity (NCP) to the wing margin and the normalized eyespot proximity (NEP) to the wing margin did not vary significantly for any wing cell between seasonal forms, but did show clear differences between clades. Finally, the normalized eyespot diameter (NED) varied significantly between seasonal forms and also showed clear differences between clades. It should also be noted that it is likely that the parafocal element also varies between seasonal forms, but difficulty in consistently defining the boundaries of this pattern element led to its omission from this study.

13.3.2 Variation by Wing Cell

Variation in pattern elements was also dependent on the wing cell in which the pattern elements are located. There was little variation in NED, NSP, NCP, or NEP for wing cells Sc+R1, M2, and M3 between seasonal forms for all clades because these eyespots are typically reduced or absent in ventral hind wings in both the wet and dry season forms. In contrast, the Rs, M1, and Cu1 wing cells showed significant differences in NED and NCP across clades, but only NED varied significantly between seasonal forms. *Junonia almana* (Fig. 13.3) provides a clear example of the differences in seasonal eyespot size variation across wing cells. In this species, the Sc+R1 eyespot is absent in both seasonal forms. The M2 and M3 eyespots are absent in the wet season form, but are present and highly reduced in the

Fig. 13.3 *Junonia almana* seasonal forms. (*Left*) Wet season form exhibiting regions of high color contrast, large well-defined eyespots, and a low-angularity apex shape. (*Right*) Dry season form exhibiting low color contrast throughout the entire wing, drastically reduced eyespots, and a high-angularity forewing apex shape

dry season form. Finally, the Rs, M1, and Cu1 eyespots are large in the wet season form and highly reduced in the dry season form (Fig. 13.3).

13.3.3 Seasonal Eyespot Variation by Clade

Parsimony analysis of seasonal eyespot variation resulted in three similar parsimony character state reconstructions for the Rs, M1, and Cu1 eyespots when mapped to the molecular phylogeny. (Figs. 13.4, 13.5, and 13.6). However, the seasonal eyespot variation for each wing cell was not consistent across clades.

All species in the upper *Junonia* clade exhibited high seasonal eyespot size variation for the Rs, M1, and Cu1 eyespots with two exceptions: *J. westermanni* has no variation in the M1 eyespot (Fig. 13.5) and *J. hierta* in the Cu1 eyespot (Fig. 13.6).

The lower *Junonia* clade varied greatly both across species and by wing cell for seasonal eyespot size. For the Rs eyespot, most species showed no seasonal size variation except slightly in *J. cytora* and *J. touhilimasa* (Fig. 13.4). However, for the M1 and Cu1 eyespots, all of the species showed some seasonal size variation, though to different degrees, with the single exception of *J. cytora* in the Cu1 cell (Figs. 13.5 and 13.6).

The *Yoma* clade also exhibited a wide range of seasonal eyespot size variation. *Yoma algina* showed large seasonal eyespot size variation for the Rs, M1, and Cu1 eyespots, while its sister taxon *Y. sabina* showed only minimal variation in the Rs and M1 eyespots and no variation in the Cu1 eyespot. *Protogoniomorpha parhassus* exhibited seasonal eyespot size variation in the M1 and Cu1 eyespots but not the Rs eyespot. Finally, *Salamis cacta* showed no seasonal eyespot size variation whatsoever (Figs. 13.4, 13.5, and 13.8).

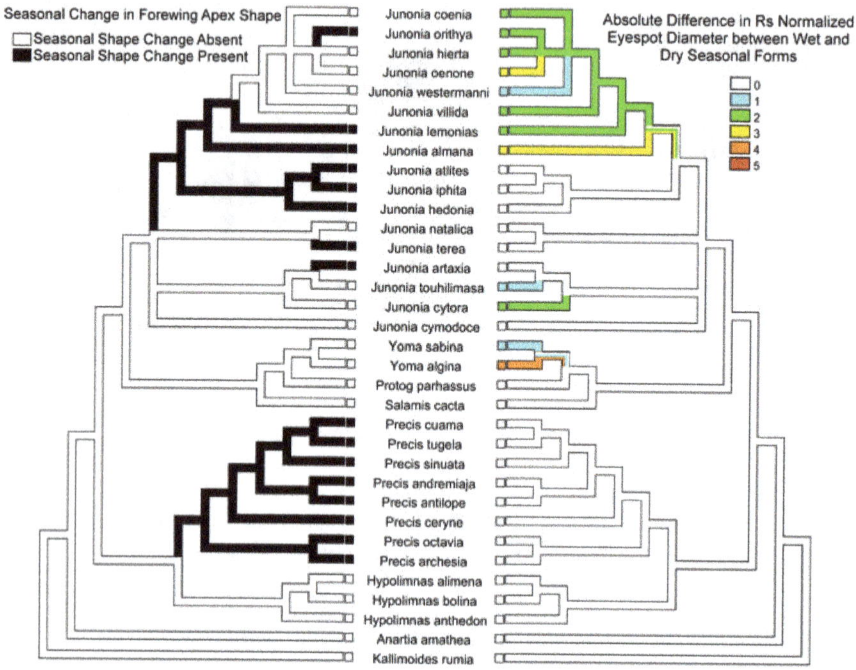

Fig. 13.4 Mirrored parsimony character state reconstructions for seasonal change in forewing apex shape class and the absolute difference in Rs normalized eyespot diameter between wet and dry season forms. Warmer colors represent a larger disparity between eyespot diameters of wet and dry forms (Tree topology from Kodandaramaiah and Wahlberg 2007)

Finally, the *Precis* clade, the *Hypolimnas* clade, and the Asian *Junonia* showed almost no seasonal eyespot size variation for any eyespot with one exception in *H. anthedon* which had the unusual characteristic of having no eyespots in (Figs. 13.4, 13.5, and 13.6).

13.3.4 Seasonal Forewing Apex Shape Change by Clade

The shape of the forewing apex varied both seasonally and by clade. The upper *Junonia* clade exhibited a mix of seasonal variation with shape change present only in *J. orithya*, *J. lemonias*, and *J. almana*. Similarly, in the lower *Junonia* clade, only two species, *J. terea* and *J. artaxia*, showed seasonal shape change of the forewing. The remaining clades, however, do not have this mix of shape change between seasonal forms. Neither the *Yoma* nor the *Hypolimnas* clades showed seasonal shape change. In contrast, both the Asian *Junonia* and *Precis* clades showed seasonal shape change in every species sampled (Fig. 13.7).

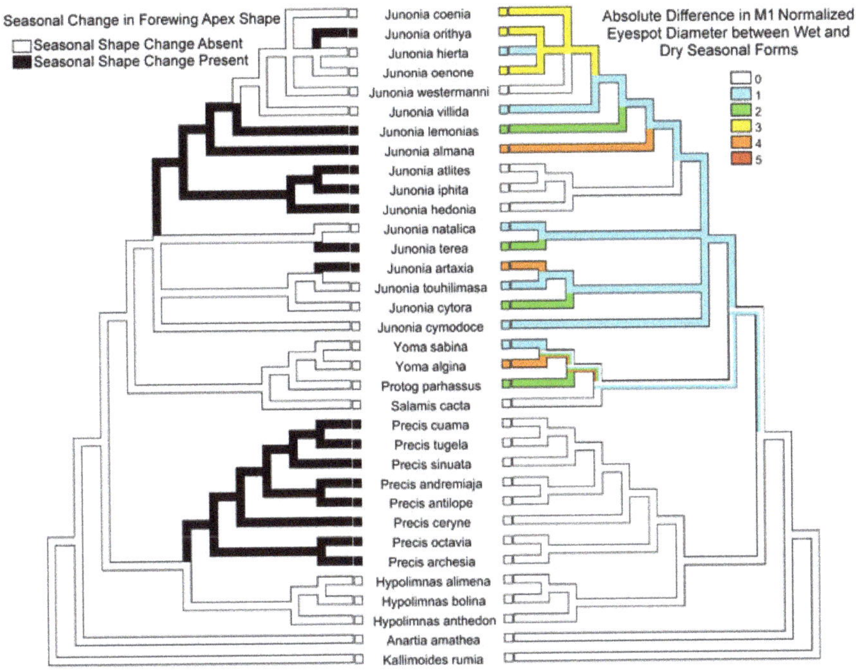

Fig. 13.5 Mirrored parsimony character state reconstructions for seasonal change in forewing apex shape class (*left*) and the absolute difference in M1 normalized eyespot diameter between wet and dry season forms (*right*). Warmer colors represent a larger disparity between eyespot diameters of wet and dry forms (Tree topology from Kodandaramaiah and Wahlberg 2007)

13.3.5 Shape Type and Shape Change

There exists an association between the shape of the forewing apex of the dry season form of a species and whether or not there was seasonal shape variation for that species. Species whose dry season form had a low angularity or leaflike forewing apex invariably did not exhibit seasonal forewing shape change (Fig. 13.7). Furthermore, species whose forewing apex varied seasonally did so according to a pattern of increasing angularity in the dry season form compared to the wet season form (Figs. 13.7 and 13.8 – bottom). Species with seasonal shape change whose wet season form had a low-angularity forewing apex had dry season forms with medium- or high-angularity forewing apex shapes. Similarly, species with seasonal shape change whose wet season form had a medium-angularity forewing apex had dry season forms with high-angularity forewing apex shapes. The result of this association is the general trend that species that exhibit seasonal change in forewing shape tend to have more high-angularity forewing morphologies, while species that do not exhibit seasonal change tend to have low-angularity forewing morphologies.

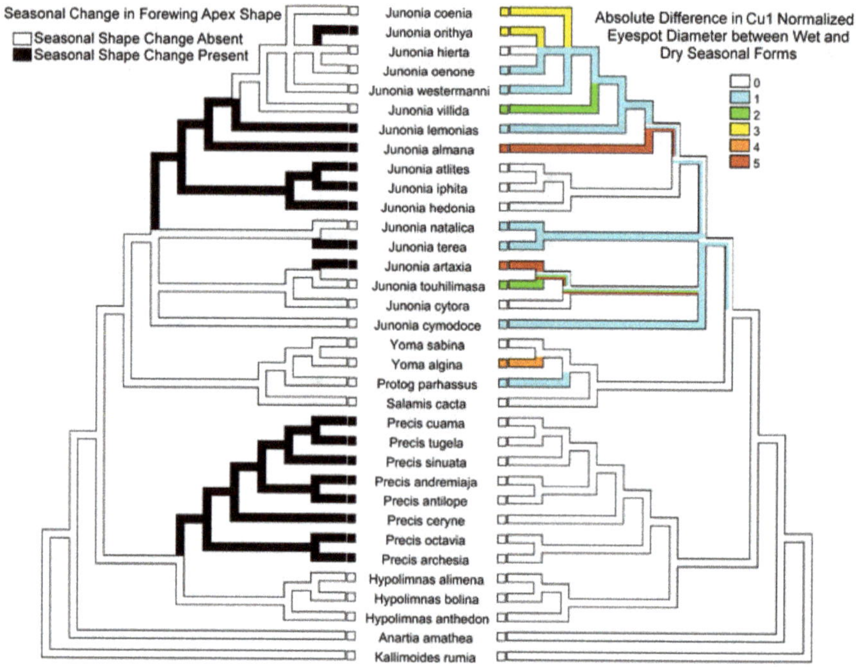

Fig. 13.6 Mirrored parsimony character state reconstructions for seasonal change in forewing apex shape class (*left*) and the absolute difference in Cu1 normalized eyespot diameter between wet and dry season forms (*right*). Warmer colors represent a larger disparity between eyespot diameters of wet and dry forms (Tree topology from Kodandaramaiah and Wahlberg 2007)

13.3.6 Discussion

The results of this investigation reveal a dynamic relationship between the components of a seasonally polyphenic phenotype and the pattern by which they evolve. They show that the phenotype elements of a seasonal form can be evolutionarily decoupled, that these plastic elements have responded to selective pressures in diverse ways, and that the diversity of these responses is reflected in the tribe's phylogenetic history. The different pattern elements of seasonally polyphenic forms do not evolve as a single cohesive unit.

The methods by which the seasonal forms have evolved in these butterflies also offer insight about how pattern elements are involved in responding to selection for crypsis in the dry season. The findings of this study suggest that the location of pattern elements on the wing does not change in the formation of alternative seasonal phenotypes. This suggests that either selection is not acting strongly on position or that element position is under some developmental constraint and therefore unable to respond to selective pressures. When considering the variation in the position of eyespots specifically, it seems more likely to be a lack of selection on position given that eyespot position is characteristically different among clades,

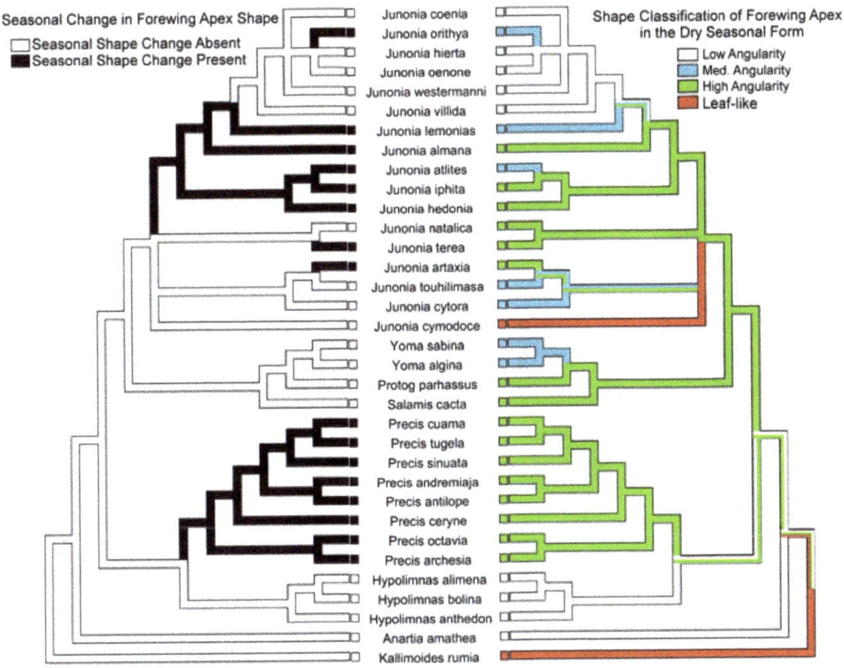

Fig. 13.7 Mirrored parsimony character state reconstructions for seasonal change in forewing apex shape class and the shape classification of the forewing apex in the dry season form. Note the association between exhibiting seasonal shape change and having higher forewing apex angularity in the dry season form (Tree topology from Kodandaramaiah and Wahlberg 2007)

implying at least some freedom from constraint during the early evolution of the major Junoniini clades.

However, the variation in eyespot size tells us a different story. In species where eyespot size is variable, the dry season form always has reduced eyespot size compared to the wet season form, suggesting a strong response to selection. At the same time, there is little variation in the wing cells in which large eyespots develop. For instance, the upper and lower *Junonia* clades all have stable large eyespots in the Rs, M1, and Cu1 wing cells and stable small eyespots in the M2 and M3 wing cells in the wet season form, whereas species in the *Precis* clade have stable small eyespots in all of the wing cells. There is no case where stable large eyespots have evolved in the M2 and M3 wing cells. Thus there appear to be constraints both on the capacity to have size-variable eyespots and on the wing cell in which the eyespot is found (Figs. 13.4 and 13.8 – bottom).

Similarly, in clades that have seasonal variation in wing shape, the dry season form always has a more angular or falcate forewing shape, again suggesting a strong response to selection (Figs. 13.7 and 13.8 – bottom). Other clades, by contrast, have evolved phenotypically stable low-angularity rounded (*Hypolimnas* clade) or high-angularity falcate (*Yoma* and Asian *Junonia* clades) wing shapes.

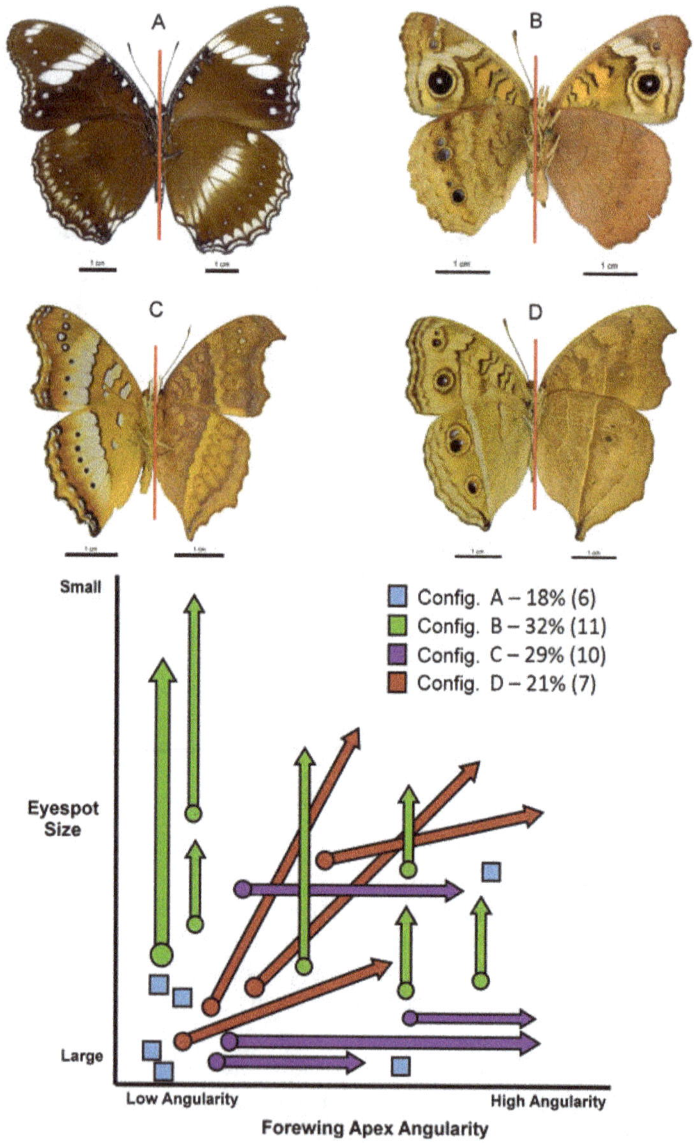

Fig. 13.8 *Top*: Plates representing four configurations found among Junoniini butterflies – (**a**) Neither forewing shape nor eyespot size vary seasonally; (**b**) Only eyespot size varies seasonally; (**c**) Only forewing shape varies seasonally; (**d**) Both forewing shape and eyespot size vary seasonally. Each plate is a composite of the wet season form (*left*) and the dry season form (*right*). *Bottom*: Schematic diagram of the relationship between seasonal changes in forewing shape and eyespot size. Wet season forms are shown as *circles*, dry season forms as *arrows*, and when the two forms overlap they are shown as *squares*. Note that the change for forewing shape is always from lower to higher angularity, and the change in eyespot size is always a reduction

Thus the Junoniini show both plastic alternative wing shape phenotypes and evolutionary fixation of the alternative shapes, and the fixation of either the low- or high-angularity wing shapes corresponds to major clade divergences in the phylogeny. This suggests that while the capacity to develop alternative wing shapes evolved early for the entire tribe, the fixation of alternative wing shapes occurred later in the establishment of the genera.

The decoupling of independent phenotype elements has allowed for the emergence of what we loosely refer to here as "phenotypic configurations" which can be thought of as arrangements of phenotype element variation. To illustrate this idea, we used only the combination of eyespot size variation and forewing apex shape variation which can be roughly categorized into the four configurations that represent butterflies whose seasonal forms exhibit no variation, size variation only, shape variation only, or both size and shape variation: Configurations A, B, C, and D, respectively (Fig. 13.8 – top).

The distribution of these configurations in the phylogeny shows some interesting patterns. First, each of the configurations corresponds to the clades described in Fig. 13.1 – top: the *Hypolimnas* clade shows almost no seasonal change in wing shape or eyespot size (Figs. 13.4 and 13.8 – top: Configuration A); the upper *Junonia* exhibit minimal seasonal change in wing shape, but exhibit great seasonal change in eyespot size (Figs. 13.4 and Fig. 13.8 – top: Configuration B); the inverse is true for the *Precis* and Asian *Junonia* clades whose members show almost no seasonal change in eyespot size, while they all exhibit seasonal change in forewing apex shape (Figs. 13.4 and Fig. 13.8 – top: Configuration C). Second, the relative frequency of these configurations and their position in the phylogeny show that species whose wing shape varies seasonally tend not to be the same species whose eyespot size varies seasonally, with a few exceptions. A highly angular wing shape in the dry season form seems to have evolved very early in the tribe but was lost independently in both *Hypolimnas* and upper *Junonia* (Fig. 13.4 right). Interestingly, the capacity to develop plastic seasonally distinct wing shapes may have evolved around the same time as the falcate wing shape but saw successive loss in *Hypolimnas*, *Yoma*, and some species in lower and upper *Junonia* (Fig. 13.4 left). An alternative interpretation might be that the capacity for wing shape change evolved multiple times – once in the *Precis* clade and once in genus *Junonia* with only the latter having some subsequent loss of the trait (Fig. 13.4 left).

Finally, the leaflike wing shape seems to have evolved independently with respect to the well-known leaf mimics of the genus *Kallimoides*. This raises some interesting questions regarding the importance of the phenotypic elements for crypsis. If a species can achieve crypsis by either reducing its eyespots or changing its wing shape from season to season, why do one or the other? Is the plasticity of one phenotype element enough to render the plasticity of another unnecessary? What, then, of species who have variation in both or neither of these elements?

Another interesting question regarding the interplay between phenotypic elements and their evolutionary trajectory is the role of color and contrast of the wing pattern. In the same way, the conspicuousness of an eyespot can be diminished by reducing its size; it can also be recolored or recontrasted to match the surrounding region of the wing, which effectively achieves the same result. This is the case in *Precis atlites*, whose eyespots remain the same in size but become less bold and more similar in color to the background of the wing rendering them more difficult to detect. Although seasonal changes in color and contrast are widespread in the tribe, that is to say all of the dry season forms become duskier in color and less striking in the boldness of their patterns, it is unclear as to what extent changing color and contrast compensate for the inability to modify either wing shape or eyespot size. A detailed analysis of these elements will be presented in a separate paper.

The seasonally polyphenic forms of butterflies are often thought of as a single trait. In reality, because butterfly wing patterns are comprised of serially homologous phenotypic elements that are developmentally semiautonomous, they can be uncoupled, modified, and reconfigured to respond to selection and produce constraints in diverse ways. The seasonal forms of Junoniini butterflies have changed over time by invoking at least three distinct developmental mechanisms, including wing shape morphogenesis, pigment synthesis pathways, and pattern element positioning mechanisms. Rather than inheriting a seasonal form, these butterflies inherit the tools to create a seasonal form, and the methods by which they have convergently evolved to become cryptic are written in their evolutionary history.

Acknowledgments I am grateful to the Smithsonian Institution National Museum of Natural History and the National History Museum in London for access to their butterfly collections, Professor H.F. Nijhout for his direction and guidance, Richard Gawne and Kenneth McKenna for their helpful discussion and criticism, Leo Kerner for assistance in image databasing, the anonymous reviewers for their criticism and feedback, and the support of the following National Science Foundation grants awarded to H.F. Nijhout: IOS-0641144, IOS-1557341, and IOS-1121065.

References

Brakefield PM, Larsen TB (1984) The evolutionary significance of dry and wet season forms in some tropical butterflies. Biol J Linnean Soc 22(1):1–12

Brakefield PM, Gates J, Keys D, Kesbeke F, Wijngaarden PJ, Monteiro A, French V, Carroll SB (1996) Development, plasticity and evolution of butterfly eyespot patterns. Nature 384 (6606):236–242

Kodandaramaiah U (2009) Eyespot evolution: phylogenetic insights from Junonia and related butterfly genera (Nymphalidae: Junoniini). Evol Dev 11(5):489–497

Kodandaramaiah U, Wahlberg N (2007) Out-of-Africa origin and dispersal-mediated diversification of the butterfly genus Junonia (Nymphalidae: Nymphalinae). J Evol Biol 20(6):2181–2191

Maddison WP, Maddison DR (2015) Mesquite: a modular system for evolutionary analysis. Version 3.04. http://mesquiteproject.org

McLeod L (1968) Controlled environment experiments with Precis octavia Cram (Nymphalidae). J Res Lepidoptera 8(2):53–54

Monteiro A, Tong X, Bear A, Liew SF, Bhardwaj S, Wasik BR, Dinwiddie A, Bastianelli C, Cheong WF, Wenk MR, Cao H (2015) Differential expression of ecdysone receptor leads to variation in phenotypic plasticity across serial homologs. PLoS Genet 11(9):e1005529

Nijhout HF (1991) The development and evolution of butterfly wing patterns, Smithsonian series in comparative evolutionary biology. Smithsonian Institution Press, Washington

Oostra V, de Jong MA, Invergo BM, Kesbeke F, Wende F, Brakefield PM, Zwaan BJ (2011) Translating environmental gradients into discontinuous reaction norms via hormone signalling in a polyphenic butterfly. Proc R Soc Lond B Biol Sci 278(1706):789–797

Prudic KL, Stoehr AM, Wasik BR, Monteiro A (2015) Eyespots deflect predator attack increasing fitness and promoting the evolution of phenotypic plasticity. Proc R Soc B 282 (1798):20141531. The Royal Society

Rountree DB, Nijhout HF (1995) Genetic control of a seasonal morph in Precis coenia (Lepidoptera: Nymphalidae). J Insect Physiol 41(12):1141–1145

Schneider CA, Rasband WS, Eliceiri KW (2012) NIH Image to ImageJ: 25 years of image analysis. Nat Method 9(7):671–675

Wahlberg N (2006) That awkward age for butterflies: insights from the age of the butterfly subfamily Nymphalinae (Lepidoptera: Nymphalidae). Syst Biol 55(5):703–714

Wahlberg N, Brower AV, Nylin S (2005) Phylogenetic relationships and historical biogeography of tribes and genera in the subfamily Nymphalinae (Lepidoptera: Nymphalidae). Biol J Linnean Soc 86(2):227–251

Win NZ, Choi EY, Park J, Park JK (2016) Taxonomic review of the tribe Junoniini (Lepidoptera: Nymphalidae: Nymphalinae) from Myanmar. J Asia-Pacific Biodiv 9(3):383–388

14

Male Pipevine Swallowtail Butterflies and their Mating Success

Nayuta Sasaki, Tatsuro Konagaya, Mamoru Watanabe and Ronald L. Rutowski

Abstract Sexual dimorphism in wing coloration is pervasive in butterflies and has been attributed to the process of sexual selection. However, this view has rarely been tested, partly owing to difficulties in estimating the mating success of males in the field. In the present study, we describe a method for assessing the mating success of male pipevine swallowtail (*Battus philenor*) butterflies, based on the appearance of their reproductive tracts. Laboratory experiments indicated that, in response to mating, components of the males' reproductive tracts become shorter, decrease in mass, and change in appearance, irrespective of age; and these changes persist for at least 2 days. Using these indicators of recent mating, we examined the reproductive tracts of 68 field-caught males and found that the color of the dorsal hindwing, a feature that females use in mate choice, was significantly greener in males that had recently mated than in males that had not.

Keywords *Battus philenor* • Ejaculate substance • Male mating success • Ornaments • Sexual selection

---

N. Sasaki (✉)
Field Science Center for Northern Biosphere, Hokkaido University, 053-0035 Takaoka, Tomakomai, Hokkaido, Japan
e-mail: nayutaSSK@gmail.com

T. Konagaya
Graduate School of Science, Kyoto University, 606-8502 Kyoto, Japan
e-mail: konagaya@ethol.zool.kyoto-u.ac.jp

M. Watanabe
Graduate School of Life and Environmental Sciences, University of Tsukuba, 305-8572 Ibaraki, Japan
e-mail: papilio-platycnemis@nifty.com

R.L. Rutowski
School of Life Sciences, Arizona State University, Tempe, AZ 85287-4501, USA
e-mail: R.RUTOWSKI@asu.edu

14.1 Introduction

There is a long history of interest in the diversity of butterfly wing pattern and coloration. Starting with work of Darwin (1874) and Wallace (1889), researchers have observed and discussed various issues related to wing pattern, such as interspecific similarity and variation (Nijhout 1990), ecological relevance (Rutowski 1997), color production mechanisms (Koch et al. 1998), evolutionary and developmental plasticity (Beldade et al. 2002), and genetics (Carroll et al. 1994), as well as intersexual differences. Because male butterflies typically exhibit brighter wing coloration and sometimes exhibit pattern elements that are not found in females, many researchers, including Darwin (1874), have speculated that the coloration of male wings results from female mating preferences associated with exaggerated visual signals (Kemp and Rutowski 2011; but see Allen et al. 2011).

A considerable amount of research has also been motivated by the intersexual variation of butterfly wing coloration; however, relative to other groups of colorful animals, such as birds, fish, and lizards (Blount et al. 2003; Grether et al. 2005; Hill and Montgomerie 1994; Keyser and Hill 1999), relatively little is known about the selective factors that promote the sexual dimorphism of wing color in butterflies (Kemp 2007). This deficit has partly stemmed from the difficulty of setting up the necessary assays of female preference. In addition, since sexual selection ultimately results from biased reproductive success, it is necessary to elucidate the relationship between male traits and reproductive success. However, the highly dispersed, cryptic, and ephemeral nature of butterfly copulation hinders the estimation of male mating success in the field (e.g., Rutowski 1997; Takeuchi 2016).

Here we report a new technique to assess the recent mating success of males in Lepidoptera that relies on changes that occur in the appearance of internal reproductive organs during mating. In some lepidopteran species, males transfer an ejaculate to females that can account for as much as 15% of the male body mass (e.g., Rutowski et al. 1983; Svärd and Wiklund 1989), and in many species, males can produce more than one spermatophore; however, it takes time for males to produce an ejaculate that is comparable in size to the one transferred during the previous mating (Bissoondath and Wiklund 1996; Watanabe and Hirota 1999). Therefore, the internal reproductive organs of mated males might differ in size, contents, or appearance from those of unmated males, at least for a few days after mating.

The typical arrangement of internal reproductive organs in male butterflies includes two fused testes that give rise to a pair of vas deferens, which are secretory ducts, that lead to the duplex, which is a pair of sperm storage organs. Two duplex ducts unite caudally to form the simplex, which is a single duct that leads to the intromittent organ, or aedeagus. Sperm move from the testes into the duplex via the vas deferens (Riemann et al. 1974; LaChance et al. 1977). Due to the arrangement of the reproductive organs in the male body (c.f. Fig. 14.1), the spermatophore materials and accessory substances in the simplex are transferred to the female body during mating before the sperm are transferred (Watanabe and Sato 1993). Thus,

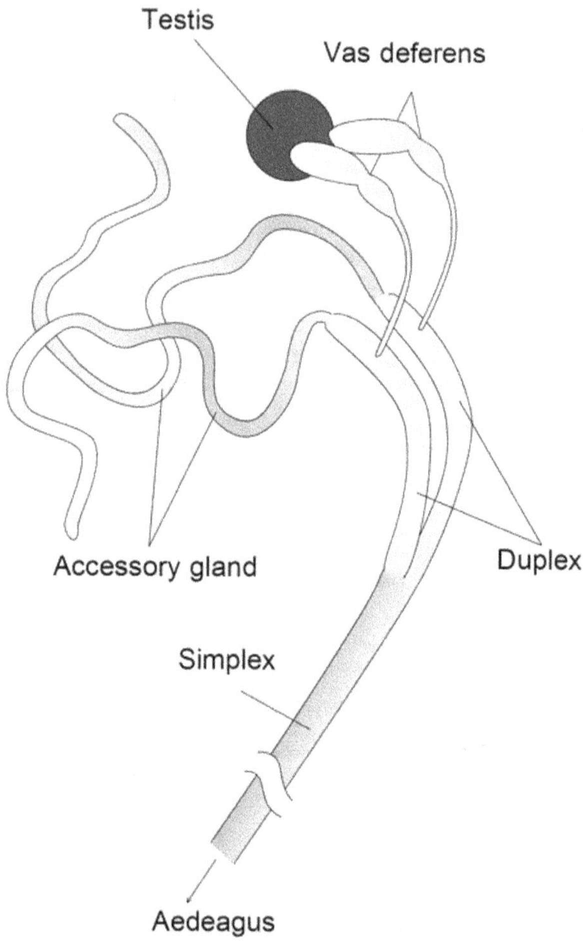

Fig. 14.1 A schematic representation of the internal reproductive organs of a *B. philenor* male (After Sasaki et al. 2015)

males might not be able to reserve spermatophore materials or accessory substance in the simplex and so the quantity or nature of these materials might be a good indicator of recent mating activity.

The aims of the present study were (1) to document any changes that occur in the appearance of reproductive structures in male pipevine swallowtail (*Battus philenor*) butterflies as a result of mating, as well as the persistence of such changes after mating, in order to develop criteria for identifying males that had recently mated and (2) to examine the reproductive tracts of field-caught *B. philenor* males, assess the variation in their recent mating history, and determine whether their recent mating success was related to their phenotypes. Rutowski and Rajyaguru (2013) have reported that, in a captivity *B. philenor*, females use the dorsal hindwing coloration of males in mate choice.

This issue is mainly reporting previously published results and ideas in Sasaki et al. (2015).

14.2 Materials and Methods

14.2.1 Source of Animals Used

All specimens were from a population of *B. philenor* that thrives near the confluence of Mesquite Wash and Sycamore Creek in the Mazatzal Mountains, Arizona (33° 43′ 50″ N, 111° 30′ 50″ W). Animals used in the mating studies were reared from eggs and early instar larvae collected in the field from early June to mid-July in 2011. All larvae were reared in a walk-in environmental chamber, programmed for 14 h of light at 30 °C and 10 h of dark at 24 °C with relative humidity held constant at 55%, and were fed ad libitum on cuttings of the local larval food plant, *Aristolochia watsonii*. On the day of eclosion, males were weighed, their forewing length measured, and given an individual number. Sexes were kept separately in small flight cages (~1 m^3) at room temperature (~24 °C) and individually fed 20% sucrose solution for about 20 min each day.

14.2.2 Examination of Reproductive Tracts of Virgin and Mated Males

To examine the effect of mating on the appearance of the male's reproductive tract, we hand-paired males with 0–3-day-old virgin females using the method of Watanabe and Hirota (1999). Then, each male was dissected and his reproductive tract examined to assess changes in the appearance of simplex with age and with mating experience. We divided males into three experimental groups: (1) males that never mated and dissected on the day of eclosion or 3 or 6 days after eclosion; (2) males that mated 1, 3, or 5 days after eclosion and dissected immediately after the mating; and (3) males mated 1 day after eclosion and dissected right after the mating or 1, 2, 3, or 5 days after mating.

Before dissection, each male was immobilized by gently pinching their thorax. Each male's abdomen was then removed from the body and placed in a petri dish filled with fresh insect Ringer's solution. The reproductive organs including the simplex and duplex were carefully removed from his abdomen. To describe the simplex of each male, we measured its length, appearance, and mass. We first imaged each simplex after removing any fat bodies attached to it and then recorded its appearance with a digital camera attached to a microscope. After capturing images, each simplex was separated from the attached duplex and aedeagus. Wet mass of each simplex was then determined to the nearest 0.01 mg.

14.2.3 Estimation of Recent Mating Success of Field-Caught Male

Sixty-eight wild males were collected from 16 July to 1 August 2011 in the morning near Sunflower, Arizona. Each captured male was scored as to his wing wear as an indicator of his age. Age-class was scored on the scale (I (least worn) to V (most worn)) described by Watanabe et al. (1986). The forewing length of each male was measured from the wing base to the wing tip. All males were dissected on the day of capture. To assess recent mating success of males, the mass, length, and transparency differences of each male's simplex were measured.

14.2.4 Spectral Analyses of Iridescent Wing Areas

In preparation for spectral measurements, the left hindwing of each butterfly was removed from the thorax and mounted dorsal side up on black card stock with spray adhesive. Reflectance spectra were collected from these wings using techniques described in Rutowski et al. (2010). Reflectance relative to a magnesium oxide white standard was measured between 300 and 700 nm from the wings. Because the reflectance spectra of these iridescent wing surfaces are unimodal, we extracted three color parameters, intensity, hue, and chroma, to describe and analyze the properties of the wing reflectance (Montgomerie 2008).

14.3 Results

14.3.1 Virgin Male Reproductive Tract

For virgin males, the simplex mass adjusted for forewing length ((simplex mass)$^{1/3}$/forewing length), simplex length adjusted for forewing length (simplex length/forewing length), and the transparency difference of simplex was approximately 0.5, 1.3, and 1.3, respectively, and these did not change with male age [mass (ANOVA, $F_{2,21}$=1.554, p = 0.237); length (ANOVA, $F_{2,21}$=2.276, p = 0.130); transparency difference (ANOVA, $F_{2,21}$=0.475, p = 0.629)]. Consequently, simplex of virgin males did not change in appearance with time since eclosion.

14.3.2 Reproductive Tract of Males Immediately After Mating

For males, immediately after the termination of copulation, the simplex mass adjusted for forewing length, the simplex length adjusted for forewing length, and the transparency difference were approximately 0.32, 0.5, and 0.3, respectively, and these did not vary with the age of the male at mating [mass (ANOVA, $F_{2,16}$=0.027, $p = 0.974$); length (ANOVA, $F_{2,16} = 1.331$, $p = 0.296$); transparency difference (ANOVA, $F_{2,16}$=0.170, $p = 0.845$)]. Although the simplex of males just after the termination of copulation was different in appearance from that of virgin males, these were not affected by age at mating.

14.3.3 Changes in the Male's Reproductive Tract with Time Since Mating

Although the simplex of males just after the termination of copulation was short, it lengthened and refilled again as time passed since mating. During this period, the color of the simplex turned from yellow to colorless, and the amount granular substances increased in the basal end of the tube. Statistically, the mass, length, and transparency difference of simplex all changed with time between mating and dissection (Fig. 14.2; mass: ANOVA: $F_{5,61} = 59.202$, $p < 0.001$; length: ANOVA: $F_{5,55} = 42.770$, $p < 0.001$; transparency difference: ANOVA: $F_{5,55} = 17.139$, $p < 0.001$). After copulation, simplex mass (A) and length (B) dropped to half their precopulatory values, but returned to precopulatory values in about 2 days. The transparency difference decreased with mating but also returned to premating values within about 2 days (C).

Using the results of these analyses, we developed criteria for assessing whether a male's reproductive tract showed evidence of recent mating. The distribution of simplex mass adjusted for forewing length of males within 1 day after mating was 0.251–0.453, whereas that of virgin males was 0.465–0.565, with no overlap in these ranges. The observed ranges of simplex length adjusted for forewing length and the transparency difference of males within 1 day after mating and virgin males also did not overlap (length, 0.401–0.940 vs 1.084–1.628; transparency difference, 0.013–0.964 vs 1.105–2.148). So, we set the lower end of the ranges of values for simplex characteristics of virgin males as the value below which would indicate that the male had recently mated. That is, a field-caught male that had a simplex of less than 0.46 in mass, less than 1.0 in length, or less than 1.0 in transparency difference was taken as indicating that the male had recently mated.

Fig. 14.2 Simplex mass (**a**), length (**b**), and transparency difference (**c**) of simplex for virgin males (V, 0, 3, 6 days old) and for mated males (1, 3, 5 days old) dissected at various number of days after mating (mean±S.D.) (After Sasaki et al. 2015). *,** and *** represent $p < 0.05$, $p < 0.01$, and $p < 0.001$ in Tukey's HSD test, respectively

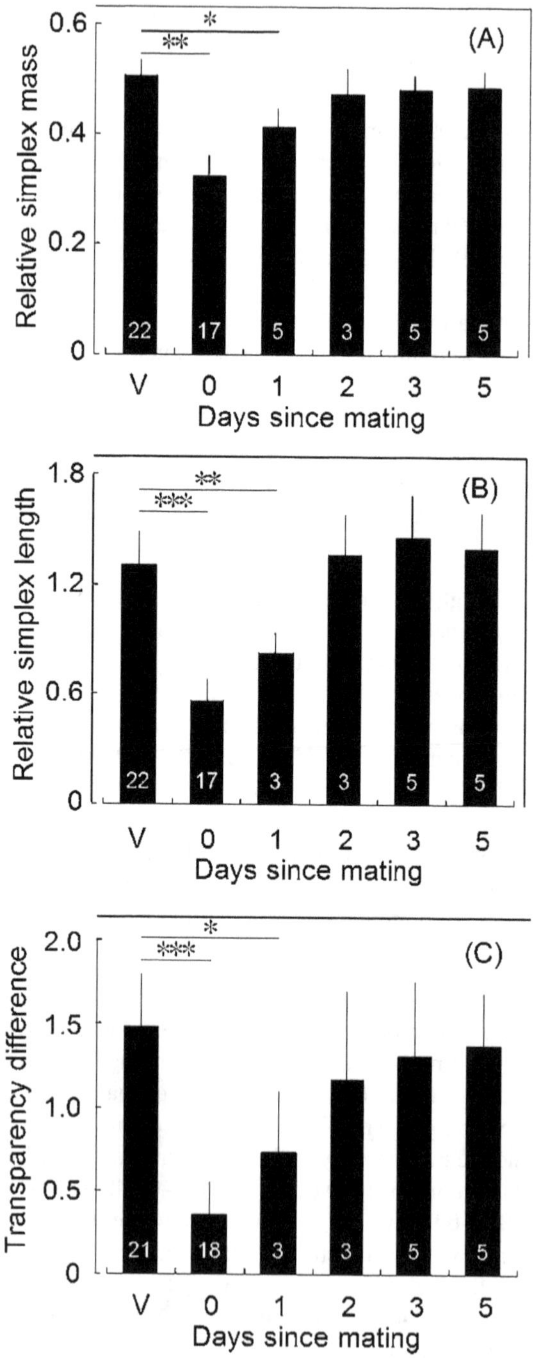

14.3.4 *Mating Success of Field-Caught Males*

All field-caught males were evaluated and placed in groups based on which of the three criteria for recent mating they met and which they did not (Table 14.1). For the eight possible groups and to maximize contrasts any group that met two or more of the criteria (Groups E to H) we labeled as showing strong evidence of recent mating. However, because there were no individuals in Group G, we regarded males in Group E, F, and H as recently mated males. Of the 68 males in the list, 12 showed this strong evidence of having mated recently. We also confidently labeled as not recently mated, males that met none of the criteria (Group A).

Using GLM with binomial errors and a logit link function, we compared the phenotypic characteristics of those that had recently mated (Groups E, F, and H) with those that had not (Group A). The characters included in the analysis were the intensity and hue of the iridescent area of male dorsal hindwing and age-class. Chroma was not included as an independent variable in the analysis because there were significant correlations between chroma and all other characteristics (Table 14.2). As shown in Table 14.3, while intensity was not related to their recent mating success, hue and age-class significantly affect their recent mating success. Recently mated males were older and had a higher hue value (were greener) than males that had not recently mated (Figs. 14.3 and 14.4).

Table 14.1 A summary of the state of reproductive tract components of 68 males caught in the field

	Group							
Criterion	A	B	C	D	E	F	G	H
Simplex mass	–	–	–	+	–	+	+	+
Simplex length	–	–	+	–	+	–	+	+
Transparency difference	–	+	–	–	+	+	–	+
N	23	30	2	1	2	5	0	5

After Sasaki et al. (2015)

A plus sign means that the state of that component met the criteria we set for indicating the male recently mated. A minus sign means it did not meet the criteria. In later analysis, any group that met two or more of the criteria (Groups E to H) and males that met none of the criteria (Group A) were used as recently mated males and as not recently mated males, respectively

Table 14.2 Spearman correlation coefficients for the relationship between male age and the various color parameters for the dorsal hindwing coloration

Measure	1	2	3	4
1. Intensity	–			
2. Chroma	0.584*	–		
3. Hue	−0.087	−0.383*	–	
4. Age-class	−0.102	−0.581*	0.211	–

After Sasaki et al. (2015)

Significant correlation between intensity and chroma, chroma and hue, and chroma and age-class were found

*$p < 0.01$

Table 14.3 Results of an ANOVA (GLM) of factors that contribute to variation in recent mating success of wild males

Effect	df	Deviance
	34	45.004
Intensity	1	0.669
Hue	1	4.744**
Age	1	1.760*

After Sasaki et al. (2015)

Effect of hue and age-class on male recent mating success were significant. * and ** represent $p < 0.05$ and $p < 0.01$, respectively

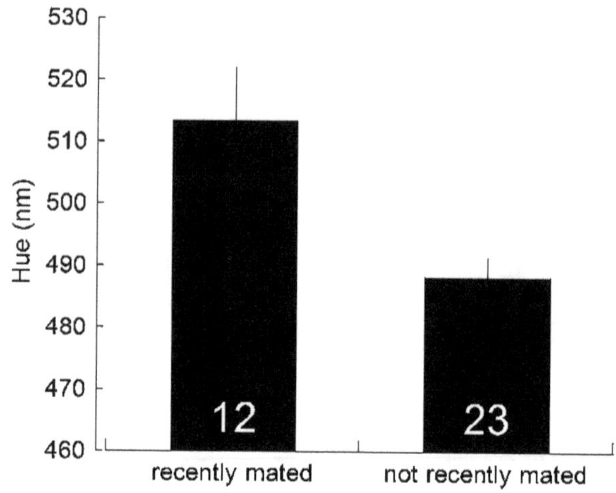

Fig. 14.3 The hue (wavelength of maximum reflectance) of the dorsal hindwing for field-caught males that met the criteria for evidence of having recently mated and those males that did not meet the criteria (±S.E.) (After Sasaki et al. 2015)

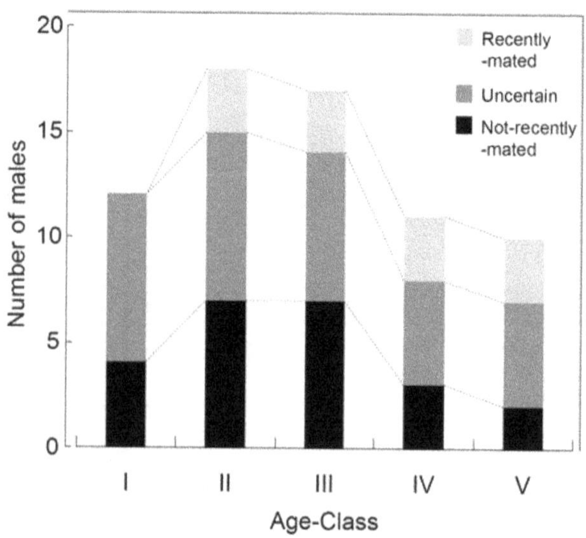

Fig. 14.4 Change with age-class in the number of clearly recently mated males (*light-gray bar*), clearly not recently mated males (*black bar*), and males of uncertain recent mating history (*dark-gray bar*) for field-caught males (After Sasaki et al. 2015)

14.4 Discussion

14.4.1 Assessing the Mating History of Male Butterflies in the Field

The most convincing demonstrations of the evolutionary significance of mating preferences are those in which the results of manipulative experiments are matched by observations in the wild or in wild-caught populations (Kemp 2007). In butterflies, many laboratory experiments have demonstrated the occurrence of female preference for particular male traits, including wing coloration (Krebs and West 1988; Robertson and Monteiro 2005; Andersson et al. 2007). However, the conclusions of these studies have rarely been validated against data obtained in more natural field-based settings (Kemp and Rutowski 2011). In other insects, the comparison of traits from copulating males and unattached males in the field has been extensively used to make inferences about population mating biology (Flecker et al. 1988; Harari et al. 1999; Alcock and Kemp 2006). However, this strategy has not been used in butterflies, except in cases of extremely high density, owing to the difficulty of observing butterfly mating in the field (Kemp and Rutowski 2011). Therefore, to determine the effect of female preference on the evolution of traits in male butterflies, it was necessary to establish an alternative method for evaluating male mating success.

Tsubaki and Matsumoto (1998) estimated the mating frequency of male *Luehdorfia japonica* by assessing the degree of scale loss from males' claspers. In this species, males consume scales and use them to form a mating plug on the female abdomen during copulation. Kazuma (1987) reported that the degree of scale loss in laboratory-reared males increased with repeated hand pairing. The degree of scale loss was scored on a scale of 0 (slight scale loss) to 3 (almost all scales were lost), and each stage corresponded to 0, 1, 2, and 3 or more matings. Since mating frequency must have a strong relationship to lifetime reproductive success, this method may provide an accurate estimation of reproductive success in wild males. However, because the mating-related loss of scales is not the rule among lepidopterans, this method is not applicable to all species.

In the present study, we reported a new technique for assessing male mating success that relies on changes that occur in the appearance of internal reproductive organs during mating. Since the internal reproductive organs of butterflies are not much different between species and it is common for males to ejaculate spermatophores and accessory substances during mating (e.g., Drummond 1984), this method can likely be applied to any butterfly species. In fact, it is known that the status of the male simplex just after mating is also different from that before mating in *Papilio xuthus* (Sasaki, personal observation), *P. machaon* (Sasaki, personal observation), *Byasa alcinous* (Sasaki, personal observation), and *Eurema hecabe* (Konagaya, personal communication). In addition, this new strategy can be used at both middle- or low-density mating sites, since it can detect the occurrence of male mating within a few days, and even though the technique only distinguishes between males that have recently mated and those that have not, it still holds promise for examining variation in male mating success and, thereby, investigating traits that correlate with mating success and the intensity of sexual selection.

We note that, when using this method, there might be some traits that are not suitable for investigating the relationship with mating history. For example, it is impossible, in principle, to investigate the relationship between recent mating success and spermatophore production capacity, which is closely related to the reproductive success of both male and female butterflies.

14.4.2 Phenotypic Correlates of Mating Success in Male **B. philenor** *in the Field*

Our results indicate that males that exhibit signs of having recently mated differ from those that do not, in that they are older and have a greener coloration. This could mean that age and coloration are important determinants of male mating success and under selection, either in the context of (1) female choice, in which females prefer older and greener males, or (2) male competition, in which older and greener males are, for some reason, more effective competitors. In general, the results of the present study support the prediction that male coloration and mating success are related. However, Rutowski and Rajyaguru (2013) reported that the dorsal hindwings of successfully mated *B. philenor* males possessed more chromatic iridescence than those that failed to mate, rather than a different hue, as reported here.

Such differences between field and laboratory study have also been reported by previous studies. For example, in-copula males of *Eurema hecabe* were reportedly older than their free-flying counterparts and possessed significantly less bright markings, while brighter males were preferred by females in the laboratory (Kemp 2008). In this case, the difference was caused by the existence of newly emerged females that could not reject mating, and since the density of individuals at the study site was high and activity is centered at localized breeding sites, males could profitably locate such females.

The reasons for difference the color parameters correlated with the mating success of male *B. philenor* have not yet been clarified. It is possible that the data reported here about the specifics of male age and color associated with recent mating success were affected by the several significant correlations among color parameters and between coloration and age. We made efforts to control for these correlations by excluding variables, such as chroma, from our analysis, but in the end, it is difficult to make conclusions with confidence about the reasons for the connections between male color, age, and male mating success suggested by our data set. In addition, we have not controlled or taken into consideration several other variables that might affect male mating history, such as body size, population density, time during the breeding season, and weather. To convincingly identify factors that determine male mating success in *B. philenor*, the experimental manipulation of candidate variables is needed.

Acknowledgments We thank Masaru Hasegawa and the members of Rutowski Laboratory, especially Sean Hannam, for assistance in the field and laboratory work as well as for valuable discussions. This study was supported by the Japan Society for the Promotion of Science for the Institutional Program for Young Researcher Overseas Visits to NS and TK and in part by JSPS KAKENHI Grant Number 24570019 (MW) and by NSF Grant IOS 1145654 (to R. L. Rutowski). We would like to thank Editage (www.editage.jp) for English language editing.

References

Alcock J, Kemp DJ (2006) The behavioral significance of male body size in the tarantula hawk wasp *Hemipepsis ustulata* (Hymenoptera: Pompilidae). Ethology 112:691–698

Allen CE, Zwaan BJ, Brakefield PM (2011) Evolution of sexual dimorphism in the Lepidoptera. Ann Rev Entomol 56:445–464

Andersson J, Karlson AKB, Vongvanich N, Wiklund C (2007) Male sex pheromone release and female mate choice in a butterfly. J Exp Biol 210:964–970

Beldade P, Koops K, Brakefield PM (2002) Developmental constraints versus flexibility in morphological evolution. Nature 416:844–847

Bissoondath CJ, Wiklund C (1996) Male butterfly investment in successive ejaculates in relation to mating system. Behav Ecol Sociobiol 39:285–292

Blount JD, Metcalfe NB, Birkhead TR, Surai PF (2003) Carotenoid modulation of immune function and sexual attractiveness in zebra finches. Science 300:125–127

Carroll SB, Gates J, Keys DN, Paddock SW, Panganiban GE, Selegue JE, Williams JA (1994) Pattern formation and eyespot determination in butterfly wings. Science 265:109–114

Darwin C (1874) The descent of man and selection in relation to sex. John Murray and Sons, London

Drummond BA (1984) Multiple mating and sperm competition in the Lepidoptera. In: Smith RL (ed) Sperm competition and the evolution of animal mating systems, pp 291–370

Flecker AS, Allan JD, McClintock NL (1988) Male body size and mating success in swarms of the mayfly *Epeorus longimanus*. Ecography 11:280–285

Grether GF, Cummings ME, Hudon J (2005) Countergradient variation in the sexual coloration of guppies (*Poecilia reticulata*): drosopterin synthesis balances carotenoid availability. Evolution 59:175–188

Harari AR, Handler AM, Landolt PJ (1999) Size-assortative mating, male choice and female choice in the curculionid beetle *Diaprepes abbreviatus*. Anim Behav 58:1191–1200

Hill GE, Montgomerie R (1994) Plumage color signals nutritional condition in the house finch. Proc R Soc Lond B 258:47–52

Kazuma M (1987) Mating patterns of a sphragis-bearing butterfly, *Luehdorfia japonica* Leech (Lepidoptera: Papilionidae), with descriptions of mating behavior. Res Popul Ecol 29:97-110

Kemp DJ (2007) Female butterflies prefer males bearing bright iridescent ornamentation. Proc R Soc London B 274:1043–1047

Kemp DJ (2008) Female mating biases for bright ultraviolet iridescence in the butterfly *Eurema hecabe* (Pieridae). Behav Ecol 19:1–8

Kemp DJ, Rutowski RL (2011) The role of coloration in mate choice and sexual interactions in butterflies. Adv Study Behav 43:55–92

Keyser AJ, Hill GE (1999) Condition-dependent variation in the blue-ultraviolet coloration of a structurally based plumage ornament. Proc R Soc Lond B 266:771–777

Koch PB, Keys DN, Rocheleau T, Aronstein K, Blackburn M, Carroll SB (1998) Regulation of dopa decarboxylase expression during colour pattern formation in wild-type and melanic tiger swallowtail butterflies. Development 125:2303–2313

Krebs RA, West AD (1988) Female mate preference and the evolution of female-limited Batesian mimicry. Evolution 42:1101–1104

LaChance LEO, Richard RD, Ruud RL (1977) Movement of eupyrene sperm bundles from the testis and storage in the ductus ejaculatoris duplex of the male pink bollworm: effects of age, strain, irradiation, and light. Ann Entomol Soc Am 70:647–651

Montgomerie R (2008) CLR, version 1.05. Queen's University, Kingston, Canada. Available as of 14 December 2011 at http://post.queensu.ca/~mont/color/analyze.html

Nijhout HF (1990) A comprehensive model for colour pattern formation in butterflies. Proc R Soc London B 239:81–113

Riemann JG, Thorson BJ, Ruud RL (1974) Daily cycle of release of sperm from the testes of the Mediterranean flour moth. J Insect Physiol 20:195–207

Robertson KA, Monteiro A (2005) Female *Bicyclus anynana* butterflies choose males on the basis of their dorsal UV-reflective eyespot pupils. Proc R Soc London B 272:1541–1546

Rutowski RL (1997) Sexual dimorphism, mating systems and ecology in butterflies. In: Choe JC, Crespi BJ (eds) The evolution of mating systems in insects and arachnids. Cambridge University Press, Cambridge, pp 257–272

Rutowski RL, Nahm A, Macedonia JM (2010) Iridescent hindwing patches in the pipevine swallowtail: differences in dorsal and ventral surfaces relate to signal function and context. Funct Ecol 24:767–775

Rutowski RL, Newton M, Schaefer J (1983) Interspecific variation in the size of the nutrient investment made by male butterflies during copulation. Evolution 34:708–713

Rutowski RL, Rajyaguru PK (2013) Male-specific iridescent coloration in the pipevine swallowtail (*Battus philenor*) is used in mate choice by females but not sexual discrimination by males. J Insect Behav 26:200–211

Sasaki N, Konagaya T, Watanabe M, Rutowski RL (2015) Indicators of recent mating success in the pipevine swallowtail butterfly (*Battus philenor*) and their relationship to male phenotype. J Insect Physiol 83:30–36

Svärd L, Wiklund C (1989) Mass and production rate of ejaculates in relation to monandry/polyandry in butterflies. Behav Ecol Sociobiol 24:395–402

Takeuchi T (2016) Agonistic display or courtship behavior? A review of contests over mating opportunity in butterflies. J Ethol:1–10

Tsubaki Y, Matsumoto K (1998) Fluctuating asymmetry and male mating success in a sphragis-bearing butterfly *Luehdorfia japonica* (Lepidoptera: Papilionidae). J Insect Behav 11:571–582

Wallace AR (1889) Darwinism: an exposition of the theory of natural selection, with some of its applications. Macmillan & Co, London

Watanabe M, Hirota M (1999) Effects of sucrose intake on spermatophore mass produced by male swallowtail butterfly *Papilio xuthus* L. Zool Sci 16:55–61

Watanabe M, Sato K (1993) A spermatophore structured in the bursa copulatrix of the small white *Pieris rapae* (Lepidoptera, Pieridae) during copulation, and its sugar content. J Res Lepid 32:26–36

Watanabe M, Nozato K, Kiritani K (1986) Studies on ecology and behavior of Japanese black swallowtail butterflies (Lepidoptera: Papilionidae): V. Fecundity in summer generations. Appl Entomol Zool 21:448–453

Part V

Larva and other Insects: A Close Study of Color Patterns

15

Larval Cuticular Pigmentation and Larval Body Color Patterns in P. xuthus and Papilio

Hongyuan Jin and Haruhiko Fujiwara

Abstract In lepidopterans (butterflies and moths), larval body color pattern, which is an important mimicry trait involved in prey–predator interactions, presents a great diversity of pigmentation and patterning. Unlike wing patterns, larval body color patterns can switch during development with larval molting. For example, in the Asian swallowtail butterfly *Papilio xuthus*, a younger larva (first–fourth instar) has a white/black color pattern that mimics bird droppings, whereas the final instar (fifth) larva drastically changes to a greenish pattern that provides camouflage on plants. Insect mimicry has interested scientists and the public since Darwin's era. Broadly, mimicry is an antipredation strategy whereby one creature's color, shape, or behavior resembles another creature or object. In this review, I address basic knowledge about larval cuticular pigmentation and advanced understanding of its regulatory mechanism in *P. xuthus*; I also discuss larval body color patterns among members of the genus *Papilio*, followed by conclusions and prospects for further research.

Keywords Larval pigmentation • Lepidoptera • Mimicry • Cuticular melanization • Ecdysteroid • Juvenile hormone • *Papilio xuthus* • *Papilio polytes* • *Papilio machaon*

15.1 Introduction

About 150 years ago, following H.W. Bates' report on mimicry in insects (Bates 1862), Charles Darwin wrote to Bates and said: "In my opinion, it is one of the most remarkable & admirable papers I ever read in my life. The mimetic cases are truly marvelous..." (Darwin 1863). Today, the mimicry phenomenon remains as an interesting evolutionary theme as ever, attracting the interest of both scientists and the public. To better understand the molecular mechanisms behind insect mimicry, we need to understand how wing and body color patterns evolve.

H. Jin (✉) • H. Fujiwara
Department of Integrated Biosciences, Graduate School of Frontier Sciences, The University of Tokyo, Kashiwa, Chiba 277-8562, Japan
e-mail: 1169399970@edu.k.u-tokyo.ac.jp

Lepidopterans (butterflies and moths) show highly diverse wing colors and patterns and are considered to be an ideal model system for examining color pattern formation and evolution. In lepidopterans, evidence for wing color and pattern evolution has been frequently reported (Nijhout 1991; Reed et al. 2011; Heliconius Genome 2012; Kunte et al. 2014), whereas our knowledge of larval body coloration and pattern formation is relatively limited.

Generally, lepidopteran larvae are soft bodied and cannot escape by flight from predators. Natural selection has led to the development of many chemical and morphological devices in larvae that aid survival in the wild (Scoble and Scoble 1992). Among those, body color pattern is particularly interesting because it is important in visual recognition. Two different strategies are commonly used for predator defense. Toxic larvae tend to warn predators with their colorful markings which act as warning signals. The majority of larvae, which are palatable and nonpoisonous, mimic an item in the surroundings (such as a bud, a twig, or even a moss) or conceal their bodies in the environmental background (Pasteur 1982).

In the case of *P. xuthus*, a larva switches its body color pattern with larval molting (Fig. 15.1). A younger larva (first–fourth instar) mimics bird droppings with a black/white body color (denoted mimetic pattern, Fig. 15.1a). The fifth (final) instar larva dramatically switches to a greenish body pattern with a pair of eyespots on the metathorax, which allows it to blend in with the color and pattern of its host plant (denoted cryptic pattern, Fig. 15.1b). A similar switching of body color pattern is observed in other *Papilio* species (Prudic et al. 2007) and is considered to be a successful survival strategy for this genus (Tullberg et al. 2005). Recent studies have reported that two critical insect hormones, ecdysone (Fig. 15.2a, b) and juvenile hormone (JH), directly regulate pigmentation and color pattern switch in the larva of *P. xuthus* (Futahashi and Fujiwara 2007, 2008a).

In this chapter, I review recent progress in understanding the molecular mechanisms underlying cuticular melanization and the hormonal regulation of pigmentation in the larva of *P. xuthus*. I also discuss possible evolutionary changes among three *Papilio* species, followed by conclusions and prospects for further research.

Fig. 15.1 (**a**) Fourth instar larva with bird-dropping body pattern; (**b**) Fifth instar larva with *green* body pattern of *P. xuthus*

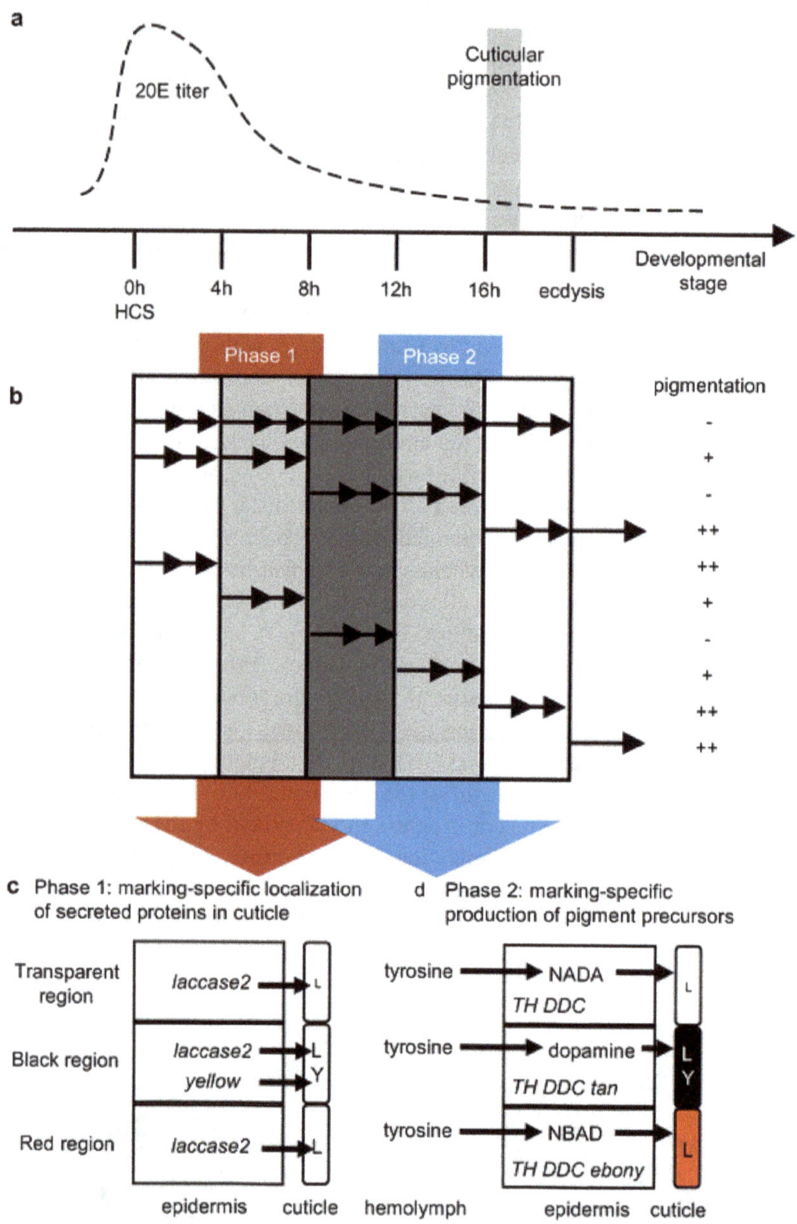

Fig. 15.2 A working model for the two-phase cuticular pigmentation in larvae of *P. xuthus*. (**a**) 20E titer in hemolymph during the fourth molt; (**b**) The timing effect of 20E on *black* pigment synthesis. The intervals of 20E applications (*arrows*, every 2 h); (**c**) phase 1; (**d**) phase 2. N-b-alanyldopamine (NBAD). N-acetyldopamine (NADA). L and Y in cuticle indicate laccase 2 and Yellow proteins (Modified from Futahashi et al. 2010)

15.2 Pigmentation of Larval Cuticle in *P. xuthus*

An insect cuticle is a hardened exoskeleton composed of chitin and proteins. In lepidopteran larvae, black cuticular pigments mainly comprise melanin, which is produced by the oxidization of dopamine or L-3,4-dihydroxyphenylalanine (DOPA) (Kramer and Hopkins 1987; Hiruma and Riddiford 2009; Wright 1987). In both *P. xuthus* and *Manduca sexta* (tobacco hornworm), the pigmentation procedure of larval cuticle can be summarized in two steps: localization of secreted proteins (Fig. 15.2c, phase 1) and production of pigment precursors (Fig. 15.2d, phase 2) (Hiruma and Riddiford 2009; Futahashi et al. 2010; Walter et al. 1991). These steps occur in the cuticle and the epidermal cell, respectively (Fig. 15.2).

In phase 1, laccase 2 (Lac2), which is a phenol oxidase (PO), and other pigment-related proteins (such as Yellow) are synthesized and deposited into the newly forming cuticle (Fig. 15.2c) (Futahashi and Fujiwara 2007; Kramer and Hopkins 1987; Hiruma and Riddiford 1988, 2009). Lac2 catalyzes the oxidation of dopamine to dopamine–melanin in many species (Hiruma and Riddiford 2009; Noh et al. 2016; Futahashi et al. 2011). In *Tribolium castaneum*, laccase 2 (coded by *TmLac2*) is the major PO involved in the tanning of larval, pupal, and adult cuticles (Arakane et al. 2005).

In *P. xuthus*, Futahashi et al. (2010) found that *Pxlaccase2* (*Pxlac2*) expression is strongly associated with the presumptive black pigment (11 h after head capsule slippage (HCS) at the fourth molt) (Futahashi et al. 2010). Typically, expression of Lac2 begins in the middle period of molting, and the deposited Lac2 is on standby until the pigment precursors reach the cuticular surface. These events precede the expression of melanin synthesis genes at mRNA levels and the production of pigment precursors (Walter et al. 1991; Hiruma and Riddiford 1988; True et al. 1999; Futahashi and Fujiwara 2005). Another pigmentation-related protein, Yellow (coded by *Pxyellow*), shows an expression pattern similar to that of Lac2 in phase 1. However, the precise function of *Pxyellow* gene remains unclear (Futahashi et al. 2010; Noh et al. 2016). It is inferred that PxYellow may be secreted into the cuticle and probably acts as a cofactor (Futahashi and Fujiwara 2005).

In phase 2, the precursors of melanin compounds are synthesized from phenolic amino acids (mainly tyrosine) (Fig. 15.2c). The dopamine–melanin synthesis pathway is conserved in many insects (Hiruma and Riddiford 2009; Noh et al. 2016; Futahashi and Fujiwara 2005; Massey and Wittkopp 2016). First, tyrosine is converted to DOPA by tyrosine hydroxylase (TH), and then dopamine is synthesized from DOPA by DOPA decarboxylase (DDC) (Futahashi and Fujiwara 2005). Dopamine is a prominent black pigment precursor in many insects (Hiruma et al. 1985). After its synthesis in an epidermal cell, dopamine is incorporated into the cuticle and converted to dopamine–melanin by PO and other proteins. However, it also can be converted to a reddish brown pigment by ebony or to a transparent pigment called N-acetyldopamine (NADA) by dopamine N-acetyltransferase (DAT) activity (Futahashi et al. 2010; Futahashi and Fujiwara 2005; Massey and Wittkopp 2016; Wittkopp et al. 2002).

In *P. xuthus*, spatially specific localization of melanin synthesis genes contributes to the color pattern (Futahashi and Fujiwara 2005). Futahashi and Fujiwara (2005) showed that the spatial expression of melanin synthesis genes (*TH*, *DDC*, and *tan*) perfectly corresponds with the presumptive black pigment (Futahashi et al. 2010; Futahashi and Fujiwara 2005) and that the expression of *ebony* is limited to the red area within the eyespot (Futahashi and Fujiwara 2005). They also demonstrated that the addition of excess tyrosine did not promote pigmentation, whereas the application of DOPA with 3-iodotyrosine (3IT, a competitive inhibitor of TH protein) led to a clear color pattern with an overall pigmentation in vitro (Futahashi and Fujiwara 2005). Their results indicate that cuticle color patterns form from spatially specific localization of melanin synthesis genes rather than the differential uptake of melanin precursors into individual epidermal cells.

Cuticular pigmentation occurs in the latter half of the molting period just before ecdysis (16–18 h after HCS during the fourth molting period). When Futahashi and Fujiwara (2005) examined the timing of expression of *PxTH*, *PxDDC*, *Pxebony*, and *Pxtan*, they noticed that the expression of these melanin synthesis genes precisely coincides with melanization onset. Therefore, cuticular pigmentation is predictably strictly controlled by ecdysteroid, the molting hormone (Futahashi and Fujiwara 2005, 2007; Futahashi et al. 2010).

15.3 Hormonal Regulation of Larval Pigmentation

Ecdysone and juvenile hormone are directly and indirectly involved in larval pigmentation in insects (Futahashi and Fujiwara 2008a; Hiruma and Riddiford 1990, 2009; Hwang et al. 2003).

15.3.1 Ecdysone-Induced Cuticular Pigmentation

Ecdysone is a steroid hormone and the central regulator in insect development and reproduction (Kopec 1926). The periodic release of ecdysone triggers larval molting and pupal metamorphosis (Yamanaka et al. 2013).

The first evidence of ecdysone-regulated pigmentation was reported by Karlson and Sekeris in 1976. They showed that ecdysone causes elevated activity of DDC in *Calliphora* (Hiruma and Riddiford 2009; Karlson and Sekeris 1976). In *M. sexta*, regulation of DDC expression requires exposure of 20-hydroxyecdysone (also known as 20E, an active form of ecdysone), followed by its withdrawal during larval molting (Hiruma and Riddiford 1986, 1990; Hiruma et al. 1995; Hiruma and Riddiford 2007). Hiruma et al. (1995) found continuous exposure of 20E insufficient for DDC expression, unless there is a 20E-free period (Hiruma et al. 1995).

In *P. xuthus*, Futahashi and Fujiwara (2007) successfully tested the effect of 20E exposure on larval pigmentation using a topical application method in vivo.

Consistent with the results in *M. sexta*, they demonstrated that cuticular melanization and epidermal pigmentation are inhibited through 20E treatment during the molt and confirmed that the removal of ecdysone is necessary for the onset of normal coloration. Moreover, they showed that 20E inhibited pigmentation if it was applied at the middle of the molt when native ecdysone titers decline (Fig. 15.2b) (Futahashi and Fujiwara 2007). As expected, the expression of melanin synthesis genes, including *TH*, *DDC*, and *ebony*, was repressed by high 20E concentration (Futahashi and Fujiwara 2007). Unexpectedly, the expression of *Pxyellow* was promoted by a high concentration of 20E. This led Futahashi and Fujiwara to hypothesize that PxYellow must function as a cofactor for other melanin synthesis enzymes since it alone is not sufficient for melanization (Futahashi and Fujiwara 2007).

Like the pigment synthesis genes, some upstream regulatory factors are also controlled by ecdysone. In the ecdysone signaling pathway, 20E acts as a hormonal signal and regulates the expression of downstream transcription factors (Yamanaka et al. 2013; Yao et al. 1992). Hiruma and Riddiford (2007) found that two nuclear transcription factors, E75B and MHR4, are 20E-induced inhibitors of *Msddc* in vitro (Hiruma and Riddiford 2007). Evidence also showed that there is at least one other suppressive protein other than E75B and MHR4 that binds to a specific sequence (GGCTTATGCGCTGCA) in the DDC promoter when the ecdysone titer decreases (Hiruma et al. 1995). In *Drosophila melanogaster*, *DmDDC* is directly modulated by an ecdysone response element (EcRE), located at position −97 to −83 bp relative to the transcription initiation site (Chen et al. 2002). In *D. melanogaster*, Yellow is known to be a prepattern factor as well as a pigmentation factor in adult body patterning (Massey and Wittkopp 2016). Recently, comprehensive yeast one-hybrid and RNAi screens were carried out by Kalay et al. (2016). They screened and identified four ecdysone-induced nuclear reporters (*Hr78*, *Hr38*, *Hr46*, and *Eip78C*) that showed a statistically significant interaction with at least one Yellow enhancer. In an RNAi experiment, all four caused altered pigmentation when knocked down (Kalay et al. 2016). In *Bombyx mori*, Yamaguchi et al. (2013) used a type of *L* (multi lunar) mutant with twin-spot markings on the sequential segments and proved that the gene responsible for this phenotype (*BmWnt1*) can be induced by high concentrations of 20E in vitro (Yamaguchi et al. 2013).

In *P. xuthus*, Futahashi et al. (2012) used a microarray EST dataset to recognize *E75A* and *E75B*, which are transcription factors involved in ecdysone signaling, as candidates involved in specific marking-specific patterning (Futahashi et al. 2012). The expression of *E75A* and *E75B* is specifically localized at the eyespot marking region, and temporal expression patterns are similar to those of *Pxyellow*, as described before. It is known that *E75* is active early in ecdysone signaling (Palli et al. 1995; Jindra et al. 1994; Jindra and Riddiford 1996). Taken together, this suggests that 20E-induced *E75A* and/or *E75B* expression may regulate both the prepattern of marking and the stage specificity of several black marking-associated genes (Futahashi et al. 2012). Interestingly, 3-dehydroecdysone 3β-reductase (coded by the *3DE 3β-reductase* gene) has a clear marking-specific expression in the presumptive black region, similar to TH or DDC. Since its function is converting inactivated 3-dehydroecdysone to ecdysone, localized marking-specific ecdysone synthesis may be critical for complex cuticular pigmentation and patterning (Futahashi et al. 2012).

The evidence above shows that there is a complicated relationship between ecdysone signaling and larval cuticular pigmentation and patterning. However, because only some of the regulatory genes have been identified, the detailed regulatory mechanisms remain to be uncovered.

15.3.2 Juvenile Hormone Directly Regulates Larval Color Pattern Switch

Juvenile hormone (JH) is a group of acyclic sesquiterpenoids secreted from the corpora allata (CA), which is an endocrine gland near the brain (Jindra et al. 2013). Like ecdysteroids, JH plays a critical role in molting, metamorphosis, reproduction, and other physiological processes in insects (Jindra et al. 2013). JH is also known as "*status quo* hormone," because the presence of JH prevents insect metamorphosis (Riddiford 1996). In a simplified model, a lepidopteran progresses through a larva-to-larva molt when JH is present and a larva-to-pupa metamorphosis when JH is absent at the final molting stage. It has been hypothesized that JH modulates the action of ecdysteroid-molting hormones, but the detailed mechanisms of the modulation are still unclear (Jindra et al. 2013; Urena et al. 2014; Kayukawa et al. 2016).

There is some evidence that JH has an effect on larval pigmentation. Lack of sufficient JH (caused by the artificial removal of the CA from the larva) causes black larvae in the tobacco hornworm, *M. sexta*. In addition, when the larvae of the black strain are treated with JH, they revert to their normal green color (Riddiford 1975).

As described above, *P. xuthus* larvae markedly switch from a black/white body pattern to a greenish one after the fourth–fifth larval ecdysis (Fig. 15.1). Futahashi (2006) found that when 20E was injected at the early fourth instar stage, precociously molted fifth larva appeared with a black/white mimetic pattern instead of the normal green pattern (Futahashi 2006). It is known and proven that JH controls the action of ecdysteroid at least through direct inhibition of *Broad–Complex* (*BR-C*) activity (Kayukawa et al. 2016; Nijhout and Wheeler 1982; Ogihara et al. 2015). To define the role of JH in facilitating larval color pattern regulation, Futahashi and Fujiwara (2008a) performed experiments using three types of JH analogs (JHA), which they artificially applied on the integument of fourth instar larvae (Fig. 15.3). Their results showed that some individuals failed to switch color patterns, either completely or partially, after the fourth molt. The larvae treated with fenoxycarb (JHA) kept a fourth instar-like black/white pattern or developed an intermediate color pattern with elements of both fourth and fifth instars (Fig. 15.3).

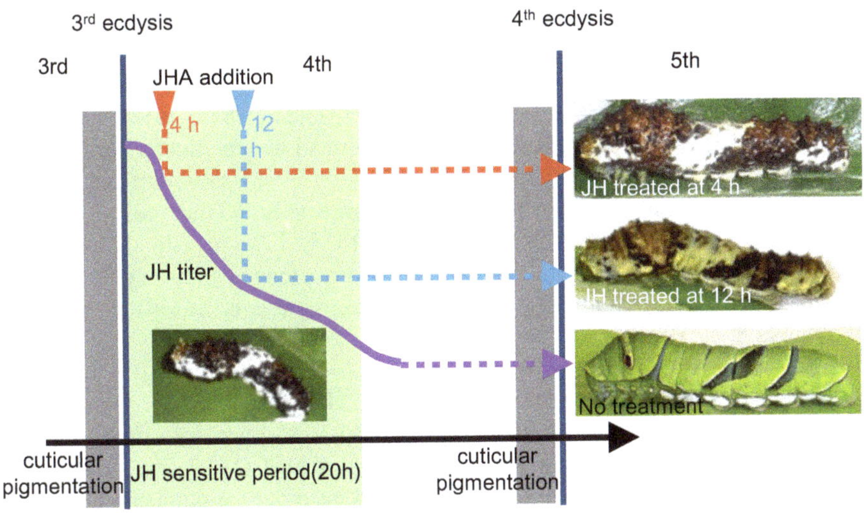

Fig. 15.3 Treatment of juvenile hormone analogs during the JH-sensitive period (Modified from Futahashi and Fujiwara 2008a)

Furthermore, they noticed that the epidermis is only sensitive to JHA during the first 20 h of the fourth instar stage. Exposure after this relatively short time frame did not prevent the color pattern switch. Hence, they named that specific time window "JH-sensitive period." In nontreated species, JH titer in the hemolymph was measured and found to be decreasing continuously during the early days of the fourth instar stage. Taken together, this evidence indicates that the decline of JH titer within a restricted developmental stage regulates the body color pattern switch in *P. xuthus* larvae (Futahashi and Fujiwara 2008a).

Because of our fragmentary knowledge of JH pathways, the molecular mechanisms underlying how JH alters color patterning and controls pigment synthesis are still under investigation (Jindra et al. 2013). Jin et al. have found some candidate genes involved in the larval color pattern switch by RNAi screening using the latest genomic information of *P. xuthus* (unpublished data). In my opinion, future studies may shed light on the downstream regulation of the JH cascade in larval pigmentation.

15.4 Species-Specific Color Patterns in the *Papilio* Genus

15.4.1 A Combination of Yellow and Blue Makes the Larval Body Green

A greenish body pattern follows the bird-dropping pattern in many *Papilio* species, making us wonder what the identity of the "green" pigment is. Green body coloration seems to be a beneficial adaptation for the final instar larvae of *Papilio*, which helps them conceal themselves in the host plant. The chemical nature of caterpillar's green pigment was once misunderstood as chlorophyll derived from the plant because of the strong color resemblance (Meldola 1873). However, studies show that larval green pigmentation is instead formed by a particular combination of yellow and blue pigments (Przibram and Lederer 1933). Przibram and Lederer (1933) proposed that the yellow pigments are carotenoids and that most of the blue pigments are biliverdins (Przibram and Lederer 1933). Later investigations led to a model that postulates that pigments are intimately associated with specific proteins and that the complex of pigment-conjugated proteins presents the visible coloration (Kawooya et al. 1985).

The blue pigment-binding protein (or bilin-binding protein, BBP) has been isolated and identified in various lepidopterans (Riley et al. 1984; Huber et al. 1987; Saito and Shimoda 1997; Kayser et al. 2009). In *M. sexta*, insecticyanin (INS) was identified to be a bilin-binding protein. Riddiford et al. (1990) found that INS is synthesized in the epidermis and is mainly stored in epidermal pigment granules or secreted into the hemolymph and cuticle (Riddiford et al. 1990). Other pigment-binding proteins are less well known. Although carotenoid-binding protein (CBP) has been well studied in vertebrates (Bhosale and Bernstein 2007), few homologs have been recognized among the Lepidoptera. In Lepidoptera, the *yellow-blood* mutant (Y) of *B. mori* (which produces yellow cocoon) was identified (Tsuchida and Sakudoh 2015); however, the expression of *BmCBP* was not detected in the epidermis. Using next-generation sequencing (NGS) technology, whole genomes of several lepidopteran species were recently released (Suetsugu et al. 2013; Li et al. 2015; Nishikawa et al. 2015; Kanost et al. 2016). Putative BmCBP homologs in other lepidopteran species can be found by BLAST search. Nonetheless, no biological experiment has been performed, and the molecular functions of these putative CBPs are largely unknown.

In *P. xuthus*, two related genes, *bilin-binding protein 1* (*BBP1*) and *yellow-related protein* (*YRG*), were identified to be associated with greenish epidermal coloration by Futahashi and Fujiwara (2008a, b) and Shirataki et al. (2010), respectively. In addition, two *putative carotenoid-binding proteins* (*PCBP1*, *PCBP2*) and other members of BBP family were later identified, which proved to be specifically expressed in the green epidermal regions during the final larval ecdysis (Futahashi et al. 2012).

15.4.2 Species-Specific Color Pattern Among Papilio *Species*

Another vital question in adaptive evolution is how the larval body pattern evolves among closely related species. There are about 200 species included in the genus *Papilio*, and these cover more than one-third of all Papilionidae (Prudic et al. 2007). In the genus *Papilio*, all the larvae share a similar bird-dropping coloration (mimetic pattern) until the fourth or fifth (final) instar (Prudic et al. 2007). The color pattern in the final instar stage is divided into three patterns: bird-dropping mimetic pattern, green cryptic pattern, and aposematic pattern with orange or black spots and black or white stripes (Prudic et al. 2007; Yamaguchi et al. 2013).

Shirataki et al. (2010) investigated the larval color pattern formation using three *Papilio* species: *P. xuthus*, *P. machaon*, and *P. polytes* (Shirataki et al. 2010). In 2015, whole-genome sequences of those three species were released and made freely accessible (Li et al. 2015; Nishikawa et al. 2015). The last instar larvae of *P. xuthus* and *P. polytes* exhibit similar green cryptic body patterns, with a pair of eyespots on the metathorax and a V-shaped marking on the abdomen, whereas the fourth and fifth instar larvae of *P. machaon* have aposematic color patterns, with a greenish epidermis covered by black bands and an orange twin-spot marking. However, *P. xuthus* and *P. machaon* are more closely related to each other than either is to *P. polytes* (Fig. 15.4) (Zakharov et al. 2004).

Shirataki et al. (2010) cloned several pigmentation-related genes, including *TH*, *DDC*, *yellow*, *BBP1*, and *YRG*, from all three species and compared their expression patterns using in situ hybridization (Fig. 15.4). The results showed a perfect correlation between gene expression and pigmentation among species. Expression of TH, DDC, and yellow matched the black regions in the eyespot, the V-shaped markings of *P. xuthus* and *P. polytes*, and the black bands of *P. machaon*. Regardless of the universal expression of BBP1 and YRG in the green regions among all the three species, BBP1 was specifically expressed in the blue spots in *P. polytes*, and YRG was tightly associated with the orange spots in *P. machaon*. Notably, a unique expression pattern of *ebony* was only detected in the red area within the eyespot region in *P. xuthus*. This work led to the model described in Fig. 15.3.

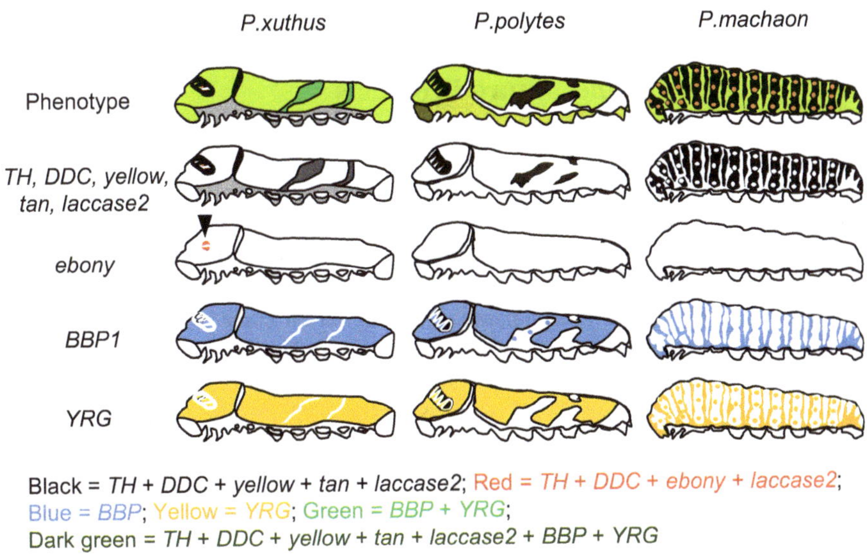

Fig. 15.4 Schema of species-specific body color pattern among three *Papilio* species (Modified from Shirataki et al. 2010)

15.4.3 Trans-regulation of YRG in the Genus Papilio

Morphological and phenotypic differences arising from evolutionary change, particularly using large-scale genetic information, have been recently identified in lepidopterans (Kunte et al. 2014; Nishikawa et al. 2015; Wallbank et al. 2016). Some studies have examined the genetic basis underlying intraspecific differences among members of the genus *Drosophila (*Massey and Wittkopp 2016; Wittkopp et al. 2009*)*. F1 hybrids allow researchers to understand regulation changes between close species (Wittkopp et al. 2003, 2008; Wittkopp and Kalay 2012).

Although hybrids of *Papilio* species are difficult to breed under laboratory conditions (Watanabe 1968), Shirataki et al. (2010) successfully bred an F1 hybrid by hand-pairing a *P. xuthus* male with a *P. polytes* female (Clarke and Sheppard 1956), and the fifth instar larvae showed intermediate characteristics between parents (Shirataki et al. 2010). One pigment-related gene, *YRG*, was selected to study expression patterns in the F1 hybrid because both the nucleotide and amino acid sequences had diverged enough to include species-specific regions. Species-specific YRG probes (PxYRG and PpYRG) were designed, and the spatial expression pattern was detected in the last instar larvae of the F1 hybrid. Both the PxYRG and PpYRG probes showed similar expression patterns, indicating that changes in expression of the *YRG* gene are mainly caused by trans-regulatory changes (Shirataki et al. 2010).

15.5 Conclusion and Future Prospects

In the swallowtail butterfly, the larval body color pattern is a vital ecological trait that affects prey–predator interactions. It is precisely regulated by ecdysteroid and juvenile hormone. In *P. xuthus*, pigmentation mechanisms and pathways have been recently elucidated. However, the details of hormonal regulation need to be understood, and the molecular mechanism underlying larval body color patterning has not been studied. New information from next-generation whole-genome sequencing projects will provide a valuable resource that can be used to gain insight into the genetic basis underlying those questions. Moreover, pioneering functional analysis methods, like electroporation-mediated transgenic methods (Ando and Fujiwara 2013) and the CRISPR/Cas9 system (Li et al. 2015), may also lead to new approaches for examining gene functions in non-model species, such as *P. xuthus*.

References

Ando T, Fujiwara H (2013) Electroporation-mediated somatic transgenesis for rapid functional analysis in insects. Development 140(2):454–458. doi:10.1242/dev.085241

Arakane Y, Muthukrishnan S, Beeman RW, Kanost MR, Kramer KJ (2005) Laccase 2 is the phenoloxidase gene required for beetle cuticle tanning. Proc Natl Acad Sci U S A 102 (32):11337–11342. doi:10.1073/pnas.0504982102

Bates HW (1862) XXXII. Contributions to an insect fauna of the Amazon valley. Lepidoptera: Heliconidæ. Trans Linn Soc Lond 23(3):495–566. doi:10.1111/j.1096-3642.1860.tb00146.x

Bhosale P, Bernstein PS (2007) Vertebrate and invertebrate carotenoid-binding proteins. Arch Biochem Biophys 458(2):121–127. doi:10.1016/j.abb.2006.10.005

Chen L, Reece C, O'Keefe SL, Hawryluk GW, Engstrom MM, Hodgetts RB (2002) Induction of the early-late Ddc gene during Drosophila metamorphosis by the ecdysone receptor. Mech Dev 114(1–2):95–107

Clarke CA, Sheppard PM (1956) Hand-pairing of butterflies. Lepid News New Haven 10:47–53

Darwin CR (1863) To H.W. Bates "Letter no.3816,". Cleveland Health Sciences Library, Cleveland

Futahashi R (2006) Molecular mechanisms of mimicry in larval body marking of the swallowtail butterfly, Papilio xuthus. The University of Tokyo, Tokyo

Futahashi R, Fujiwara H (2005) Melanin-synthesis enzymes coregulate stage-specific larval cuticular markings in the swallowtail butterfly, Papilio xuthus. Dev Genes Evol 215 (10):519–529. doi:10.1007/s00427-005-0014-y

Futahashi R, Fujiwara H (2007) Regulation of 20-hydroxyecdysone on the larval pigmentation and the expression of melanin synthesis enzymes and yellow gene of the swallowtail butterfly, Papilio xuthus. Insect Biochem Mol Biol 37(8):855–864. doi:10.1016/j.ibmb.2007.02.014

Futahashi R, Fujiwara H (2008a) Juvenile hormone regulates butterfly larval pattern switches. Science 319(5866):1061. doi:10.1126/science.1149786

Futahashi R, Fujiwara H (2008b) Identification of stage-specific larval camouflage associated genes in the swallowtail butterfly, Papilio xuthus. Dev Genes Evol 218(9):491–504. doi:10.1007/s00427-008-0243-y

Futahashi R, Banno Y, Fujiwara H (2010) Caterpillar color patterns are determined by a two-phase melanin gene prepatterning process: new evidence from tan and laccase2. Evol Dev 12 (2):157–167. doi:10.1111/j.1525-142X.2010.00401.x

Futahashi R, Tanaka K, Matsuura Y, Tanahashi M, Kikuchi Y, Fukatsu T (2011) Laccase2 is required for cuticular pigmentation in stinkbugs. Insect Biochem Mol Biol 41(3):191–196. doi:10.1016/j.ibmb.2010.12.003

Futahashi R, Shirataki H, Narita T, Mita K, Fujiwara H (2012) Comprehensive microarray-based analysis for stage-specific larval camouflage pattern-associated genes in the swallowtail butterfly, Papilio xuthus. BMC Biol 10:46. doi:10.1186/1741-7007-10-46

Heliconius Genome C (2012) Butterfly genome reveals promiscuous exchange of mimicry adaptations among species. Nature 487(7405):94–98. doi:10.1038/nature11041

Hiruma K, Riddiford LM (1986) Inhibition of dopa decarboxylase synthesis by 20-hydroxyecdysone during the last larval moult of Manduca sexta. Insect Biochem 16 (1):225–231. doi:10.1016/0020-1790(86)90100-9

Hiruma K, Riddiford LM (1988) Granular phenoloxidase involved in cuticular melanization in the tobacco hornworm: regulation of its synthesis in the epidermis by juvenile hormone. Dev Biol 130(1):87–97

Hiruma K, Riddiford LM (1990) Regulation of dopa decarboxylase gene expression in the larval epidermis of the tobacco hornworm by 20-hydroxyecdysone and juvenile hormone. Dev Biol 138(1):214–224

Hiruma K, Riddiford LM (2007) The coordination of the sequential appearance of MHR4 and dopa decarboxylase during the decline of the ecdysteroid titer at the end of the molt. Mol Cell Endocrinol 276(1–2):71–79. doi:10.1016/j.mce.2007.07.002

Hiruma K, Riddiford LM (2009) The molecular mechanisms of cuticular melanization: the ecdysone cascade leading to dopa decarboxylase expression in Manduca sexta. Insect Biochem Mol Biol 39(4):245–253. doi:10.1016/j.ibmb.2009.01.008

Hiruma K, Riddiford LM, Hopkins TL, Morgan TD (1985) Roles of dopa decarboxylase and phenoloxidase in the melanization of the tobacco hornworm and their control by 20-hydroxyecdysone. J Comp Physiol B 155(6):659–669

Hiruma K, Carter MS, Riddiford LM (1995) Characterization of the dopa decarboxylase gene of Manduca sexta and its suppression by 20-hydroxyecdysone. Dev Biol 169(1):195–209. doi:10.1006/dbio.1995.1137

Huber R, Schneider M, Mayr I, Muller R, Deutzmann R, Suter F, Zuber H, Falk H, Kayser H (1987) Molecular structure of the bilin binding protein (BBP) from Pieris brassicae after refinement at 2.0 A resolution. J Mol Biol 198(3):499–513

Hwang JS, Kang SW, Goo TW, Yun EY, Lee JS, Kwon OY, Chun T, Suzuki Y, Fujiwara H (2003) cDNA cloning and mRNA expression of L-3,4-dihydroxyphenylalanine decarboxylase gene homologue from the silkworm, Bombyx mori. Biotechnol Lett 25(12):997–1002

Jindra M, Riddiford LM (1996) Expression of ecdysteroid-regulated transcripts in the silk gland of the wax moth, Galleria mellonella. Dev Genes Evol 206(5):305–314. doi:10.1007/s004270050057

Jindra M, Sehnal F, Riddiford LM (1994) Isolation, characterization and developmental expression of the ecdysteroid-induced E75 gene of the wax moth Galleria mellonella. Eur J Biochem 221(2):665–675

Jindra M, Palli SR, Riddiford LM (2013) The juvenile hormone signaling pathway in insect development. Annu Rev Entomol 58:181–204. doi:10.1146/annurev-ento-120811-153700

Kalay G, Lusk R, Dome M, Hens K, Deplancke B, Wittkopp PJ (2016) Potential direct regulators of the Drosophila yellow gene identified by yeast one-hybrid and RNAi screens. G3 (Bethesda) 6(10):3419–3430. doi:10.1534/g3.116.032607

Kanost MR, Arrese EL, Cao X, Chen YR, Chellapilla S, Goldsmith MR, Grosse-Wilde E, Heckel DG, Herndon N, Jiang H, Papanicolaou A, Qu J, Soulages JL, Vogel H, Walters J, Waterhouse RM, Ahn SJ, Almeida FC, An C, Aqrawi P, Bretschneider A, Bryant WB, Bucks S, Chao H, Chevignon G, Christen JM, Clarke DF, Dittmer NT, Ferguson LC, Garavelou S, Gordon KH, Gunaratna RT, Han Y, Hauser F, He Y, Heidel-Fischer H, Hirsh A, Hu Y, Jiang H, Kalra D, Klinner C, Konig C, Kovar C, Kroll AR, Kuwar SS, Lee SL, Lehman R, Li K, Li Z, Liang H, Lovelace S, Lu Z, Mansfield JH, McCulloch KJ, Mathew T, Morton B, Muzny DM, Neunemann D, Ongeri F, Pauchet Y, Pu LL, Pyrousis I, Rao XJ, Redding A, Roesel C, Sanchez-Gracia A, Schaack S, Shukla A, Tetreau G, Wang Y, Xiong GH, Traut W, Walsh TK, Worley KC, Wu D, Wu W, Wu YQ, Zhang X, Zou Z, Zucker H, Briscoe AD, Burmester T, Clem RJ, Feyereisen R, Grimmelikhuijzen CJ, Hamodrakas SJ, Hansson BS, Huguet E, Jermiin LS, Lan Q, Lehman HK, Lorenzen M, Merzendorfer H, Michalopoulos I, Morton DB, Muthukrishnan S, Oakeshott JG, Palmer W, Park Y, Passarelli AL, Rozas J, Schwartz LM, Smith W, Southgate A, Vilcinskas A, Vogt R, Wang P, Werren J, Yu XQ, Zhou JJ, Brown SJ, Scherer SE, Richards S, Blissard GW (2016) Multifaceted biological insights from a draft genome sequence of the tobacco hornworm moth, Manduca sexta. Insect Biochem Mol Biol. doi:10.1016/j.ibmb.2016.07.005

Karlson P, Sekeris CE (1976) Control of tyrosine metabolism and cuticle sclerotization by ecdysone. The insect integument. 1-571. Elsevier Scientific Publishing Company, Amsterdam/Oxford/New York

Kawooya JK, Keim PS, Law JH, Riley CT, Ryan RO, Shapiro JP (1985) Why are green caterpillars green. ACS Symp Ser 276:511–521

Kayser H, Mann K, Machaidze G, Nimtz M, Ringler P, Muller SA, Aebi U (2009) Isolation, characterisation and molecular imaging of a high-molecular-weight insect biliprotein, a member of the Hexameric Arylphorin protein family. J Mol Biol 389(1):74–89. doi:10.1016/j.jmb.

2009.03.075
Kayukawa T, Nagamine K, Ito Y, Nishita Y, Ishikawa Y, Shinoda T (2016) Kruppel homolog 1 inhibits insect metamorphosis via direct transcriptional repression of broad-complex, a pupal specifier gene. J Biol Chem 291(4):1751–1762. doi:10.1074/jbc.M115.686121
Kopec S (1926) Experiments on metamorphosis of insects. Bull Int Acad Polon Sci Cracow B 1917:57–60
Kramer KJ, Hopkins TL (1987) Tyrosine metabolism for insect cuticle tanning. Arch Insect Biochem Physiol 6(4):279–301. doi:10.1002/arch.940060406
Kunte K, Zhang W, Tenger-Trolander A, Palmer DH, Martin A, Reed RD, Mullen SP, Kronforst MR (2014) Doublesex is a mimicry supergene. Nature 507(7491):229–232. doi:10.1038/nature13112
Li X, Fan D, Zhang W, Liu G, Zhang L, Zhao L, Fang X, Chen L, Dong Y, Chen Y, Ding Y, Zhao R, Feng M, Zhu Y, Feng Y, Jiang X, Zhu D, Xiang H, Feng X, Li S, Wang J, Zhang G, Kronforst MR, Wang W (2015) Outbred genome sequencing and CRISPR/Cas9 gene editing in butterflies. Nat Commun 6:8212. doi:10.1038/ncomms9212
Massey JH, Wittkopp PJ (2016) The genetic basis of pigmentation differences within and between Drosophila species. Curr Top Dev Biol 119:27–61. doi:10.1016/bs.ctdb.2016.03.004
Meldola R (1873) On a certain class of cases of variable protective colouring in insects. Proc Zool Soc:153–162
Nijhout HF (1991) The development and evolution of butterfly wing patterns. Smithsonian series in comparative evolutionary biology: the development and evolution of butterfly wing patterns. Smithsonian Institution Press, Washington, DC/London
Nijhout HF, Wheeler DE (1982) Juvenile-hormone and the physiological-basis of insect polymorphisms. Q Rev Biol 57(2):109–133
Nishikawa H, Iijima T, Kajitani R, Yamaguchi J, Ando T, Suzuki Y, Sugano S, Fujiyama A, Kosugi S, Hirakawa H, Tabata S, Ozaki K, Morimoto H, Ihara K, Obara M, Hori H, Itoh T, Fujiwara H (2015) A genetic mechanism for female-limited Batesian mimicry in Papilio butterfly. Nat Genet. doi:10.1038/ng.3241
Noh MY, Muthukrishnan S, Kramer KJ, Arakane Y (2016) Cuticle formation and pigmentation in beetles. Curr Opin Insect Sci 17:1–9. doi:10.1016/j.cois.2016.05.004
Ogihara MH, Hikiba J, Iga M, Kataoka H (2015) Negative regulation of juvenile hormone analog for ecdysteroidogenic enzymes. J Insect Physiol. doi:10.1016/j.jinsphys.2015.03.012
Palli SR, Sohi SS, Cook BJ, Lambert D, Ladd TR, Retnakaran A (1995) Analysis of ecdysteroid action in Malacosoma disstria cells: cloning selected regions of E75- and MHR3-like genes. Insect Biochem Mol Biol 25(6):697–707
Pasteur G (1982) A classificatory review of mimicry systems. Annu Rev Ecol Syst 13:169–199
Prudic KL, Oliver JC, Sperling FA (2007) The signal environment is more important than diet or chemical specialization in the evolution of warning coloration. Proc Natl Acad Sci U S A 104 (49):19381–19386. doi:10.1073/pnas.0705478104
Przibram H, Lederer E (1933) Das Tiergrun der Heuschrecken als Mischung aus Farbstoffen. Anz Akad Wiss Wien 70:163–165
Reed RD, Papa R, Martin A, Hines HM, Counterman BA, Pardo-Diaz C, Jiggins CD, Chamberlain NL, Kronforst MR, Chen R, Halder G, Nijhout HF, McMillan WO (2011) Optix drives the repeated convergent evolution of butterfly wing pattern mimicry. Science 333 (6046):1137–1141. doi:10.1126/science.1208227
Riddiford LSLM (1975) The biology of the black larval mutant of the tobacco hornworm, Manduca sexta. J Insect Physiol 21(12):1931–1933, 1935–1938
Riddiford LM (1996) Juvenile hormone: the status of its "status quo" action. Arch Insect Biochem Physiol 32(3–4):271–286. doi:10.1002/(SICI)1520-6327(1996)32:3/4<271::AID-ARCH2>3.0.CO;2-W
Riddiford LM, Palli SR, Hiruma K, Li W, Green J, Hice RH, Wolfgang WJ, Webb BA (1990) Developmental expression, synthesis, and secretion of insecticyanin by the epidermis of the tobacco hornworm, Manduca sexta. Arch Insect Biochem Physiol 14(3):171–190. doi:10.1002/

arch.940140305
Riley CT, Barbeau BK, Keim PS, Kezdy FJ, Heinrikson RL, Law JH (1984) The covalent protein structure of insecticyanin, a blue biliprotein from the hemolymph of the tobacco hornworm, Manduca sexta L. J Biol Chem 259(21):13159–13165
Saito H, Shimoda M (1997) Insecticyanin of Agrius convolvuli: purification and characterization of the biliverdin-binding protein from the larval hemolymph. Zool Sci 14(5):777–783. doi:10.2108/zsj.14.777
Scoble MJ, Scoble MJ (1992) The Lepidoptera. British Museum/Oxford University Press, London/Oxford
Shirataki H, Futahashi R, Fujiwara H (2010) Species-specific coordinated gene expression and trans-regulation of larval color pattern in three swallowtail butterflies. Evol Dev 12 (3):305–314. doi:10.1111/j.1525-142X.2010.00416.x
Suetsugu Y, Futahashi R, Kanamori H, Kadono-Okuda K, Sasanuma S, Narukawa J, Ajimura M, Jouraku A, Namiki N, Shimomura M, Sezutsu H, Osanai-Futahashi M, Suzuki MG, Daimon T, Shinoda T, Taniai K, Asaoka K, Niwa R, Kawaoka S, Katsuma S, Tamura T, Noda H, Kasahara M, Sugano S, Suzuki Y, Fujiwara H, Kataoka H, Arunkumar KP, Tomar A, Nagaraju J, Goldsmith MR, Feng Q, Xia Q, Yamamoto K, Shimada T, Mita K (2013) Large scale full-length cDNA sequencing reveals a unique genomic landscape in a lepidopteran model insect, Bombyx mori. G3 (Bethesda) 3(9):1481–1492. doi:10.1534/g3.113.006239
True JR, Edwards KA, Yamamoto D, Carroll SB (1999) Drosophila wing melanin patterns form by vein-dependent elaboration of enzymatic prepatterns. Curr Biol 9(23):1382–1391
Tsuchida K, Sakudoh T (2015) Recent progress in molecular genetic studies on the carotenoid transport system using cocoon-color mutants of the silkworm. Arch Biochem Biophys 572:151–157. doi:10.1016/j.abb.2014.12.029
Tullberg BS, Merilaita S, Wiklund C (2005) Aposematism and crypsis combined as a result of distance dependence: functional versatility of the colour pattern in the swallowtail butterfly larva. Proc Biol Sci 272(1570):1315–1321. doi:10.1098/rspb.2005.3079
Urena E, Manjon C, Franch-Marro X, Martin D (2014) Transcription factor E93 specifies adult metamorphosis in hemimetabolous and holometabolous insects. Proc Natl Acad Sci U S A 111 (19):7024–7029. doi:10.1073/pnas.1401478111
Wallbank RW, Baxter SW, Pardo-Diaz C, Hanly JJ, Martin SH, Mallet J, Dasmahapatra KK, Salazar C, Joron M, Nadeau N, McMillan WO, Jiggins CD (2016) Evolutionary novelty in a butterfly wing pattern through enhancer shuffling. PLoS Biol 14(1):e1002353. doi:10.1371/journal.pbio.1002353
Walter MF, Black BC, Afshar G, Kermabon AY, Wright TR, Biessmann H (1991) Temporal and spatial expression of the yellow gene in correlation with cuticle formation and dopa decarboxylase activity in Drosophila development. Dev Biol 147(1):32–45
Watanabe K (1968) A study of interspecific hybrids between Papilio machaon hippocrates and P. xuthus. Trans Lep Soc Jpn 19(1&2):3
Wittkopp PJ, Kalay G (2012) Cis-regulatory elements: molecular mechanisms and evolutionary processes underlying divergence. Nat Rev Genet 13(1):59–69. doi:10.1038/nrg3095
Wittkopp PJ, Vaccaro K, Carroll SB (2002) Evolution of yellow gene regulation and pigmentation in Drosophila. Curr Biol 12(18):1547–1556
Wittkopp PJ, Carroll SB, Kopp A (2003) Evolution in black and white: genetic control of pigment patterns in Drosophila. Trends Genet 19(9):495–504. doi:10.1016/S0168-9525(03)00194-X
Wittkopp PJ, Haerum BK, Clark AG (2008) Regulatory changes underlying expression differences within and between Drosophila species. Nat Genet 40(3):346–350. doi:10.1038/ng.77
Wittkopp PJ, Stewart EE, Arnold LL, Neidert AH, Haerum BK, Thompson EM, Akhras S, Smith-Winberry G, Shefner L (2009) Intraspecific polymorphism to interspecific divergence: genetics of pigmentation in Drosophila. Science 326(5952):540–544. doi:10.1126/science.1176980
Wright TR (1987) The genetics of biogenic amine metabolism, sclerotization, and melanization in Drosophila melanogaster. Adv Genet 24:127–222

Yamaguchi J, Banno Y, Mita K, Yamamoto K, Ando T, Fujiwara H (2013) Periodic Wnt1 expression in response to ecdysteroid generates twin-spot markings on caterpillars. Nat Commun 4:1857. doi:10.1038/ncomms2778

Yamanaka N, Rewitz KF, O'Connor MB (2013) Ecdysone control of developmental transitions: lessons from Drosophila research. Annu Rev Entomol 58:497–516. doi:10.1146/annurev-ento-120811-153608

Yao TP, Segraves WA, Oro AE, McKeown M, Evans RM (1992) Drosophila ultraspiracle modulates ecdysone receptor function via heterodimer formation. Cell 71(1):63–72

Zakharov EV, Caterino MS, Sperling FA (2004) Molecular phylogeny, historical biogeography, and divergence time estimates for swallowtail butterflies of the genus Papilio (Lepidoptera: Papilionidae). Syst Biol 53(2):193–215. doi:10.1080/10635150490423403

16

Pigmentation Pattern Formation in Drosophila Guttifera

Shigeyuki Koshikawa, Yuichi Fukutomi and Keiji Matsumoto

Abstract A polka-dotted fruit fly, *Drosophila guttifera*, has a unique pigmentation pattern made of black melanin and serves as a good model system to study color pattern formation. Because of its short generation time and the availability of transgenics, it is suitable for dissecting the genetic mechanisms of color pattern formation. While the ecology and life history of *D. guttifera* in the wild are not well understood, it is known to be resistant to a mushroom toxin, and this physiological trait is under molecular scrutiny. Pigmentation around crossveins and longitudinal vein tips is common in closely related species of the *quinaria* group, in addition to which *D. guttifera* has evolved species-specific pigmentation spots around the campaniform sensilla. Regulatory evolution of the Wnt signaling ligand Wingless, which locally induces pigmentation in the developing wing epithelium, has driven the evolution of distinct aspects of wing and body pigmentation. A melanin biosynthesis pathway gene, *yellow*, is also involved in the elaboration of these traits, downstream of *wingless*. Unraveling the detailed mechanism of pigmentation pattern formation of this species sheds light on the general principles of morphological evolution and foreshadows potential parallels with other systems, such as the pigmented wings of butterflies.

Keywords *Drosophila guttifera* • Pigmentation • Color pattern • Evolution • Development • Transgenic • *Cis*-regulatory element • Phylogeny • Ecology • Life history • Taxonomy

---

S. Koshikawa (✉)
The Hakubi Center for Advanced Research, Kyoto University, Sakyo-ku, Kyoto 606-8501, Japan

Graduate School of Science, Kyoto University, Sakyo-ku, Kyoto 606-8501, Japan
e-mail: koshikawa@mdb.biophys.kyoto-u.ac.jp

Y. Fukutomi
Graduate School of Science, Kyoto University, Sakyo-ku, Kyoto 606-8501, Japan

K. Matsumoto
Graduate School of Science, Kyoto University, Sakyo-ku, Kyoto 606-8501, Japan

Graduate School of Science, Osaka City University, Sumiyoshi-ku, Osaka 558-8585, Japan

16.1 Introduction

Research on butterfly color patterns has greatly advanced in recent years. Knowledge of the characteristics of the genome, mechanisms of pattern formation, and the function and evolutionary mode of the pattern is rapidly growing. This was enabled by utilization of multiple model species, including species of *Bicyclus*, *Heliconius*, *Junonia*, *Vanessa*, *Papilio*, and others, and by the best use of characteristics of materials (Nijhout 1991; Carroll et al. 1994; Brakefield et al. 1996; Joron et al. 2011; Reed et al. 2011; The Heliconius Genome Consortium 2012; Martin et al. 2012; Kunte et al. 2014; Monteiro 2015; Nishikawa et al. 2015; Beldade and Peralta 2017).

In vertebrates, zebrafish (*Danio rerio*) has been a model of color pattern formation, and recently, domestic and wild cats and a four-striped mouse (*Rhabdomys pumilio*) were also used for research, making this an exciting time for color pattern studies (Singh and Nüsslein-Volhard 2015; Kaelin et al. 2012; Mallarino et al. 2016).

We have been using a dipteran insect, *Drosophila guttifera*, to study a mechanism of color pattern formation (Fig. 16.1). *D. guttifera* has a pattern on its wings, which is a commonality with butterflies; however, there are also some important differences. In contrast with the pigmented scales of butterflies and moths (as an exception, see Stavenga et al. 2010), *Drosophila* pigmentation is embedded in the cuticle layers of the wing membrane. This pigmentation is believed to be made of black melanin. A congeneric species, *Drosophila melanogaster*, is a model organism widely used in genetics and various biological researches, and we can utilize its knowledge, techniques, and resources to study *D. guttifera*. This phylogenetic proximity is an asset, as it is possible to transfer a part of the genetic system, such as an enhancer involved in pattern formation, into *D. melanogaster* and analyze its function in a heterologous context. *D. guttifera* has the potential to

Fig. 16.1 Adult male of *Drosophila guttifera*. The pigmentation pattern is very similar between the sexes

approach the same problem of color pattern as in butterflies but from a different angle. It also enables a good comparison, since its complex pigmentation patterns evolved independently from the ones seen in butterflies.

In this chapter, we present an overview of the biology of *D. guttifera*. Then we discuss differences in pattern formation between *D. guttifera* and butterflies and the advantage and potential of *D. guttifera* to contribute to the general understanding of animal color pattern formation.

16.2 Phylogenetic Position of *D. guttifera*

Fruit flies (drosophilid flies) belong to family Drosophilidae, order Diptera, and consist of 72 genera and more than 4000 described species (Yassin 2013). Among them, genus *Drosophila* includes more than 1160 described species (Markow and O'Grady 2006; Toda 2017). The best-studied species, *D. melanogaster*, also belongs to this genus. It should be noted, however, that the genus *Drosophila* is not monophyletic and potentially includes multiple genera within this clade, and there is ongoing debate on the proper taxonomic treatment of this genus (O'Grady 2010).

D. guttifera was described by an English entomologist, Francis Walker, based on a specimen collected in Florida (Walker 1849). This description consisted of 4 lines in Latin and 21 lines in English with no illustration and was one of many descriptions of a museum collection of the British Museum. In his taxonomic revision of North American drosophilids, Sturtevant (1921) examined multiple specimens of *D. guttifera* and redescribed the morphological features. Sturtevant (1942) established "species groups" to classify species within the genus *Drosophila*. *D. guttifera* was assigned to a monospecific *guttifera* group. He also established the *quinaria* group, which includes 11 species (*D. quinaria*, *deflecta*, *palustris*, *subpalustris*, *occidentalis*, *suboccidentalis*, *munda*, *subquinaria*, *transversa*, and possibly *phalerata* and *nigromaculata*). Patterson (1943) revised drosophilids of the Southwestern United States and Northern Mexico and redescribed many species with beautiful illustrations. *D. guttifera* was redescribed with illustrations of a pupa and internal organs of reproduction and a color illustration of the whole body. Patterson also described three new species in the *quinaria* group (*D. suffusca*, *tenebrosa*, and *innubila*). After that, many species were described in the *quinaria* group, and currently it includes 31 species (Markow and O'Grady 2006, Toda 2017).

The close relationship between *D. guttifera* and the *quinaria* group is almost certain at this time, based on molecular genetic evidence (Perlman et al. 2003; Izumitani et al. 2016). Morphological similarity between *D. guttifera* and the *quinaria* group was also noticed (Patterson and Stone 1952), and some authors even placed *D. guttifera* in the *quinaria* group (Throckmorton 1962, 1975; Markow and O'Grady 2006). Species-level relationships among *D. guttifera* and species of the *quinaria* group are not completely resolved; however, the commonly supported result is bifurcation into two clades, one including mostly North American species and one including mostly Eurasian species (Perlman et al. 2003, Markow and O'Grady 2006, Izumitani et al. 2016).

There are species with pigment patterns on the thorax, abdomen, and wings to various degrees in the *quinaria* group (Patterson 1943; Werner and Jaenike 2017), but *D. guttifera* has distinctive vertical stripes on the thorax and a polka dot pattern on the abdomen and wings. Even when compared with the *quinaria* group species, *D. guttifera* has the most prominently pigmented appearance.

16.3 Food Habits, Poison Resistance, and Behavioral Ecology of *D. guttifera*

The life history and ecology of *D. guttifera* in the wild have not been well studied. There are many species of the *quinaria* group that utilize mushrooms as a food source. Sturtevant (1921) assumed *D. guttifera* is also a mushroom feeder based on the facts that *D. guttifera* was found around mushrooms and that he could rear *D. guttifera*, from eggs to adults, with mushrooms (he noted that both gill fungi and pore fungi can be utilized, but he did not describe mushroom species). Bunyard and Foote (1990a) studied what kind of dipteran insects emerged from mushrooms collected in the state of Ohio and reported that *D. guttifera* emerged from two mushroom species, *Psilocybe polytrichophila* and *Collybia dryophila*. They tested oviposition site preference among commercial *Agaricus bisporus*, banana, tomato, lettuce, and agar and found that *Agaricus* was the most preferred site (Bunyard and Foote 1990b). They also confirmed that *D. guttifera* can grow from eggs to adults with *Agaricus*. In laboratory conditions, however, we can keep strains of *D. guttifera* with artificial food containing sugar/corn meal/yeast/agar (sugar food) or molasses/corn meal/yeast/agar (molasses food) without adding mushrooms.

Some fungus-feeding drosophilids are known to have high tolerance to a mushroom toxin, alpha-amanitin, which is highly poisonous to most animals (Spicer and Jaenike 1996). *D. guttifera* has the potential to be a model system to study this phenomenon. Alpha-amanitin exerts its toxicity by binding to RNA polymerase II, an enzyme essential for transcription. A mutant strain of *D. melanogaster* with high alpha-amanitin tolerance had an amino acid substitution in RNA polymerase II (Chen et al. 1993). However, *D. guttifera* and other species with the tolerance do not have the same substitution, indicating that other mechanisms are involved (Stump et al. 2011). There are other strains of *D. melanogaster* with alpha-amanitin tolerance but without RNA polymerase II mutation. The responsible locus was mapped, and gene expression profiles were analyzed in these strains (Begum and Whitley 2000; Mitchell et al. 2014, 2015).

There are some other studies of *D. guttifera* behavior. Oviposition site preference of *D. guttifera* was affected by larval food condition, and this is known as a classic example of olfactory conditioning of animals (Cushing 1941). The mating behavior of *D. guttifera* was also studied (Grossfield 1977). The ecological significance and function of pigmentation patterns of *D. guttifera* is not well understood. Some drosophilids are known to use wing pigmentation in courtship displays (Ringo and Hodosh 1978; Yeh et al. 2006; Fuyama 1979). Dombeck and Jaenike (2004) analyzed fitness effects of abdominal spot number in *D. falleni*.

16.4 The Evolution of Wing Pigmentation Pattern

Dombeck and Jaenike (2004) analyzed and illustrated the evolutionary path of wing and abdominal pigmentations of *D. guttifera* and seven species of the *quinaria* group. We summarize here the evolution of wing pigmentation pattern of *D. guttifera* and the *quinaria* group species based on molecular phylogenetics (Fig. 16.2). As previously explained, the *quinaria* group is divided into two major clades (Perlman et al. 2003; Markow and O'Grady 2006, Izumitani et al. 2016). We defined the clade with mostly North American species as "clade A" and the clade with mostly Eurasian species as "clade B." Species in clade A have relatively simple patterns; pigmentations are formed only around crossveins except in *D. innubila*, which has no pigmentation. The evolution of patterns in clade B is rather complicated. The relationships among basal species of clade B [*D. guttifera*, *nigromaculata*, and (*deflecta* + *palustris* + *subpalustris*)] have not been completely resolved, because the topologies of the phylogenetic trees depend on the analytical methods. These four species have pigmentations around crossveins and longitudinal vein tips. In addition, *D. guttifera* has pigmentations around the campaniform sensilla, which is unique to this species [at least unique among the clade of (*quinaria* group + *D. guttifera*) and probably among the genus *Drosophila*]. Among the rest of the species in clade B, *D. quinaria* has weak pigmentations on the tips of longitudinal veins in addition to crossveins. *D. recens* and many other species within this cluster have pigmentations around crossveins. *D. kuntzei*, which has a similar pattern to *D. quinaria*, branches from the most basal position of clade B according to Perlman et al. (2003), although the statistical support for this topology was low. Due to the lack of a robust phylogeny, it would be premature to propose a simple scenario stepwise pattern of gain and loss within the *quinaria* group. It is plausible that the instances of longitudinal vein tip pigmentation are the result of convergent evolution, perhaps via parallel mechanisms, although we cannot exclude the possibility of a single gain of the longitudinal vein tip pigmentation and a secondary loss in derived species of clade B. Nevertheless, the other dot-like patterns of *D. guttifera*, which overlap in position with innervated cupules known as campaniform sensilla (see below), are unique to this species and are assumed to form a true evolutionary novelty.

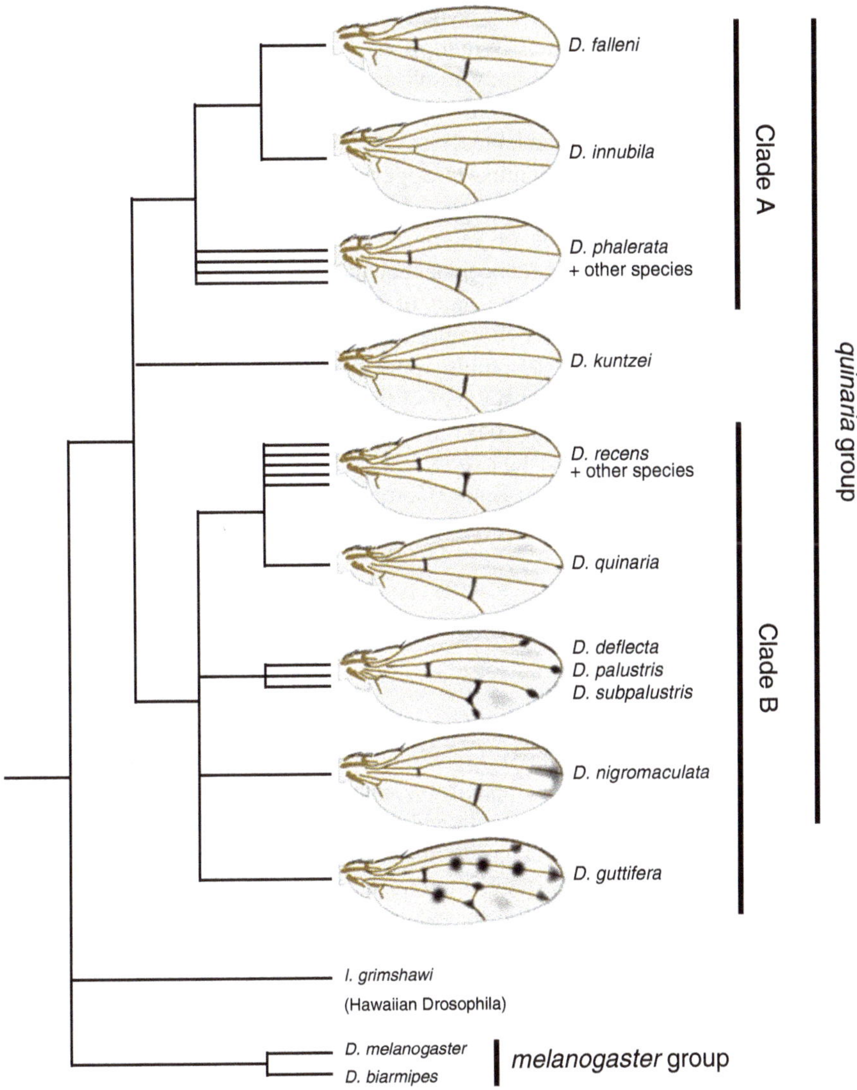

Fig. 16.2 Phylogenetic relationships of *D. guttifera* and species in the *quinaria* group. The topology was drawn from a consensus between Perlman et al. (2003) and Izumitani et al. (2016). See also Fig. 16.3 for interpretation of pigmentation

16.5 Wing Pigmentation Pattern Formation in *Drosophila*

The initial study of the mechanism of wing pigmentation pattern formation was done by True et al. (1999). They argued that patterns are formed through patterning by gene expression and subsequent elaboration by precursor trafficking through wing veins, based on experiments using *Drosophila grimshawi* (synonym of

Idiomyia grimshawi), *D. rajasekari* (synonym of *D. biarmipes*), and mutants and transgenics of *D. melanogaster*. Wittkopp et al. (2002) studied the function of *yellow* and *ebony* genes in the body trunk and wings of *D. melanogaster*. They also showed that the future spot position had more Yellow protein and less Ebony protein. Yellow is known to enhance black melanin synthesis, and Ebony is an enzyme that conjugates beta-alanine to dopamine and produces NBAD (N-beta-alanyldopamine) resulting in repression of black melanin synthesis. Gompel et al. (2005) analyzed the regulation of *yellow* gene expression in *D. biarmipes* and showed that evolution of an enhancer (a sequence that enhances expression of a nearby gene) was involved in the gain of pigmentation. In *D. biarmipes* and *D. guttifera*, they showed that Yellow protein was localized in future black spots and Ebony protein was localized in future transparent (no pigmentation) places. The *yellow* expression in the anteriodistal part of the wing in *D. biarmipes* results from regulation by at least two factors: posterior expression of *engrailed* repressing the *yellow* expression and anteriodistal expression of *Distal-less* enhancing expressions of *yellow* and other pigmentation genes (Gompel et al. 2005; Arnoult et al. 2013).

16.6 Features of Wing Pigmentation Pattern in *D. guttifera*

D. guttifera has prominent black polka dots on its wings, and these are believed to be made with melanin (Fig. 16.3). Pigmentations are formed around crossveins, longitudinal vein tips, and the campaniform sensilla. Weak pigmentations are also formed in intervein regions. As mentioned previously, crossvein pigmentation is widely observed in the *quinaria* group and also found in many species in other species groups. The crossvein pigmentation in *D. guttifera* is constricted in the center, forming an hourglass shape (or calabash shape), and this is unique to this species. Longitudinal vein tip pigmentations are observed in a few species, but the pigmentation area is largest in *D. guttifera*. Campaniform sensilla pigmentation is a trait unique to *D. guttifera*, although some species, such as a Hawaiian species, *Idiomyia grimshawi* (synonym of *Drosophila grimshawi*), have dappled spots all over the wings. The campaniform sensilla are lined on the third longitudinal vein in the same way as in other drosophilids, but in *D. guttifera*, one campaniform sensillum is also found on the fifth longitudinal vein, which is unique to this species. This campaniform sensillum is also surrounded by pigmentation (Sturtevant 1921; Werner et al. 2010). The wing pigmentation of *D. guttifera* starts to form in the pupal period, and it continues until one day old adult (Fukutomi et al. 2017).

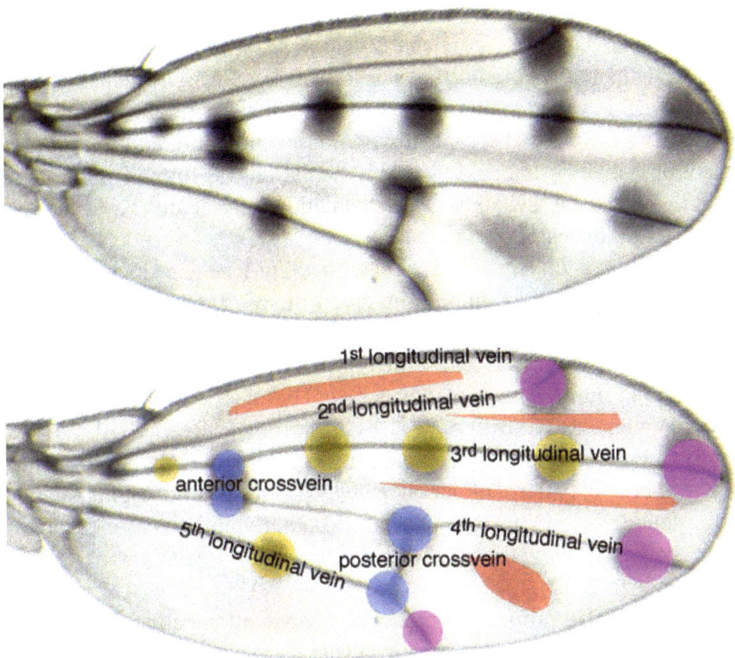

Fig. 16.3 *Top* Wing pigmentation of *D. guttifera*. *Bottom* Interpretation of the pigmentation pattern. Blue marks pigmentations around crossveins, purple marks longitudinal vein tips, yellow marks campaniform sensilla, and red marks intervein shading

16.7 *Wingless* Gene Induces Pigmentation Pattern Formation in *D. guttifera*

Werner et al. (2010) analyzed the *cis*-regulatory region of the *yellow* gene and identified *vein spot* CRE, which is an enhancer driving expression in all the polka dots, and *intervein shade* CRE, which is an enhancer driving expression in the intervein region. *Vein spot* CRE drove polka dots in *D. guttifera* but drove around crossveins and longitudinal vein tips if introduced in *D. melanogaster*. This difference means there is a difference in localization of a *trans*-regulatory factor that has an input to *vein spot* CRE. Gene expression patterns were known for several genes in *D. melanogaster*, and therefore they found candidate genes from genes showing similar expression with the *vein spot* CRE pattern. Among the candidate genes, *wingless*, a gene encoding a ligand of the Wnt signaling pathway, showed expression in the center of future spot positions (crossveins, longitudinal vein tips, and campaniform sensilla) in *D. guttifera*. There was no *wingless* expression in the campaniform sensilla in a closely related species, *D. deflecta*, which does not have pigmentation around them. A spontaneous mutant line of *D. guttifera*, *schwarzvier*, has additional pigmentation on the fourth longitudinal vein. In this mutant line, *wingless* was ectopically expressed on the fourth longitudinal vein. To obtain direct functional evidence, they tried to make ectopic expressions of *wingless* by construction of the GAL4/UAS system in *D. guttifera*. Although they did not obtain optimal GAL4 lines, they found that one of the UAS-*wingless* lines had ectopic expression of *wingless*, probably caused by the enhancer trap mechanism. In this line, *wingless* was expressed ectopically on the second, third, and fourth longitudinal veins of pupal wings, and additional pigmentation was formed on these veins in adult wings. With these evidences, they concluded that *wingless* is the upstream *trans*-factor that induces pigmentation.

In *Heliconius* and *Limenitis* butterflies, the *WntA* gene, which also seems to encode a ligand of Wnt signaling, is involved in specifying wing pattern shapes, including in melanic elements (Martin et al. 2012; Gallant et al. 2014; Martin and Reed 2014). In *Junonia coenia* and some other butterfly species, *wingless* is known to be expressed in future pattern elements called basal (B), discal (DI and DII), and marginal (EI) elements (Carroll et al. 1994; Martin and Reed 2010, 2014; Huber et al. 2015) and was also identified at the center of eyespot patterns (Monteiro et al. 2006). The thoracic pattern of larval *Bombyx mori* is also regulated by *Wnt1* (homolog of *wingless*) (Yamaguchi et al. 2013). Evolutionary roles of secreted ligand genes such as *wingless* are reviewed in chapter 4 of this book (Martin and Courtier-Orgogozo 2017).

Werner et al. (2010) proposed a model of pigmentation pattern formation based on the assumption that Wingless protein diffuses from the source and serves as a long-range signal. There are a limited number of cells expressing *wingless*, and they are located in centers of future pigmented spots. In their model, secreted Wingless protein is diffused or transported to wider regions and transduces the signal. The signal is probably mediated by an unknown transcription factor and activates transcription of melanin synthesis-related genes, including *yellow*. Melanin should be synthesized by products of these genes and wings are consequently pigmented. This model should be validated by future research.

16.8 *Cis*-Regulatory Evolution of *Wingless*

The expression pattern of *wingless* evolved uniquely in *D. guttifera*. To examine how this unique expression pattern evolved, the genomic region around *wingless* was analyzed using a fluorescent reporter assay. As a result, three novel enhancer activities (in longitudinal vein tips, campaniform sensilla, and thoracic stripes) were found (Fig. 16.4). These novel enhancer activities are thought to have been involved in the evolution of the novel pigmentation pattern (Koshikawa et al. 2015). This study provided unique insights into the evolution of novel traits, illustrating how gains of novel enhancer activities at developmental regulatory gene were associated with derived expression domains and the emergence of novel traits (Rebeiz et al. 2011; Koshikawa et al. 2015; Rebeiz and Williams 2017).

We can generalize this concept as follows. In many organisms, gains of novel expression domains by gains of enhancer activities for a developmental regulatory

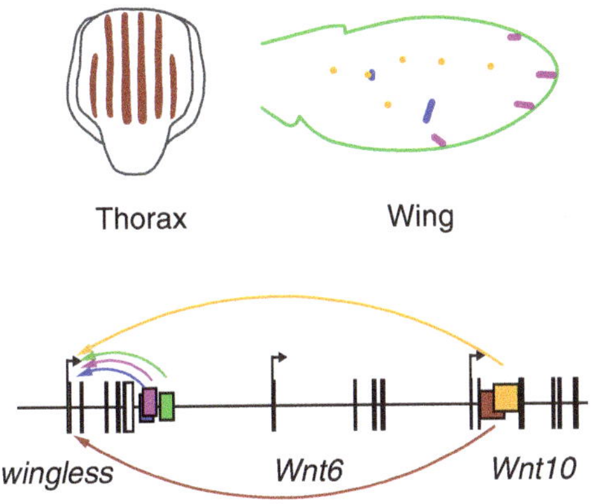

Fig. 16.4 Enhancers driving pupal wing and thoracic expressions of *wingless* in *D. guttifera*. Color code indicates correspondence of enhancer positions and expression domains. *Green*: wing margin. *Blue*: crossveins. *Purple*: longitudinal vein tips. *Yellow*: campaniform sensilla. *Brown*: thoracic stripes. Expressions in the wing margin and crossveins are ancestral (common in *D. melanogaster* and *D. guttifera*), and the longitudinal vein tips, campaniform sensilla, and thoracic stripes are novel (found in *D. guttifera* but not in *D. melanogaster*) (Modified from Koshikawa et al. (2015) and Koshikawa (2015))

gene could be a part of possible mechanisms of heterotopy (evolutionary duplication of a pre-existing trait in a different place on the body) (Gould 1977; West-Eberhard 2003; Rubinstein and de Souza 2013; Rebeiz et al. 2015; for more discussion see Koshikawa 2015).

16.9 Trials of Artificial Production of Pigmentation on *D. melanogaster* Wings

For now, only two genes, the upstream pattern inducer *wingless* and the melanin synthesis-related gene *yellow*, have been identified in the machinery required for pigmentation pattern formation in *D. guttifera*. In many cases, the Wingless signal is transduced through the so-called canonical pathway, where Pangolin/dTCF is an effector transcription factor regulating transcriptions of downstream genes. There were consensus sequences of Pangolin/dTCF binding sites in *vein spot* CRE, but replacement of these sequences by nonsense sequences did not change the expression pattern of the reporter gene (Werner et al. 2010). This means that the positional information of *wingless* does not directly regulate *yellow* through the canonical Wnt pathway. Involvement of another transcription factor is assumed, but so far it has not been identified. Furthermore, we know *yellow* is involved in pigmentation, but overexpression of *yellow* alone does not cause additional pigmentation in *D. melanogaster* (Gompel et al. 2005; Riedel et al. 2011). Proper expression or repression of melanin synthesis-related genes and/or proper supply of melanin precursors, such as dopa and dopamine, could be required for artificial production of pigmentation in *D. melanogaster* wings.

16.10 Diversity and Generality in Color Pattern Formation

We summarized above what was revealed by studies of *D. guttifera*, but will it apply to pattern formation in other organisms? Due to the experimental strengths of this system, we can be optimistic that we will reach an integrated model for pigmentation pattern formation in *Drosophila*. Butterflies show interesting parallels with the *Drosophila* wing patterning genes, as *Wnt* genes and *Distal-less* are key players in both lineages (Werner et al. 2010; Martin et al. 2012; Brakefield et al. 1996; Arnoult et al. 2013). If we expand the comparison to vertebrates, there are large differences in genes involved in pattern formation and melanin synthesis (Kopp 2009; Kronforst et al. 2012; Kaelin et al. 2012; Mallarino et al. 2016). Still we assume we can find some common mechanisms, such as a way of measuring distance in a tissue, and a hierarchical regulatory architecture. Comparing comprehensive datasets will be instrumental in answering this question of fundamental interest for our understanding of the mechanisms that generate biodiversity on Earth.

Acknowledgments We thank Toshiro Sekimura, Frederik H. Nijhout, and persons involved in the meeting at Chubu University in 2016 for stimulating us to write this chapter. We also thank Arnaud Martin and Takao K. Suzuki for reviewing this chapter, Masanori J. Toda for advice on taxonomy, Elizabeth Nakajima for English editing, and Noriko Funayama for hosting our research. A part of the writing was supported by KAKENHI (15K18586) and the Sumitomo Foundation.

References

Arnoult L, Su KF, Manoel D, Minervino C, Magriña J, Gompel N, Prud'homme B (2013) Emergence and diversification of fly pigmentation through evolution of a gene regulatory module. Science 339(6126):1423–1426. doi:10.1126/science.1233749

Begun DJ, Whitley P (2000) Genetics of alpha-amanitin resistance in a natural population of *Drosophila melanogaster*. Heredity 85(2):184–190. doi:10.1046/j.1365-2540.2000.00729.x

Beldade P, Peralta CM (2017) Developmental and evolutionary mechanisms shaping butterfly eyespots. Curr Opin Insect Sci 19:22–29. doi:10.1016/j.cois.2016.10.006

Brakefield PM, Gates J, Keys D, Kesbeke F, Wijngaarden PJ, Monteiro A, French V, Carroll SB (1996) Development, plasticity and evolution of butterfly eyespot patterns. Nature 384 (6606):236–242. doi:10.1038/384236a0

Bunyard B, Foote BA (1990a) Acalyptrate Diptera reared from higher fungi in northeastern Ohio. Entomol News 101(2):117–121

Bunyard B, Foote BA (1990b) Biological notes on *Drosophila guttifera* (Diptera: Drosophilidae), a consumer of mushrooms. Entomol News 101(3):161–163

Carroll SB, Gates J, Keys DN, Paddock SW, Panganiban GE, Selegue JE, Williams JA (1994) Pattern formation and eyespot determination in butterfly wings. Science 265(5168):109–114. doi:10.1126/science.7912449

Chen Y, Weeks J, Mortin MA, Greenleaf AL (1993) Mapping mutations in genes encoding the two large subunits of *Drosophila* RNA polymerase II defines domains essential for basic transcription functions and for proper expression of developmental genes. Mol Cell Biol 13 (7):4214–4222. doi:10.1128/MCB.13.7.4214

Cushing JE (1941) An experiment in olfactory conditioning in *Drosophila guttifera*. Proc Natl Aca Sci U S A 27(11):496–499

Dombeck I, Jaenike J (2004) Ecological genetics of abdominal pigmentation in *Drosophila falleni*. Evolution 58(3):587–596. doi:10.1554/03-299

Fukutomi Y, Matsumoto K, Agata K, Funayama N, Koshikawa S (2017) Pupal development and pigmentation process of a polka-dotted fruit fly, *Drosophila guttifera* (Insecta, Diptera). Dev Genes Evol 227(3):171–180. doi:10.1007/s00427-017-0578-3

Fuyama Y (1979) A visual stimulus in the courtship of *Drosophila suzukii*. Experientia 35 (10):1327–1328

Gallant JR, Imhoff VE, Martin A, Savage WK, Chamberlain NL, Pote BL, Peterson C, Smith GE, Evans B, Reed RD, Kronforst MR, Mullen SP (2014) Ancient homology underlies adaptive mimetic diversity across butterflies. Nat Commun 5:4817. doi:10.1038/ncomms5817

Gompel N, Prud'homme B, Wittkopp PJ, Kassner VA, Carroll SB (2005) Chance caught on the wing: cis-regulatory evolution and the origin of pigment patterns in *Drosophila*. Nature 433 (7025):481–487. doi:10.1038/nature03235

Gould SJ (1977) Ontogeny and phylogeny. Harvard University Press, Cambridge

Grossfield J (1977) *Drosophila* courtship: decapitated quinaria group females. J NY Entomol Soc 85(3):119–126

Huber B, Whibley A, Poul YL, Navarro N, Martin A, Baxter S, Shah A, Gilles B, Wirth T, McMillan WO, Joron M (2015) Conservatism and novelty in the genetic architecture of adaptation in *Heliconius* butterflies. Heredity 114(5):515–524. doi:10.1038/hdy.2015.22

Izumitani HF, Kusaka Y, Koshikawa S, Toda MJ, Katoh T (2016) Phylogeography of the subgenus *Drosophila* (Diptera: Drosophilidae): evolutionary history of faunal divergence between the old and the new worlds. PLoS One 11(7), e0160051. doi:10.1371/journal.pone.0160051

Joron M, Frezal L, Jones RT, Chamberlain NL, Lee SF, Haag CR, Whibley A, Becuwe M, Baxter SW, Ferguson L, Wilkinson PA, Salazar C, Davidson C, Clark R, Quail MA, Beasley H, Glithero R, Lloyd C, Sims S, Jones MC, Rogers J, Jiggins CD, ffrench-Constant RH (2011) Chromosomal rearrangements maintain a polymorphic supergene controlling butterfly mimicry. Nature 477(7363):203–206. doi:10.1038/nature10341

Kaelin CB, Xu X, Hong LZ, David VA, McGowan KA, Schmidt-Küntzel A, Roelke ME, Pino J, Pontius J, Cooper GM, Manuel H, Swanson WF, Marker L, Harper CK, van Dyk A, Yue B, Mullikin JC, Warren WC, Eizirik E, Kos L, O'Brien SJ, Barsh GS, Menotti-Raymond M (2012) Specifying and sustaining pigmentation patterns in domestic and wild cats. Science 337 (6101):1536–1541. doi:10.1126/science.1220893

Kopp A (2009) Metamodels and phylogenetic replication: a systematic approach to the evolution of developmental pathways. Evolution 63(11):2771–2789. doi:10.1111/j.1558-5646.2009.00761.x

Koshikawa S (2015) Enhancer modularity and the evolution of new traits. Fly (Austin) 9 (4):155–159. doi:10.1080/19336934.2016.1151129

Koshikawa S, Giorgianni MW, Vaccaro K, Kassner VA, Yoder JH, Werner T, Carroll SB (2015) Gain of *cis*-regulatory activities underlies novel domains of *wingless* gene expression in *Drosophila*. Proc Natl Acad Sci U S A 112(24):7524–7529. doi:10.1073/pnas.1509022112

Kronforst MR, Barsh GS, Kopp A, Mallet J, Monteiro A, Mullen SP, Protas M, Rosenblum EB,

Schneider CJ, Hoekstra HE (2012) Unraveling the thread of nature's tapestry: the genetics of diversity and convergence in animal pigmentation. Pigment Cell Melanoma Res 25 (4):411–433. doi:10.1111/j.1755-148X.2012.01014.x

Kunte K, Zhang W, Tenger-Trolander A, Palmer DH, Martin A, Reed RD, Mullen SP, Kronforst MR (2014) *Doublesex* is a mimicry supergene. Nature 507(7491):229–232. doi:10.1038/nature13112

Mallarino R, Henegar C, Mirasierra M, Manceau M, Schradin C, Vallejo M, Beronja S, Barsh GS, Hoekstra HE (2016) Developmental mechanisms of stripe patterns in rodents. Nature 539 (7630):518–523. doi:10.1038/nature20109

Martin A, Reed RD (2010) *Wingless* and *aristaless2* define a developmental ground plan for moth and butterfly wing pattern evolution. Mol Biol Evol 27(12):2864–2878. doi:10.1093/molbev/msq173

Martin A, Papa R, Nadeau NJ, Hill RI, Counterman BA, Halder G, Jiggins CD, Kronforst MR, Long AD, McMillan WO, Reed RD (2012) Diversification of complex butterfly wing patterns by repeated regulatory evolution of a Wnt ligand. Proc Natl Aca Sci U S A 109 (31):12632–12637. doi:10.1073/pnas.1204800109

Martin A, Reed RD (2014) Wnt signaling underlies evolution and development of the butterfly wing pattern symmetry systems. Dev Biol 395(2):367–378. doi:10.1016/j.ydbio.2014.08.031

Martin A, Courtier-Orgogozo V (2017) Morphological evolution repeatedly caused by mutations in signaling ligand genes. In: Diversity and evolution of butterfly wing patterns: an integrative approach. Springer, New York

Markow TA, O'Grady PM (2006) Drosophila: a guide to species identification and use. Academic Press, New York

Mitchell CL, Saul MC, Lei L, Wei H, Werner T (2014) The mechanisms underlying α-amanitin resistance in *Drosophila melanogaster*: a microarray analysis. PLoS One 9(4):e93489. doi:10.1371/journal.pone.0093489

Mitchell CL, Yeager RD, Johnson ZJ, D'Annunzio SE, Vogel KR, Werner T (2015) Long-term resistance of *Drosophila melanogaster* to the mushroom toxin alpha-amanitin. PLoS One 10 (5):e0127569. doi:10.1371/journal.pone.0127569

Monteiro A (2015) Origin, development, and evolution of butterfly eyespots. Annu Rev Entomol 60:253–271. doi:10.1146/annurev-ento-010814-020942

Monteiro A, Glaser G, Stockslager S, Glansdorp N, Ramos D (2006) Comparative insights into questions of lepidopteran wing pattern homology. BMC Dev Biol 6:52. doi:10.1186/1471-213X-6-52

Nijhout HF (1991) The development and evolution of butterfly wing patterns. Smithsonian Institution Press, Washington, DC

Nishikawa H, Iijima T, Kajitani R, Yamaguchi J, Ando T, Suzuki Y, Sugano S, Fujiyama A, Kosugi S, Hirakawa H, Tabata S, Ozaki K, Morimoto H, Ihara K, Obara M, Hori H, Itoh T, Fujiwara H (2015) A genetic mechanism for female-limited Batesian mimicry in *Papilio* butterfly. Nat Genet 47(4):405–409. doi:10.1038/ng.3241

O'Grady PM (2010) Whither Drosophila? Genetics 185(2):703–705. doi:10.1534/genetics.110.118232

Patterson JT (1943) The Drosophilidae of the Southwest. University of Texas Publication 4313:7–216

Patterson JT, Stone WS (1952) Evolution in the genus *Drosophila*. Macmillan Company, New York

Perlman SJ, Spicer GS, Shoemaker DD, Jaenike J (2003) Associations between mycophagous *Drosophila* and their *Howardula* nematode parasites: a worldwide phylogenetic shuffle. Mol Ecol 12(1):237–249. doi:10.1046/j.1365-294X.2003.01721.x

Rebeiz M, Jikomes N, Kassner VA, Carroll SB (2011) Evolutionary origin of a novel gene expression pattern through co-option of the latent activities of existing regulatory sequences. Proc Natl Acad Sci U S A 108(25):10036–10043. doi:10.1073/pnas.1105937108

Rebeiz M, Patel NH, Hinman VF (2015) Unraveling the tangled skein: the evolution of transcriptional regulatory networks in development. Annu Rev Genomics Hum Genet 16:103–131. doi:10.1146/annurev-genom-091212-153423

Rebeiz M, Williams TM (2017) Using *Drosophila* pigmentation traits to study the mechanisms of *cis*-regulatory evolution. Curr Opin Insect Sci 19:1–7. doi:10.1016/j.cois.2016.10.002

Reed RD, Papa R, Martin A, Hines HM, Counterman BA, Pardo-Diaz C, Jiggins CD, Chamberlain NL, Kronforst MR, Chen R, Halder G, Nijhout HF, McMillan WO (2011) *optix* drives the repeated convergent evolution of butterfly wing pattern mimicry. Science 333 (6046):1137–1141. doi:10.1126/science.1208227

Riedel F, Vorkel D, Eaton S (2011) Megalin-dependent *yellow* endocytosis restricts melanization in the *Drosophila* cuticle. Development 138(1):149–158. doi:10.1242/dev.056309

Ringo JM, Hodosh RJ (1978) A multivariate analysis of behavioral divergence among closely related species of endemic Hawaiian *Drosophila*. Evolution 32(2):389–397. doi:10.1111/j.1558-5646.1978.tb00654.x

Rubinstein M, de Souza FSJ (2013) Evolution of transcriptional enhancers and animal diversity. Philos Trans R Soc B 368(1632):20130017–20130017. doi:10.1098/rstb.2013.0017

Singh AP, Nüsslein-Volhard C (2015) Zebrafish stripes as a model for vertebrate colour pattern formation. Curr Biol 25(2):R81–R92. doi:10.1016/j.cub.2014.11.013

Spicer GS, Jaenike J (1996) Phylogenetic analysis of breeding site use and α-amanitin tolerance within the *Drosophila quinaria* species group. Evolution 50(6):2328–2337. doi:10.2307/2410701

Stavenga DG, Giraldo MA, Leertouwer HL (2010) Butterfly wing colors: glass scales of *Graphium sarpedon* cause polarized iridescence and enhance blue/green pigment coloration of the wing membrane. J Exp Biol 213(10):1731–1739. doi:10.1242/jeb.041434

Stump AD, Jablonski SE, Bouton L, Wilder JA (2011) Distribution and mechanism of α-amanitin tolerance in mycophagous *Drosophila* (Diptera: Drosophilidae). Environ Entomol 40 (6):1604–1612. doi:10.1603/EN11136

Sturtevant AH (1921) The North American species of *Drosophila*. Carnegie Institution of Washington, Washington, DC

Sturtevant AH (1942) The classification of the genus *Drosophila*, with the description of nine new species. University of Texas Publication 4213:5–51

The Heliconius Genome Consortium (2012) Butterfly genome reveals promiscuous exchange of mimicry adaptations among species. Nature 487(7405):94–98. doi:10.1038/nature11041

Throckmorton LH (1962) The problem of phylogeny in the genus *Drosophila*. University of Texas Publication 6205:207–343

Throckmorton LH (1975) The phylogeny, ecology, and geography of *Drosophila*. In: King RC (ed) Handbook of genetics, vol 3. Plenum Press, New York, pp 421–469

Toda MJ (2017) DrosWLD-Species: taxonomic information database for world species of Drosophilidae. Available at: http://bioinfo.lowtem.hokudai.ac.jp/db/. Accessed 25 Jan 2017

True JR, Edwards KA, Yamamoto D, Carroll SB (1999) *Drosophila* wing melanin patterns form by vein-dependent elaboration of enzymatic prepatterns. Curr Biol 9(23):1382–1391. doi:10.1016/S0960-9822(00)80083-4

Walker F (1849) List of specimens of dipterous insects of the collection of the British Museum. Part 4. British Museum (N.H.), London, pp 689–1172

Werner T, Jaenike J (2017) Drosophilids of the Midwest and Northeast. River Campus Libraries, University of Rochester, Rochester

Werner T, Koshikawa S, Williams TM, Carroll SB (2010) Generation of a novel wing colour pattern by the Wingless morphogen. Nature 464(7292):1143–1148. doi:10.1038/nature08896

West-Eberhard MJ (2003) Developmental plasticity and evolution. Oxford University Press, Oxford

Wittkopp PJ, True JR, Carroll SB (2002) Reciprocal functions of the Drosophila yellow and ebony proteins in the development and evolution of pigment patterns. Development 129 (8):1849–1858

Yamaguchi J, Banno Y, Mita K, Yamamoto K, Ando T, Fujiwara H (2013) Periodic *Wnt1* expression in response to ecdysteroid generates twin-spot markings on caterpillars. Nat Commun 4:1857. doi:10.1038/ncomms2778

Yassin A (2013) Phylogenetic classification of the Drosophilidae Rondani (Diptera): the role of morphology in the postgenomic era. Syst Entomol 38(2):349–364. doi:10.1111/j.1365-3113.2012.00665.x

Yeh SD, Liou SR, True JR (2006) Genetics of divergence in male wing pigmentation and courtship behavior between *Drosophila elegans* and *D. gunungcola*. Heredity 96(5):383–395. doi:10.1038/sj.hdy.6800814

17

Color Vision and Color Formation in Dragonflies: Present and Future Studies

Ryo Futahashi

Abstract Dragonflies are colorful diurnal insects with large compound eyes. Because they visually recognize conspecific and heterospecific individuals, their body color plays essential roles in ecology and reproductive biology. Here I introduce the recent topics of molecular mechanisms underlying color vision and color formation in dragonflies. Complex wing color polymorphism is recognized among the two closely related Japanese *Mnais* species, presumably due to stepwise character displacement to avoid interspecific mating. We discovered an extraordinary large number of visual opsin genes by RNA sequencing of 12 dragonfly species. Manual correction after de novo assembly was crucial for determining the exact number and sequence of opsin genes. Each opsin gene was differentially expressed between the adult and larva, as well as between dorsal and ventral regions of adult compound eyes, highlighting the behavior, ecology, and adaptation of aquatic larva to terrestrial adult. The repertoire of opsin genes differed among dragonfly species, plausibly involved in the diversity of the habitat and behavior of each species. We also found that sex-specific yellow-red color transition in red dragonflies is regulated by redox changes in ommochrome pigments, which unveils a previously unknown molecular mechanism underlying body color change in animals. Establishment of the methods of gene functional analyses in dragonflies is desired for future studies.

Keywords Dragonfly • Color polymorphism • Character displacement • Opsin • Color vision • Pigment • Redox • Ommochrome

17.1 Introduction

Like butterflies, dragonflies (including damselflies, Insecta: Odonata) are one of the most colorful insects, and their color patterns have been focused from ecological and evolutionary aspects for a long time (Tillyard 1917; Corbet 1999; Bybee et al.

R. Futahashi (✉)
Bioproduction Research Institute, National Institute of Advanced Industrial Science and Technology (AIST), Central 6, Tsukuba, Ibaraki 305-8566, Japan
e-mail: ryo-futahashi@aist.go.jp

2016). Dragonflies are well-known insects, and almost all Japanese people know the songs Aka-tombo (= red dragonflies; a symbol of autumn in Japan) and Tombo-no-megane (= eye glasses of dragonflies; a metaphor for the colorful compound eyes of dragonflies) (Ueda 2004; Inoue and Tani 2010). Despite the fact that the detailed genetic analyses of pattern formation in butterfly adult wing and larval body have progressed greatly in recent years as described in this book, the molecular biological study of dragonfly's color pattern formation has just started.

17.2 Important Role of Color Pattern for Partner Recognition in Dragonflies

Unlike most insects, drastic adult color transitions are widely recognized among dragonflies, resulting in conspicuous sexual dimorphism. In red dragonfly species, body colors of males turn from yellow to red in the course of sexual maturation, whereas females are yellowish throughout their adult lives in general (Fig. 17.1a–c). Moreover, many dragonfly species have color polymorphism even in the same sex (Fig. 17.1c–f, i–n), which is genetically controlled at least in several species (Futahashi 2016a). Considering that gynandromorph specimens display discontinuous male/female mosaicism in their coloration (Fig. 17.1g, h), sex-specific color formation is regulated cell-autonomously in dragonflies.

In many dragonfly species, adult body color plays important roles in partner recognition (Corbet 1999; Svensson et al. 2007; Córdoba-Aguilar 2008; Svensson et al. 2014; Takahashi et al. 2014; Beatty et al. 2015; Drury et al. 2015). Interspecific or male-male connection has sometimes been observed in the field between similar-colored individuals (Fig. 17.2a–c), and interspecific hybrids has been reported occasionally (Fig. 17.2d–f) (Corbet 1999; Futahashi 1999; Futahashi and Futahashi 2007; Moriyasu and Sugimura 2007; Ozono et al. 2012; Sánchez-Guillén et al. 2014; Futahashi 2016a). Parent combination of hybrid specimen can be determined by biparentally inherited nuclear DNA and maternally inherited mitochondrial DNA analyses (Fig. 17.2g), and it has been reported that males of *Sympetrum eroticum* are apt to catch females of other species (Futahashi 1999; Futahashi and Hayashi 2004a), suggesting that the direction of gene flow with hybridization is nonreciprocal in some cases. These misidentifications in dragonflies may be attributed to their poor sense of audition and olfaction; dragonflies lack the auditory organs, and their antennae are less developed (Yager 1999; Cocroft and Rodríguez 2005).

Fig. 17.1 Intraspecific color pattern diversity in dragonflies. (**a–d**) Sexual dimorphism, male color transition, and female color polymorphism of *Sympetrum cordulegaster*. (**e–f**) Female color polymorphism of *Ischnura senegalensis*. *Arrow* indicates a blue spot existed in males and androchrome females. (**g**) Gynandromorph of *I. senegalensis* showing the main region with female coloration and the posterior left side (*arrow*) with male coloration and appendage (Photo courtesy of Mitsutoshi Sugimura). (**h**) Gynandromorph of *Crocothemis servilia* showing the main region with male coloration and the anterior right side with female coloration (Photo courtesy of Kohji Tanaka). (**i–n**) Wing color polymorphism of *Mnais* species. Territorial males have orange (**i**) or brown (**l**) wings, whereas female mimicking males have hyaline (**j**) or pale orange (**m**) wings. Females have hyaline (**k**) or pale orange (**n**) wings. (**i–k**, **m**, **n**) *M. costalis*. (**l**) *M. pruinosa* (See also Fig. 17.3)

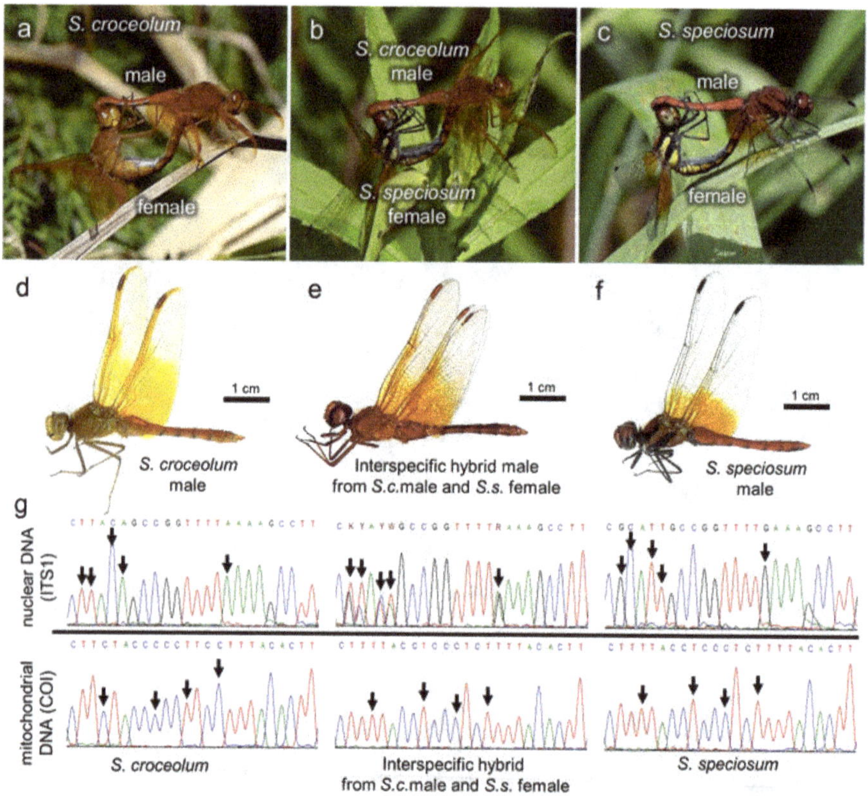

Fig. 17.2 Interspecific copulation and hybrid of dragonflies. (**a**) Normal copulation of *Sympetrum croceolum*. (**b**) Interspecific copulation between *S. croceolum* male and *Sympetrum speciosum* female (Figure modified from Ozono et al. 2012). (**c**) Normal copulation of *S. speciosum*. (**d**) Male of *S. croceolum*. (**e**) Interspecific hybrid male from *S. croceolum* male and *S. speciosum* female. (**f**) Male of *S. speciosum*. (**g**) Nuclear and mitochondrial DNA analyses of *S. croceolum, S. speciosum*, and interspecific hybrid between *S. croceolum* male and *S. speciosum* female. The internal transcribed spacer 1 (ITS1) or cytochrome c oxidase subunit I (COI) region were used for nuclear or mitochondrial DNA marker, respectively. *Arrows* indicate species specific nucleotides

17.3 Wing Color Polymorphism and Presumptive Character Displacement in Japanese *Mnais* Species

In order to avoid interspecific mating or aggression, presumptive character displacement has been reported in some species, wherein interspecific color differences are larger in sympatric populations than in allopatric populations (Waage 1975; Suzuki 1984; Tynkkynen et al. 2004; Hayashi et al. 2004b; Hassall 2014; Drury and Grether 2014; Tsubaki and Okuyama 2016). Here I introduce an interesting example of wing color polymorphism in the two closely related *Mnais* species, *M. costalis* and *M. pruinosa*, in Japan (Hayashi et al. 2004a, b; Ozono et al. 2012). These two species can be distinguished by nuclear ITS1 sequences, relative length of wing to head in adult males, shape of adult wing pterostigma, and the shape of larval caudal gill (Hayashi et al. 2004a, b). On the other hand, interspecific hybrids have been discovered occasionally and multiregional introgression of mitochondrial DNA is recognized between these two species (Hayashi et al. 2004a, 2005; Futahashi and Hayashi 2004b). Both species exhibit complex wing color polymorphism (Figs. 17.1i–n and 17.3), in which orange-winged males, hyaline-winged males, and hyaline-winged females appear widely in Japan (Fig. 17.1i–k, Asahina 1976; Hayashi et al. 2004b; Ozono et al. 2012). Male orange/hyaline wing polymorphism of *M. costalis* can be explained by a single autosomal locus, and the orange-winged phenotype is dominant (Tsubaki 2003). Previous ecological studies have shown that orange-winged males are territorial, whereas hyaline-winged males are female-mimics and usually non-territorial sneakers (Nomakuchi et al. 1984; Tsubaki et al. 1997; Hayashi et al. 2004b). In addition to these three major phenotypes, the following three phenotypes are recognized in some populations: brown-winged males of *M. pruinosa*, pale orange-winged males and females of *M. costalis* (Fig. 17.1l–n, Asahina 1976; Hayashi et al. 2004b; Ozono et al. 2012). Thus, in *M. costalis*, there are three (orange, pale orange, and hyaline) and two (pale orange and hyaline) wing color forms for males and females, respectively, whereas in *M. pruinosa,* three (brown, orange, and hyaline) and one (hyaline) wing color forms exist in males and females, respectively. Interestingly, wing polymorphism is associated with abdominal body coloration: whitish in mature territorial males (orange or brown wings) (Fig. 17.1i, l) and metallic green in female-mimicking males and females (hyaline or pale orange wings) (Fig. 17.1j, k, m, n, Asahina 1976; Hayashi et al. 2004b; Ozono et al. 2012). Exceptional untransparent white-winged phenotypes have been reported in the Boso Peninsula population of *M. pruinosa* (Asahina 1976), although these white-winged phenotypes are now almost extinct (Futahashi and Hayashi 2004b; Ozono et al. 2012).

Geographic variation of wing color polymorphism is associated with cohabitation (Suzuki 1984; Hayashi et al. 2004b; Tsubaki and Okuyama 2016). In allopatric regions, males of both *M. costalis* and *M. pruinosa* show orange/hyaline wing color polymorphism, whereas females are all hyaline-winged and monomorphic in both species (Fig. 17.3a). In central Japan where both species cohabit, males of *M. costalis* show only orange wings, while males of *M. pruinosa* show only hyaline wings in general (Fig. 17.3a). In *M. costalis*, pale orange-winged females appear prominently in the southern area, where both males and females can be distinguished solely by wing coloration (Fig. 17.3a). In addition to wing color polymorphism, sympatric populations of *M. costalis* show larger body size and prefer sunnier habitats than *M. pruinosa* and allopatric populations of *M. costalis*, suggesting that multiple character displacements emerged in central Japan to avoid interspecific mating (Tsubaki and Okuyama 2016). In western Japan, however, males of both species are polymorphic even in sympatric regions. It should be noted that two forms of female-mimicking males (pale orange and hyaline) appear in accordance with female color polymorphism in *M. costalis* (Fig. 17.3). In the southwestern area, territorial males of *M. pruinosa* have brown wings instead of orange, where both males and females can be distinguished solely by wing coloration (Fig. 17.3b, rightmost). Although it is not clear why wing color polymorphism is maintained in western Japan even in sympatric region, it has been reported that

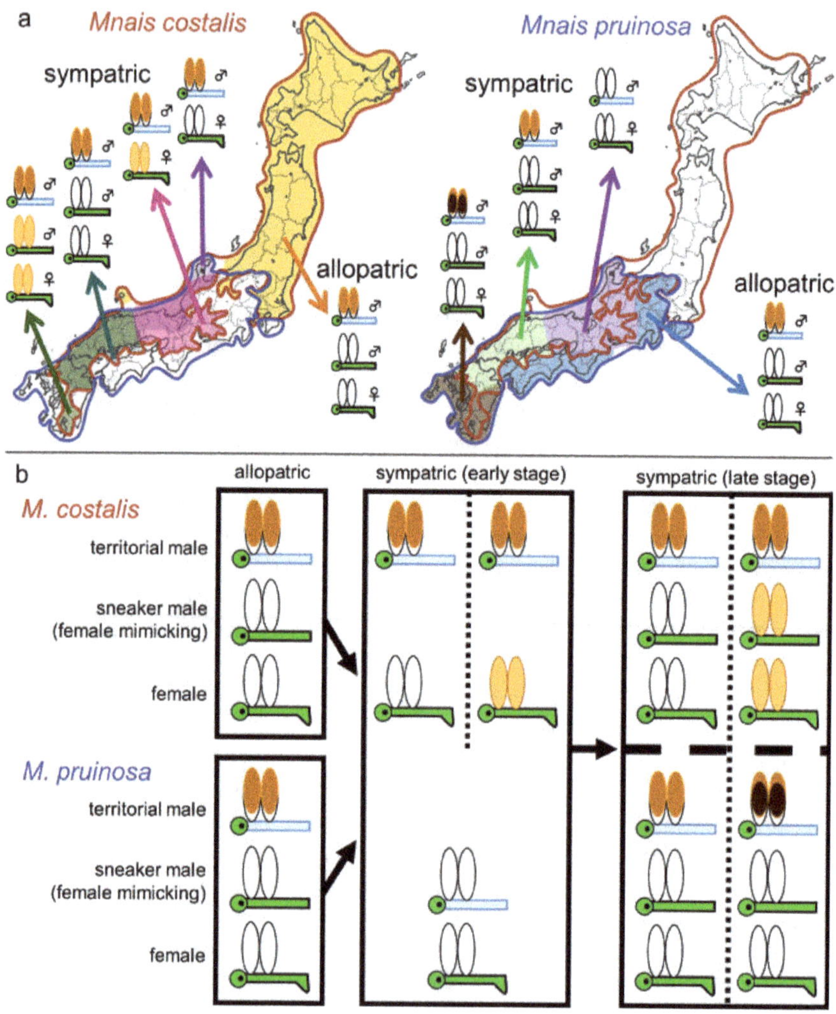

Fig. 17.3 Wing color polymorphism of two Japanese *Mnais* species. (**a**) Geographical variation of *M. costalis* and *M. pruinosa*. (**b**) Hypothetical evolutionary model of wing color polymorphism. The photos of each form are shown in Fig. 17.1i–n. As described Fig. 17.3a, wing color polymorphism varied among populations in sympatric region. Pale orange-winged males and females often appear together with hyaline-winged males and females, respectively. For example, three male forms (orange, pale orange, and hyaline) and two female forms (pale orange and hyaline) emerge simultaneously in some populations of *M. costalis*. Figure modified from Hayashi et al. 2004b; Ozono et al. 2012

degrees of mitochondrial introgression are smaller in western Japan than in central Japan (Hayashi et al. 2005), suggesting that reproductive isolating mechanism between the two *Mnais* species is more robust in western Japan. Hayashi et al. (2004b) proposed the evolutionary model of *Mnais* wing polymorphism, in which the following three stages are hypothesized (Fig. 17.3b):

1. In allopatric populations (eastern or southeastern Japan), males exhibit orange (territorial) and hyaline (female-mimicking) wing color polymorphism, while females are monomorphic (hyaline) in both species.
2. In the early stage of cohabitation (central Japan), males become monomorphic (only orange in *M. costalis* and only hyaline in *M. pruinosa*), and pale orange-winged females of *M. costalis* emerge in some places whereas all females are hyaline-winged in *M. pruinosa*. It should be noted that hyaline-winged males of *M. pruinosa* in central Japan show whitish abdomen like territorial males and have flexible territorial strategy (Fig. 17.3b middle, Siva-Jothy and Tsubaki 1989; Hayashi et al. 2004b).
3. In the late stage of cohabitation (western Japan), both territorial (orange or brown) and female-mimicking males (hyaline or pale orange) appear in both species once again.

According to this scenario, pale orange-winged males and brown males are likely to appear secondarily. The brown/orange color difference may have occurred because the wing color was lost in early sympatry and was reconstructed without models.

Orange wing color of *M. costalis* is derived from tyrosine, suggesting that pigments of orange wing are kinds of melanin (Hooper et al. 1999). Considering that dopamine, a melanin precursor, is also known as a neurotransmitter, genes involved in melanin synthesis pathway are strong candidates for analysis of pleiotropic effects on wing color formation and territoriality. Genetic mechanisms underlying wing color polymorphism deserve future studies.

17.4 Identification of Remarkable Number of Opsin Genes in Dragonflies

Because dragonflies visually recognize environment, foods, enemies, rivals, and mates, their sense of vision has been studied using electrophysiological approach. Critical flicker frequency test has revealed that dragonflies can discriminate beyond 300 Hz, suggesting that they have keen dynamic vision (McFarland and Lowe 1983). Meanwhile, based on anatomical studies, it has been hypothesized that dragonflies have approximately 20/2000 vision (Kirschfeld 1976). In addition to temporal and spatial resolution, wavelength discrimination capability (i.e., color vision) of dragonflies has been investigated, and previous studies have shown that they have three to five classes of photoreceptors (Autrum and Kolb 1968; Eguchi

1971; Meinertzhagen et al. 1983; Yang and Osorio 1991; Bybee et al. 2012; Huang et al. 2014).

Evolution of animal color vision is strongly correlated with the diversity of opsin genes (Briscoe and Chittka 2001; Terakita 2005; Briscoe 2008, Shichida and Matsuyama 2009; Hering et al. 2012; Cronin et al. 2014). Specific types of opsin gene produce light sensors sensitive to specific wavelength light. For example, human beings possess three opsin genes sensitive to blue, green, or red light and can perceive light from purple to red but not ultraviolet (UV). On the other hand, honey bees and fruit flies possess an opsin gene for UV light but not for red light, which allows them to recognize UV light, instead of discriminating red from gray. It has been thought that 2–5 opsin proteins are involved in color vision in most animals (Cronin et al. 2014).

Recently, we discovered that dragonflies possess surprisingly many opsin genes by RNA sequencing (RNA-seq) analyses using adult and larval visual organs (Futahashi et al. 2015). First we surveyed the visual transcriptomics of the red dragonfly *Sympetrum frequens* (Libellulidae). After de novo assembly using Trinity software, we obtained 60 contigs with high similarity to insect opsin proteins. When we aligned these contigs, many of them seemed to be partial or chimeric (gray and blue arrows in Fig. 17.4a). We also obtained 144 opsin gene-like contigs from the white-tailed skimmer dragonfly *Orthetrum albistylum* (Libellulidae) and found that chimeric pattern was different between these two species (Fig. 17.4a). We often encountered similar problem of chimeric contigs in de novo assembly when we focused on paralogous genes. To overcome this problem, we carefully checked and manually corrected each of the contig sequences using Integrative Genomics Viewer (Thorvaldsdóttir et al. 2013) (Fig. 17.4b). Through this manual correction, we also found that partial sequence information is often lost in automatically assembled contigs among highly paralogous genes, due to merging of several similar sequences into one (Futahashi 2016b). We verified the revised sequences by RT-PCR and DNA sequencing. Consequently, we obtained the presumably full length sequences of 20 opsin genes, consisting of 4 nonvisual opsin genes and 16 visual opsin genes of 1 UV, 5 short wavelength (SW), and 10 long wavelength (LW) type from both *S. frequens* and O. *albistylum* (Fig. 17.4c). Next we inspected the draft genome data of the scarce chaser dragonfly *Ladona fulva* (Libellulidae) and identified the same set of 20 opsin genes (Fig. 17.4c). No other opsin genes could be found in the genome. Molecular phylogenetic analysis revealed that the 20 opsin genes of these three species formed distinct 20 monophyletic clusters (Fig. 17.4c), indicating that the common ancestor of the libellulid dragonflies had these 20 genes.

Opsin genes of dragonflies are extraordinarily large in number compared with other insects (Fig. 17.5a). Why do dragonflies have so many opsin genes? In dragonflies, the structure and function of compound eyes are markedly different between not only adult and larva but also dorsal and ventral regions of adult eyes (Fig. 17.5b–c) (Labhart and Nilsson 1995). Electrophysiological analysis of *S. frequens* revealed that the dorsal eye region was sensitive to a short wavelength range from UV (300 nm) to blue-green light (500 nm), whereas the ventral eye

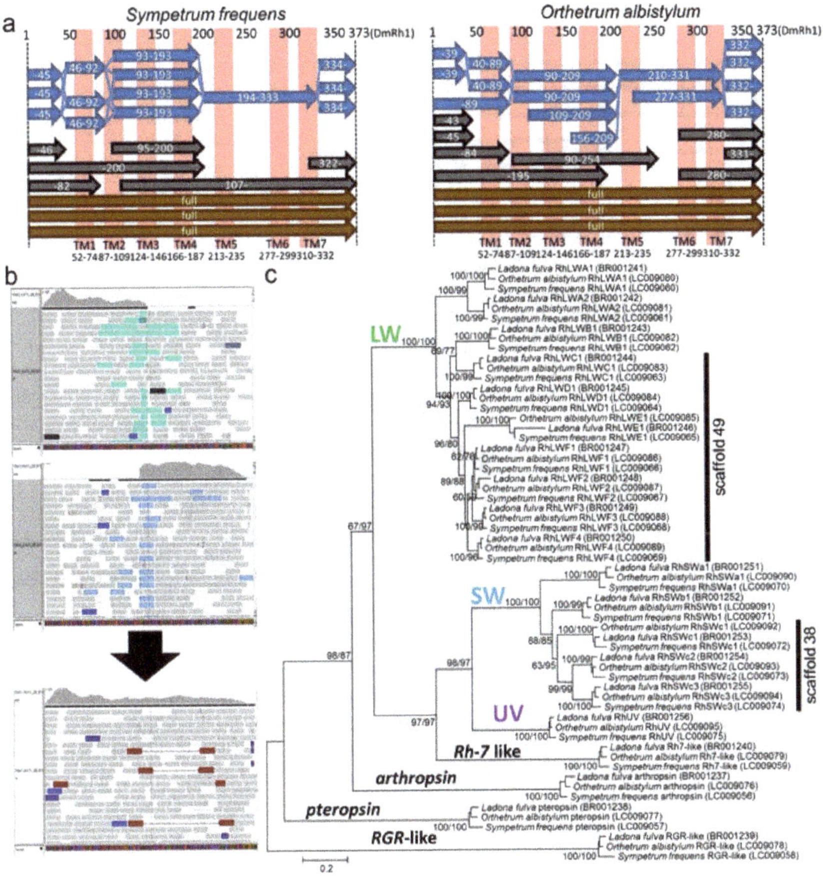

Fig. 17.4 Identification and manual assembly of 20 opsin genes in three libellulid dragonflies. (**a**) Results of de novo assembly by Trinity before manual correction. LW opsin gene-like contigs were aligned based on the similarity of *ninaE*/*Rh1* gene of *Drosophila melanogaster*. The presumptive seven transmembrane regions are shaded by *red*. *Blue* and *gray arrows* indicate chimeric and partial contigs, respectively. (**b**) Manual correction of the contig sequences using Integrative Genomics Viewer. *Cyan* and *sky-blue colors* mean paired end reads mapped on the different contigs. (**c**) Molecular phylogeny of 20 opsin genes of three libellulid dragonflies inferred from 795 aligned amino acid sites. On each node, bootstrap values are indicated in the order of neighbor-joining method/maximum-likelihood method. Accession numbers are shown in *parentheses*. On the genome of *L. fulva*, seven LW opsin genes (*LWC1*, *LWD1*, *LWE1*, and *LWF1–F4*) and three SW opsin genes (*SWc1-c3*) were located in tandem, respectively (Figure modified from Futahashi 2016b)

region was sensitive to a broader wavelength range from UV to red light (620 nm) (Fig. 17.5d). Interestingly, most of the opsin genes were expressed only at a specific life stage and in a specific region (Fig. 17.6). Although many opsin genes were expressed in adults, relatively small number of opsin genes was expressed in larvae, reflecting their lifestyle under water with less visual dependence. In adult

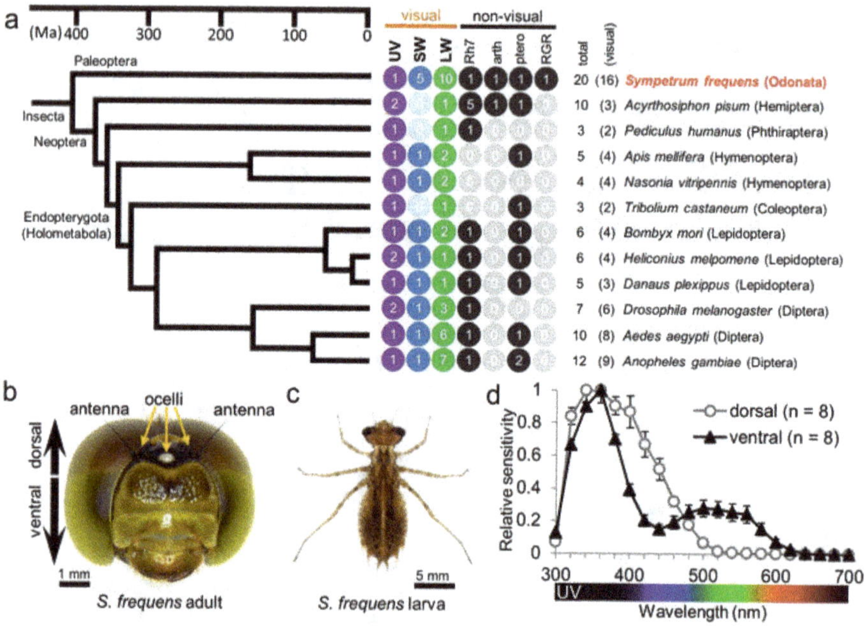

Fig. 17.5 Insect opsin genes and spectral sensitivity of adult compound eyes of *Sympetrum frequens*. (**a**) Numbers of opsin genes of ultraviolet type (UV), short wavelength type (SW), long wavelength type (LW), *rhodopsin7*-like (Rh7), *arthropsin* type (arth), *pteropsin* type (ptero), and *retinal G protein-coupled receptor*-like (RGR) are mapped on the insect phylogeny (Misof et al. 2014). (**b**) Frontal view of adult head of *S. frequens*. (**c**) Larva of *S. frequens* (Photo courtesy of Akira Ozono). (**d**) Spectral sensitivity of the dorsal and ventral regions of adult eyes of *S. frequens* measured by electroretinography (Figure modified from Futahashi et al. 2015)

compound eyes, most SW and LW opsin genes were, respectively, expressed in the dorsal and ventral regions in accordance with their spectral sensitivity, reflecting that dorsal eyes mainly perceive the SW-rich light directly from the sky, whereas the ventral eyes perceive reflected light from objects on the ground.

17.5 Diversity of Opsin Genes among Dragonflies

Body and wing color pattern, behavior, and microhabitats of dragonflies are variable among the families (Corbet 1999; Ozono et al. 2012). To investigate the opsin gene repertoire across dragonflies, comparative RNA-seq analyses were performed in additional 10 species representing 10 different dragonfly families: *Somatochlora uchidai* (Corduliidae), *Macromia amphigena* (Macromiidae), *Anotogaster sieboldii* (Cordulegastridae), *Tanypteryx pryeri* (Petaluridae), *Asiagomphus melaenops* (Gomphidae), *Anax parthenope* (Aeshnidae), *Epiophlebia superstes* (Epiophlebiidae), *Ischnura asiatica* (Coenagrionidae), *Mnais costalis*

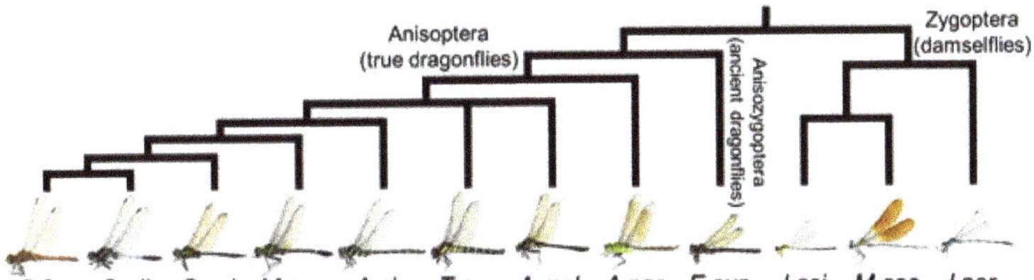

		S.fre.	O.alb.	S.uch.	M.amp.	A.sie.	T.pry.	A.mel.	A.par.	E.sup.	I.asi.	M.cos.	I.per.
UV		1 (DV)	1 (DV)	1 (DV)	1 (DV)	1 (DVO)	1 (DV)	1 (DV)	1 (DVO)	1 (DV)	1 (DV)	1 (DV)	1 (DVO)
SW	a	1 (L)	1 (L)	1 (L)	1 (L)	0	0	0	2 (VL)	1 (VL)	0	0	1 (L)
	b	1 (V)	1 (V)	1 (V)	1 (V)	1 (V)	1 (DV)	1 (DV)	1 (V)	0	2 (DVL)	1 (DVL)	1 (DV)
	c	3 (D)	3 (D)	4 (D)	4 (DO)	3 (D)	2 (D)	2 (D)	5 (DO)	6 (DV)	0	0	0
LW	A	2 (VL)	2 (VL)	2 (VL)	2 (VL)	2 (DV)	2 (VL)	2 (DV)	2 (VL)	2 (VL)	2 (DVL)	2 (DVL)	2 (DVL)
	B	1 (L)	1 (L)	1 (L)	1 (L)	1 (L)	0	0	2 (L)	1 (L)	0	0	0
	C	1 (L)	1 (L)	1 (L)	1 (L)	2 (OL)	2 (L)	1 (L)	1 (OL)	1 (OL)	1 (L)	1 (L)	1 (L)
	D	1 (O)	1 (O)	1 (O)	1 (O)	0	1 (O)	0	0	1 (O)	1 (O)	1 (O)	2 (O)
	E	1 (D)	1 (D)	3 (D)	4 (D)	1 (DO)	0	1 (DO)	16 (DVO)	1 (D)	1 (O)	1 (DO)	2 (DV)
	F	4 (V)	4 (V)	5 (V)	5 (V)	5 (DV)	5 (DV)	6 (DV)	0	5 (DV)	4 (DV)	4 (DV)	1 (DV)
Non-visual		4	4	4	4	4	4	4	3	4	4	4	4
Total		20	20	24	25	20	18	18	33	23	16	15	15

Fig. 17.6 Numbers and expression patterns of each type of opsin gene of 12 dragonfly species. Phylogenetic relationship of the dragonflies (Futahashi 2014) is shown on the *top*: *S.fre., Sympetrum frequens; O.alb., Orthetrum albistylum; S.uchi., Somatochlora uchidai; M.amp., Macromia amphigena; A.sie., Anotogaster sieboldii; T.pry., Tanypteryx pryeri; A.mel., Asiagomphus melaenops; A.par., Anax parthenope; E.sup., Epiophlebia superstes; I.asi., Ischnura asiatica; M.cos., Mnais costalis; I.per., Indolestes peregrinus.* SW and LW opsin genes are categorized into three (**a–c**) and six (**A–F**) groups, respectively. In expression pattern, major tissues and stages expressing each group of opsin genes are shown in parentheses, wherein D, V, O, and L indicate dorsal region of adult eyes, ventral region of adult eyes, adult head region containing ocelli, and larval whole head, respectively (Figure modified from Futahashi et al. 2015)

(Calopterygidae), and *Indolestes peregrinus* (Lestidae) (Fig. 17.6) (Futahashi et al. 2015). The former six species belong to true dragonflies (suborder Anisoptera), while the latter three species belong to damselflies (suborder Zygoptera). *E. superstes* belongs to ancient dragonflies (suborder Anisozygoptera, sometimes including into Anisoptera) (Ozono et al. 2012; Futahashi 2014). Among dragonfly families, the total number of opsin genes varied widely from 15 to 33 (Fig. 17.6).

One of the significant advantages of RNA-seq analyses is that the gene expression information of different developmental stages in multiple species could be efficiently obtained. Based on molecular phylogeny and expression pattern, SW and LW opsin genes were categorized into three (a, b, and c) and six (A, B, C, D, E, and

F) groups, respectively (Fig. 17.6). Nonvisual opsin genes were scarcely expressed in the larval and adult visual organs of all examined species (Futahashi et al. 2015). Stage- and region-specific expressions of opsin genes were widely conserved across dragonfly species as follows:

1. The group-a SW, group-B LW, and group-C LW opsin genes were predominantly expressed in larvae.
2. The group-b SW and group-F LW opsin genes were mainly expressed in the ventral region of adult compound eyes.
3. The group-c SW and group-E LW opsin genes were primarily expressed in the dorsal region of adult compound eyes.
4. The group-D LW opsin genes were specifically expressed in the adult ocelli (Fig. 17.6).

The dorsoventrally differentiated expression patterns were obscure in the three damselfly species (Fig. 17.6). It should be noted that compensational expression patterns associated with losses of some visual opsin genes were observed (e.g., loss of the ocellus-specific group-D LW opsin gene entailed ocellus-associated expression of the group-C or group-E genes (Fig. 17.6)). Given that the group -C, -D, -E, and -F genes were located in tandem on the genome of *L. fulva*, the rearrangement among these genes may have occurred in the course of evolution, resulting in lineage-specific expression pattern changes of these genes. Thus, dragonflies may utilize different sets of opsin genes depending on types of light environment, which can be achieved by an extraordinary increase in the number of opsin genes.

The repertoire of opsin genes differed among dragonfly species, suggesting that the opsin genes may have evolved according to the habitat or behavior of each species. For example, the absence of the SW opsin genes at larval stage coincided with their sand- or pit-dwelling behaviors in *A. sieboldii*, *T. pryeri*, and *A. melaenops*, whereas the multitude of SW and/or LW opsin gene numbers in the dorsal region of adult compound eyes are correlated with twilight flying activity for predation in *S. uchidai*, *M. amphigena*, *A. parthenope*, and *E. superstes* (Fig. 17.6; Futahashi et al. 2015). Plausibly, although speculative, the large variation of opsin genes is associated with the evolution of diverse color pattern in dragonflies.

The variety and beauty of color pattern are also prominent in Lepidopteran and Coleopteran insects, although they have only a few opsin genes (Fig. 17.5a). The small numbers of opsin genes in these insects may be attributed to nocturnal lifestyle of their ancestors like mammals (Briscoe and Chittka 2001; Feuda et al. 2016). By contrast, almost all dragonfly species are diurnal, and they diverged from other insects over 350 million years ago (Fig. 17.5a, Misof et al. 2014).

17.6 Molecular Mechanisms Underlying Color Changes in Red Dragonflies

Dragonflies display a wide variety of coloration such as red, yellow, blue, and green. Most of animal colors are derived from structural colors and/or pigment colors. The mechanisms of structural coloration of several dragonfly species have been recently investigated, wherein multilayer structures are generally involved in iridescent coloration (Vukusic et al. 2004; Hariyama et al. 2005; Schultz and Fincke 2009; Stavenga et al. 2012; Nixon et al. 2013, 2015; Guillermo-Ferreira et al. 2015). Non-iridescent blue color is also structural, attributed to coherent light scattering from the quasi-ordered nanostructures within pigment cells (Prum et al. 2004). By contrast, information on pigments in dragonflies is still limited.

We analyzed the red epidermal pigments from three species of red dragonfly, namely, the autumn darter *Sympetrum frequens*, the summer darter *Sympetrum darwinianum*, and the scarlet skimmer *Crocothemis servilia* (Futahashi et al. 2012). Two ommochrome pigments, xanthommatin (vivid red color in reduced form) and decarboxylated xanthommatin (dull red color in reduced form) were consistently identified in all these species (Fig. 17.7a), in which the ratio of xanthommatin is higher in vivid red color species. Previous studies have shown that the color of ommochrome pigments changes reversibly by redox reactions in vitro (Linzen 1974). By injecting a reductant (vitamin C) solution, we confirmed that the yellowish body color of both immature males and mature females changed into red as observed in mature males (Fig. 17.7b). Redox conditions of the extracted ommochrome pigments were measured electrochemically, and the relative abundance of the oxidized and reduced forms of pigments were evaluated. In all three species, only the mature males exhibited very high proportions of the reduced ommochrome pigments (Fig. 17.7c), indicating that sex-specific color change in mature red dragonflies is primarily attributed to redox states of the ommochrome pigments (Futahashi et al. 2012).

Pigment-based color changes in animals are mainly attributed to the following three mechanisms: synthesis and degradation of pigments, changes in localization of pigments, and accumulation of pigments from food (Stevens and Merilaita 2011). Red dragonflies adopt a previously unknown mechanism, namely, a body color change by redox reaction of the pigments. Male-specific color change of dragonflies has been considered as an ecologically important trait for reproductive success. Considering that mature males exhibit territorial behavior under the scorching sun and the reduced pigments show antioxidant abilities (Futahashi et al. 2012), male-specific red pigments may have additional role in preventing oxidative stress from UV radiation.

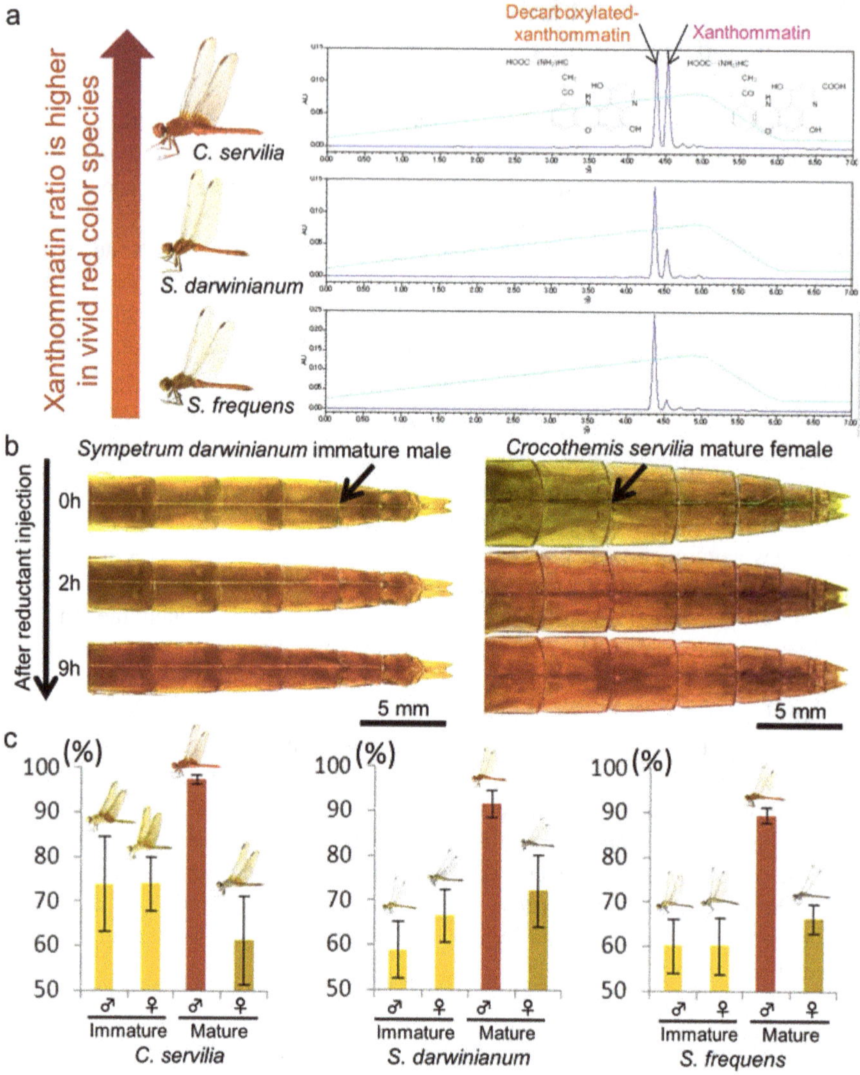

Fig. 17.7 Redox-dependent color change of the ommochrome pigments in red dragonflies. (a) Chromatograms of ommochrome pigments from males of three red dragonflies. Blue lines denote the acetonitrile gradient. (b) Reductant-induced yellow/red color change. Arrows indicate the injection sites. (c) Reduced form ratios of the extracted ommochrome pigments. Means and standard deviation are shown (n = 10 ~ 12) (Figure modified from Futahashi et al. 2012)

17.7 Conclusion and Perspective

The well-developed sense of sight and the great variety of color pattern in dragonflies have been already pointed out a century ago (Tillyard 1917). Recent progress on molecular mechanisms of color vision and color formation unveiled the outstanding diversity of visual opsin genes and the unique mechanisms of body color changes in dragonflies. Meanwhile, only limited information is available on genes potentially involved in color formation (Chauhan et al. 2014, 2016). Rapid spread of next-generation sequencing technology makes it easier than ever to analyze non-model organisms, although careful evaluation for de novo assembly is still important as described above, especially without genomic information. Molecular bases underlying color pattern formation and its evolution in dragonflies are just as fascinating and challenging as in butterflies. Recently, effective RNAi- and genome-editing methods have been developed for gene functional analyses in butterflies (Ando and Fujiwara 2013; Nishikawa et al. 2015; Li et al. 2015; Perry et al. 2016; Zhang and Reed 2016; Beldade and Peralta 2017). Applying these methods to dragonflies will be an important step toward future studies in this field (Okude et al. 2017).

Acknowledgments I would like to thank Mitsutoshi Sugimura and Kohji Tanaka for photos of gynandromorphic dragonflies, Akira Ozono for photo of *S. frequens* larva, and Genta Okude and Mizuko Osanai-Futahashi for helpful comments of the manuscript. The author's work was supported by JSPS KAKENHI Grant Numbers 23780058, 26660276, 26711021.

References

Ando T, Fujiwara H (2013) Electroporation-mediated somatic transgenesis for rapid functional analysis in insects. Development 140:454–458

Asahina S (1976) A revisional study of the genus *Mnais* (Odonata, Calopterygidae) VIII. A proposed taxonomy of Japanese *Mnais*. Tombo 19:2–16

Autrum H, Kolb G (1968) Spektrale Empfindlichkeit einzelner Sehzellen der Aeschniden. Z Vgl Physiol 60:450–477

Beatty CD, Andrés JA, Sherratt TN (2015) Conspicuous coloration in males of the damselfly *Nehalennia irene* (Zygoptera: Coenagrionidae): do males signal their unprofitability to other males? PLoS One 10:e0142684

Beldade P, Peralta CM (2017) Developmental and evolutionary mechanisms shaping butterfly eyespots. Curr Opin Insect Sci 19:22–29

Briscoe AD (2008) Reconstructing the ancestral butterfly eye: focus on the opsins. J Exp Biol 211:1805–1813

Briscoe AD, Chittka L (2001) The evolution of color vision in insects. Annu Rev Entomol 46:471–510

Bybee SM, Johnson KK, Gering EJ, Whiting MF, Crandall KA (2012) All the better to see you with: a review of odonate color vision with transcriptomic insight into the odonate eye. Org Divers Evol 12:241–250

Bybee S, Córdoba-Aguilar A, Duryea MC, Futahashi R, Hansson B, Lorenzo-Carballa MO, Schilder R, Stoks R, Suvorov A, Svensson EI, Swaegers J, Takahashi Y, Watts PC,

Wellenreuther M (2016) Odonata (dragonflies and damselflies) as a bridge between ecology and evolutionary genomics. Front Zool 13:46

Chauhan P, Hansson B, Kraaijeveld K, de Knijff P, Svensson EI, Wellenreuther M (2014) De novo transcriptome of *Ischnura elegans* provides insights into sensory biology, colour and vision genes. BMC Genomics 15:808

Chauhan P, Wellenreuther M, Hansson B (2016) Transcriptome profiling in the damselfly *Ischnura elegans* identifies genes with sex-biased expression. BMC Genomics 17(1):985

Cocroft RB, Rodríguez RL (2005) The behavioral ecology of insect vibrational communication. Bioscience 55:323–334

Corbet PS (1999) Dragonflies, behavior and ecology of odonata. Cornell University Press

Córdoba-Aguilar A (2008) Dragonflies and damselflies: model organisms for ecological and evolutionary research. Oxford, Oxford University Press

Cronin TW, Johnsen S, Marshall NJ, Warrant EJ (2014) Visual ecology. PrincetonUniversity Press, Princeton

Drury JP, Grether GF (2014) Interspecific aggression, not interspecific mating, drives character displacement in the wing coloration of male rubyspot damselflies (*Hetaerina*). Proc Biol Sci 281:20141737

Drury JP, Anderson CN, Grether GF (2015) Seasonal polyphenism in wing coloration affects species recognition in rubyspot damselflies (*Hetaerina* spp.) J Evol Biol 28:1439–1452

Eguchi E (1971) Fine structure and spectral sensitivities of retinular cells in the dorsal sector of compound eyes in the dragonfly, *Aeschna*. Z Vgl Physiol 71:201–218

Feuda R, Marlétaz F, Bentley MA, Holland PW (2016) Conservation, duplication, and divergence of five opsin genes in insect evolution. Genome Biol Evol 8:579–587

Futahashi R (1999) Notes on unusual connection and copulation in some species of dragonflies. Aeschna 36:47–55

Futahashi R (2014) A revisional study of Japanese dragonflies based on DNA analysis (2). Tombo 56:57–59

Futahashi R (2016a) Color vision and color formation in dragonflies. Curr Opin Insect Sci 17:32–39

Futahashi R (2016b) RNAseq analyses and opsin genes of dragonflies: the importance of manual assembly. Sanshi-Konchu Biotec 85(1):13–18

Futahashi R, Futahashi H (2007) A record of a black mutant of *Nannophya pygmaea* Rambur, 1842. Tombo 50:73–74

Futahashi R, Hayashi F (2004a) DNA analysis of hybrids between *Sympetrum eroticum eroticum* and *S. baccha matutinum*. Tombo 47:31–36

Futahashi R, Hayashi F (2004b) Distribution patterns of two damselfly species, *Mnais costalis* and *M. strigata*, in the Boso Peninsula, Chiba Prefecture. Tombo 47:41–46

Futahashi R, Kurita R, Mano H, Fukatsu T (2012) Redox alters yellow dragonflies into red. Proc Natl Acad Sci U S A 109:12626–12631

Futahashi R, Kawahara-Miki R, Kinoshita M, Yoshitake K, Yajima S, Arikawa K, Fukatsu T (2015) Extraordinary diversity of visual opsin genes in dragonflies. Proc Natl Acad Sci U S A 112:E1247–E1256

Guillermo-Ferreira R, Bispo PC, Appel E, Kovalev A, Gorb SN (2015) Mechanism of the wing colouration in the dragonfly *Zenithoptera lanei* (Odonata: Libellulidae) and its role in intra-specific communication. J Insect Physiol 81:129–136

Hariyama T, Hironaka M, Horiguchi H, Stavenga DG (2005) The leaf beetle, the jewel beetle, and the damselfly; insects with a multilayered show case. In: Shimozawa T, Hariyama T (eds) Structural colors in biological systems: principles and applications. University Press, Osaka, pp 153–176

Hassall C (2014) Continental variation in wing pigmentation in *Calopteryx* damselflies is related to the presence of heterospecifics. PeerJ 2:e438

Hayashi F, Dobata S, Futahashi R (2004a) Macro- and microscale distribution patterns of two closely related Japanese *Mnais* species inferred from nuclear ribosomal DNA, ITS sequences and morphology (Zygoptera: Odonata). Odonatologica 33(4):399–412

Hayashi F, Dobata S, Futahashi R (2004b) A new approach to resolve the taxonomic and ecological problems of Japanese *Mnais* damselflies (Odonata: Calopterygidae) (1). General remarks Aeschna 41:1–14

Hayashi F, Dobata S, Futahashi R (2005) Disturbed population genetics: suspected introgressive hybridization between two *Mnais* damselfly species (Odonata). Zool Sci 22(8):869–881

Hering L, Henze MJ, Kohler M, Kelber A, Bleidorn C, Leschke M, Nickel B, Meyer M, Kircher M, Sunnucks P, Mayer G (2012) Opsins in Onychophora (velvet worms) suggest a single origin and subsequent diversification of visual pigments in arthropods. Mol Biol Evol 29 (11):3451–3458

Hooper RE, Tsubaki Y, Siva-Jothy MT (1999) Expression of a costly, plastic secondary sexual trait is correlated with age and condition in a damselfly with two male morphs. Physiol Entomol 24:364–369

Huang SC, Chiou TH, Marshall J, Reinhard J (2014) Spectral sensitivities and color signals in a polymorphic damselfly. PLoS One 9:e87972

Inoue K, Tani K (2010) All about red dragonflies. Tombow, Osaka

Kirschfeld K (1976) The resolution of lens and compound eyes. In: Neural principles in vision. Springer, Berlin, pp 354–370

Labhart T, Nilsson DE (1995) The dorsal eye of the dragonfly *Sympetrum*: specializations for prey detection against the blue sky. J Comp Physiol A Neuroethol Sens Neural Behav Physiol 1995 (176):437–453

Li X, Fan D, Zhang W, Liu G, Zhang L, Zhao L, Fang X, Chen L, Dong Y, Chen Y, Ding Y, Zhao R, Feng M, Zhu Y, Feng Y, Jiang X, Zhu D, Xiang H, Feng X, Li S, Wang J, Zhang G, Kronforst MR, Wang W (2015) Outbred genome sequencing and CRISPR/Cas9 gene editing in butterflies. Nat Commun 6:8212

Linzen B (1974) The tryptophan to ommochrome pathway in insects. Adv Insect Physiol 10:117–246

McFarland WN, Lowe ER (1983) Wave produced changes in underwater light and their relations to vision. Environ Biol Fish 8:173–184

Meinertzhagen IA, Menzel R, Kahle G (1983) The identification of spectral receptor types in the retina and lamina of the dragonfly *Sympetrum rubicundulum*. J Comp Physiol 151:295–310

Misof B, Liu S, Meusemann K, Peters RS, Donath A, Mayer C, Frandsen PB, Ware J, Flouri T, Beutel RG, Niehuis O, Petersen M, Izquierdo-Carrasco F, Wappler T, Rust J, Aberer AJ, Aspöck U, Aspöck H, Bartel D, Blanke A, Berger S, Böhm A, Buckley TR, Calcott B, Chen J, Friedrich F, Fukui M, Fujita M, Greve C, Grobe P, Gu S, Huang Y, Jermiin LS, Kawahara AY, Krogmann L, Kubiak M, Lanfear R, Letsch H, Li Y, Li Z, Li J, Lu H, Machida R, Mashimo Y, Kapli P, McKenna DD, Meng G, Nakagaki Y, Navarrete-Heredia JL, Ott M, Ou Y, Pass G, Podsiadlowski L, Pohl H, von Reumont BM, Schütte K, Sekiya K, Shimizu S, Slipinski A, Stamatakis A, Song W, Su X, Szucsich NU, Tan M, Tan X, Tang M, Tang J, Timelthaler G, Tomizuka S, Trautwein M, Tong X, Uchifune T, Walzl MG, Wiegmann BM, Wilbrandt J, Wipfler B, Wong TK, Wu Q, Wu G, Xie Y, Yang S, Yang Q, Yeates DK, Yoshizawa K, Zhang Q, Zhang R, Zhang W, Zhang Y, Zhao J, Zhou C, Zhou L, Ziesmann T, Zou S, Li Y, Xu X, Zhang Y, Yang H, Wang J, Wang J, Kjer KM, Zhou X (2014) Phylogenomics resolves the timing and pattern of insect evolution. Science 346(6210):763–767

Moriyasu A, Sugimura M (2007) An interspecific hybrid between *Sympetrum croceolum* and *S. speciosum*. Mon J Entomol 433:44–45

Nishikawa H, Iijima T, Kajitani R, Yamaguchi J, Ando T, Suzuki Y, Sugano S, Fujiyama A, Kosugi S, Hirakawa H, Tabata S, Ozaki K, Morimoto H, Ihara K, Obara M, Hori H, Itoh T, Fujiwara H (2015) A genetic mechanism for female-limited Batesian mimicry in *Papilio* butterfly. Nat Genet 47(4):405–409

Nixon MR, Orr AG, Vukusic P (2013) Subtle design changes control the difference in colour reflection from the dorsal and ventral wing-membrane surfaces of the damselfly *Matronoides cyaneipennis*. Opt Express 21:1479–1488

Nixon MR, Orr AG, Vukusic P (2015) Wrinkles enhance the diffuse reflection from the dragonfly *Rhyothemis resplendens*. J R Soc Interface 12:20140749

Nomakuchi S, Higashi K, Harada M, Maeda M (1984) An experimental study of the territoriality in *Mnais pruinosa pruinosa* Selys (Zygoptera: Calopterygidae). Odonatologica 13:259–267

Okude G, Futahashi R, Kawahara-Miki R, Yoshitake K, Yajima S, Fukatsu T (2017) Electroporation-mediated RNA interference reveals a role of multicopper oxidase 2 gene in dragonfly's cuticular pigmentation. Appl Entomol Zool. doi:10.1007/s13355-017-0489-9

Ozono A, Kawashima I, Futahashi R (2012) Dragonflies of Japan, Bunichi-Sogo Syuppan., Co. Ltd

Perry M, Kinoshita M, Saldi G, Huo L, Arikawa K, Desplan C (2016) Molecular logic behind the three-way stochastic choices that expand butterfly colour vision. Nature 535(7611):280–284

Prum RO, Cole JA, Torres RH (2004) Blue integumentary structural colours in dragonflies (Odonata) are not produced by incoherent Tyndall scattering. J Exp Biol 207:3999–4009

Sánchez-Guillén RA, Córdoba-Aguilar A, Cordero-Rivera A, Wellenreuther M (2014) Genetic divergence predicts reproductive isolation in damselflies. J Evol Biol 27:76–87

Schultz TD, Fincke OM (2009) Structural colours create a flashing cue for sexual recognition and male quality in a neotropical giant damselfly. Funct Ecol 23:724–732

Shichida Y, Matsuyama T (2009) Evolution of opsins and phototransduction. Philos Trans R Soc Lond Ser B Biol Sci 364:2881–2895

Siva-Jothy MT, Tsubaki Y (1989) Variation in copulation duration in *Mnais pruinosa pruinosa* Selys (Odonata: Calopterygidae). I. Alternative mate securing tactics and sperm precedence. Behav Ecol Sociobiol 24:39–45

Stavenga DG, Leertouwer HL, Hariyama T, De Raedt HA, Wilts BD (2012) Sexual dichromatism of the damselfly *Calopteryx japonica* caused by a melanin-chitin multilayer in the male wing veins. PLoS One 7:e49743

Stevens M, Merilaita S (2011) Animal camouflage: mechanisms and function. Cambridge University Press, Cambridge

Suzuki K (1984) Character displacement and evolution of the Japanese *Mnais* damselflies (Zygoptera: Calopterygidae). Odonatologica 13:287–300

Svensson EI, Karlsson K, Friberg M, Eroukhmanoff F (2007) Gender differences in species recognition and the evolution of asymmetric sexual isolation. Curr Biol 17:1943–1947

Svensson EI, Runemark A, Verzijden MN, Wellenreuther M (2014) Sex differences in developmental plasticity and canalization shape population divergence in mate preferences. Proc Biol Sci 281:20141636

Takahashi Y, Kagawa K, Svensson EI, Kawata M (2014) Evolution of increased phenotypic diversity enhances population performance by reducing sexual harassment in damselflies. Nat Commun 5:4468

Terakita A (2005) The opsins. Genome Biol 6:213

Thorvaldsdóttir H, Robinson JT, Mesirov JP (2013) Integrative genomics Viewer (IGV): high-performance genomics data visualization and exploration. Brief Bioinform 14:178–192

Tillyard RJ (1917) The biology of dragonflies. Cambridge University Press, Cambridge

Tsubaki Y (2003) The genetic polymorphism linked to mate-securing strategies in the male damselfly *Mnais costalis* Selys (Odonata: Calopterygidae). Popul Ecol 45:263–266

Tsubaki H, Okuyama H (2016) Adaptive loss of color polymorphism and character displacements in sympatric *Mnais* damselflies. Evol Ecol 30:811

Tsubaki Y, Hooper RE, Siva-Jothy MT (1997) Differences in adult and reproductive lifespan in the two male forms of *Mnais pruinosa costalis* selys (Odonata: Calopterygidae). Res Popul Ecol 39:149–155

Tynkkynen K, Rantala MJ, Suhonen J (2004) Interspecific aggression and character displacement in the damselfly Calopteryx splendens. J Evol Biol 17:759–767

Ueda T (2004) How do the Japanese see dragonflies? Kyoto University Press, Kyoto

Vukusic P, Wootton RJ, Sambles JR (2004) Remarkable iridescence in the hindwings of the damselfly *Neurobasis chinensis chinensis* (Linnaeus) (Zygoptera: Calopterygidae). Proc Biol

Sci 271:595–601

Waage JK (1975) Reproductive isolation and the potential for character displacement in the damselflies, *Calopteryx maculata* and *C. aequabilis* (Odonata: Calopterygidae). Syst Zool 24:24–36

Yager DD (1999) Structure, development, and evolution of insect auditory systems. Microsc Res Tech 47:380–400

Yang EC, Osorio D (1991) Spectral sensitivities of photoreceptors and lamina monopolar cells in the dragonfly, *Hemicordulia tau*. J Comp Physiol A 169:663–669

Zhang L, Reed RD (2016) Genome editing in butterflies reveals that *spalt* promotes and *Distal-less* represses eyespot colour patterns. Nat Commun 7:11769

Permissions

All chapters in this book were first published by Springer; hereby published with permission under the Creative Commons Attribution License or equivalent. Every chapter published in this book has been scrutinized by our experts. Their significance has been extensively debated. The topics covered herein carry significant information for a comprehensive understanding. They may even be implemented as practical applications or may be referred to as a beginning point for further studies.

The contributors of this book come from diverse backgrounds, making this book a truly international effort. We would like to thank all the contributing authors for lending their expertise to make the book truly unique. They have played a crucial role in the development of this book. Without their invaluable contributions this book wouldn't have been possible. They have made vital efforts to compile up to date information on the varied aspects of this subject to make this book a valuable addition to the collection of many professionals and students.

This book was conceptualized with the vision of imparting up-to-date and integrated information in this field. To ensure the same, a matchless editorial board was set up. Every individual on the board went through rigorous rounds of assessment to prove their worth. After which they invested a large part of their time researching and compiling the most relevant data for our readers.

The editorial board has been involved in producing this book since its inception. They have spent rigorous hours researching and exploring the diverse topics which have resulted in the successful publishing of this book. They have passed on their knowledge of decades through this book. To expedite this challenging task, the publisher supported the team at every step. A small team of assistant editors was also appointed to further simplify the editing procedure and attain best results for the readers.

Apart from the editorial board, the designing team has also invested a significant amount of their time in understanding the subject and creating the most relevant covers. They scrutinized every image to scout for the most suitable representation of the subject and create an appropriate cover for the book.

The publishing team has been an ardent support to the editorial, designing and production team. Their endless efforts to recruit the best for this project, has resulted in the accomplishment of this book. They are a veteran in the field of academics and their pool of knowledge is as vast as their experience in printing. Their expertise and guidance has proved useful at every step. Their uncompromising quality standards have made this book an exceptional effort. Their encouragement from time to time has been an inspiration for everyone.

The publisher and the editorial board hope that this book will prove to be a valuable piece of knowledge for students, practitioners and scholars across the globe.

Index

Printed in the USA
CPSIA information can be obtained
at www.ICGtesting.com
LVHW010821221023
761596LV00005B/578

9 781639 877317